F. Tanneberger & W. Wichtmann (eds.)

Carbon credits from peatland rewetting

Franziska Tanneberger & Wendelin Wichtmann (eds.)

# Carbon credits from peatland rewetting

*Climate – biodiversity – land use*

Science, policy, implementation and recommendations of a pilot project in Belarus

with 44 contributions (see list of contributors on page 217)

Schweizerbart Science Publishers · Stuttgart, 2011

# Carbon credits from peatland rewetting

## Climate – biodiversity – land use

Science, policy, implementation and recommendations of a pilot project in Belarus

Editors:   Franziska Tanneberger & Wendelin Wichtmann
Michael Succow Stiftung
Ellernholzstraße 1/3
17489 Greifswald
Germany
info@succow-stiftung.de

The project 'Restoring peatlands and applying concepts for sustainable management in Belarus – climate change mitigation with economic and biodiversity benefits' is financed by the Federal Republic of Germany through KfW Entwicklungsbank in the framework of the International Climate Initiative of the German Federal Ministry for the Environment, Nature Conservation and Nuclear Safety (BMU). It is co-ordinated by the Royal Society for the Protection of Birds, UK, in collaboration with APB – BirdLife Belarus and the Michael Succow Foundation, Germany. It is carried out with support of the United Nations Development Programme (UNDP) in Belarus and the Ministry of Natural Resources and Environmental Protection of the Republic of Belarus.

**Disclaimer:** The authors take full responsibility for the technical and scientific content in this publication, and opinions expressed are attributed to the authors alone, and not to the institutions or personages supporting the publication.

ISBN 978-3-510-65271-6
Information on the English title: www.schweizerbart.com/9783510652716

This title is also available in Russian, information on the Russian title:
www.schweizerbart.com/9783510652723

© 2011 E. Schweizerbart'sche Verlagsbuchhandlung (Nägele u. Obermiller), Stuttgart, Germany

All rights reserved. No part of this publication may be reproduced, stored in a retrieval system, or transmitted, in any form or by any means, electronic, mechanical photocopying, recording, or otherwise, without the prior written permission of E. Schweizerbart'sche Verlagsbuchhandlung, Stuttgart.

Publisher:   E. Schweizerbart'sche Verlagsbuchhandlung (Nägele u. Obermiller)
Johannesstr. 3A, 70176 Stuttgart, Germany
mail@schweizerbart.de
www.schweizerbart.de

∞ Printed on permanent paper conforming to ISO 9706-1994

Cover: Photo by Annett Thiele

Layout: DTP + TEXT Eva Burri, Stuttgart
Printed in Germany by Gulde Druck, Tübingen

# Foreword by the United Nations Environment Programme

Since the UN Climate Conference in Bali in 2007, the global community has become increasingly aware of the important role of peatlands in regulating atmospheric greenhouse gas concentrations. Covering only 3% of the land surface, they store in their peat twice as much carbon as the total forest biomass of the world. The 'time bomb' character of this concentrated below-ground stock is illustrated by the fact that drained and degraded peatlands (covering less than 0.5% of the land surface) are responsible for a disproportional 6% of the global anthropogenic $CO_2$ emissions.

The main source of peatland emissions is Southeast Asia, where recent and continuing peat swamp conversion leads to enormous greenhouse gas emissions (GHG) from microbial peat oxidation and peat fires. The second, less well-known hotspot is in temperate Europe. Here the majority of peatlands have been drained and converted decades ago and the GHG emissions continue to date. Within Europe, Belarus has a special position as a small country with a large peatland area and belongs to the world's top 10 with respect to peatland $CO_2$ emissions, and even to the top three with respect to peatland emissions per unit land area. It is therefore laudable that Belarus has been exploring the opportunity for mitigating and adapting to climate change by peatland rewetting.

This best practice guidebook reports the diverse outcomes of the 'Restoring Peatlands and applying Concepts for Sustainable Management in Belarus, a Climate Change Mitigation Project with Economic and Biodiversity Benefits', and draws on the results of the UNDP/GEF funded project 'Renaturalization and Sustainable Management of Peatlands in Belarus to Combat Land Degradation, Ensure Conservation of Globally Valuable Biodiversity and Mitigate Climate Change', and the UNEP/GEF global programme on 'Integrated Management of Peatlands for Biodiversity and Climate Change'.

In addition to developing rewetting and management techniques and applying them on 36,000 ha of drained peatland, these initiatives describe a new peatland standard for the global voluntary carbon market, which makes appropriate methodologies available for monitoring and assessing peatland emissions and biodiversity, working out the juridical technicalities for trading carbon credits from rewetted peatland, and contributing to bringing peatlands on the agenda of the Kyoto Protocol.

The experiences gained will support similar initiatives, in Belarus, in the region, and beyond and furthermore help to strengthen the political agenda to effectively utilize the cost effective climate change mitigation options peatlands have to offer.

Achim Steiner
United Nations Under-Secretary-General
Executive Director, UNEP

# Foreword by the Minister of Natural Resources and Environmental Protection of the Republic of Belarus

Belarusian mires are unique natural ecosystems having high value for biological diversity, climate regulation and man's well-being.

In the 'National Strategy of Sustainable Social and Economic Development of the Republic of Belarus up to 2020' restoration of degraded lands and natural ecosystems is recognized as an important task for the country. In order to restore natural and economic potential of damaged lands, the Strategy foresees implementation of a complex of re-cultivation activities. In the years 2011–2020 it is expected that the area of restored lands will reach up to 15–17% of the total area of damaged lands. The Strategy also states that it is important to conserve untouched natural mire ecosystems through implementation of a complex of measures on rational utilization, restoration and protection of mires and peatlands, as well as implementation of international agreements signed by our country.

Mires are the most important long-term depositories for carbon. The idea for financing the restoration of mires and their management in order to preserve biodiversity and climate protection through the sale of carbon credits was announced for the first time at the 3rd International Conference of the Michael Otto Foundation held in June 2007 in Minsk.

Large-scale drainage in the middle of the 20th century resulted in the reduction or even complete destruction of a whole range of representatives of mire flora and fauna. Restoration of depleted peat deposits facilitates recovery of habitats of mire flora and fauna.

The Ministry of Natural Resources and Environmental Protection of the Republic of Belarus evaluates highly implementation of the project 'Restoring Peatlands and Applying Concepts for Sustainable Management in Belarus – Climate Change Mitigation with Economic and Biodiversity Benefits' aimed at the restoration of previously drained mires, their effective management with regard to economics and biodiversity, and also the development of a scheme of sustainable financing for rewetting projects within the possibilities and financial mechanisms of the world carbon market.

Significant tasks of the project required the joint efforts of local participants, national ministries and institutions and international organisations, such as United Nations Development Programme (UNDP), Global Environmental Facility (GEF), the RSPB (Royal Society for the Protection of Birds (Great Britain)), Michael Succow Foundation (Germany), Kreditanstalt für Wiederaufbau (KfW) Entwicklungsbank (Germany), APB-BirdLife Belarus. In many respects this cooperation was one of the key factors in successful project implementation and in the recognition of the importance of practical results achieved for the Republic of Belarus as well as for the world community.

Based on national and international experience the project made it possible: to test methods for rehabilitation of damaged mires and depleted peat deposits; to demonstrate the potential for management of degraded peatlands; to globally achieve valuable results. It also increased understanding of rewetting practices as an effective mechanism for sustainable management and utilization of land resources by local land users and authorities at different levels.

The practical experience, on rewetting of depleted peatlands and rehabilitation of damaged mires, gained during the implementation of this international project is very valuable to Belarus. It could be used in the countries of Western Europe where the restoration of similar territories is also a real issue. Furthermore, restoration of the project sites allows evaluation of the practical role of peatlands for climate change mitigation and analysis of the possibilities for the utilization of greenhouse gas emission reductions within the Kyoto Protocol, which the Republic of Belarus joined in 2005, and further international climatic agreements.

V. G. Tsalko
Minister of Natural Resources and Environmental Protection of the Republic of Belarus

# Foreword by the Michael Otto Foundation

The protection of our most valuable benefit – biodiversity – is a complex issue. A healthy biodiversity delivers ecosystem services that are indispensable for human well being and is the basis of all economic activity. Notwithstanding the importance of ecosystems, nature protection is all too often leading to conflicts between environmentalists and business interests.

To arbitrate this difficult dialogue and thereby to protect and conserve the vital natural resource of water and its underlying ecosystems, Dr. Michael Otto founded the Michael Otto Foundation for Environmental Protection in 1993, an organisation that comes from trade and industry while having a sense of responsibility for the environment. The foundation intends to help achieve a breakthrough for decisions which are right for the protection of nature, and which are balanced for trade and industry.

In doing so, the Otto Foundation also has a long record of cooperation in nature conservation in Belarus. In 1995 we received the report of a Belarusian/German expedition to the Prypiac Floodplains and Lowland Mires. It contained an impressive description of the rich resources of nature, especially of the large number of Aquatic Warblers breeding there. But it also indicated that this unique natural environment was endangered. Widespread projects on land reclamation and peat cutting were threatening the most valuable riverine meadows, wetland meadows and riverside moorlands of Europe. Rapid help was needed, since intact peatlands are of outmost importance as a carbon sink and as a habitat for various flagship species such as the Aquatic Warbler.

The foundation thus supported three International Prypiac Conferences. In 1997 a joint plan of action was drawn up in close cooperation with the Belarusian Ministry of the Environment. The second conference in 2003 led to an international agreement on the protection of the Aquatic Warbler in its natural habitat, while the third conference resulted in plans on rewetting up to 500,000 ha drained peatland – thereby preventing the release of damaging greenhouse gases. The conference outcome laid the foundation of the project 'Restoring peatlands and applying concepts for sustainable management in Belarus' and is outlined in this publication.

By trading the generated carbon credits on international markets, new income for water management and biodiversity conservation as well as local economic benefits are created and harmful peat fires are reduced. The project thus rests on a financially sustainable basis, indicating that business, biodiversity and climate mitigation are no contradictory issues.

The Otto Foundation strongly believes that designing and implementing best practices like the project outlined here will motivate further exemplary initiatives, ultimately leading to the fulfilment of the goal of preserving the environment for future generations.

Johannes Merck
Chair of the Michael Otto Foundation for Environmental Protection

# Table of Content

Foreword by the United Nations Environment Programme ................................................. V
Foreword by the Minister of Natural Resources and Environmental Protection
    of the Republic of Belarus ................................................................................................ VII
Foreword by the Michael Otto Foundation........................................................................... IX

## 1 Introduction .................................................................................................................. 1

## 2 Peatlands in Belarus .................................................................................................... 3
2.1 Extent and types of mires and peatlands in Belarus ................................................. 3
2.2 Investigation and drainage of peatlands ................................................................... 5
2.3 Use of peatlands and peat ........................................................................................ 7
2.4 Rewetting of peatland ............................................................................................... 9

## 3 Peatlands and climate ................................................................................................ 13
3.1 Peatlands and greenhouse gases ........................................................................... 13
3.2 The global peatland $CO_2$ picture ............................................................................ 20
3.3 Measuring GHG emissions from peatlands............................................................. 30
3.4 Vegetation as a proxy for greenhouse gas fluxes – the GEST approach ............... 37
3.5 Prediction of vegetation development with and without rewetting .......................... 42

## 4 Peatlands and biodiversity ........................................................................................ 61
4.1 Biodiversity values of Belarusian peatlands ........................................................... 61
4.2 Relationship between peatland condition and biodiversity values.......................... 68
4.3 Target and indicator species ................................................................................... 77
4.4 Peatland rewetting and biodiversity management .................................................. 81

## 5 Driving forces and funding options ......................................................................... 89
5.1 Legal obligations for the restoration of degraded peatlands in Belarus.................. 89
5.2 Sensitising global conventions for climate change mitigation by peatlands ........... 90
5.3 Selling peatland rewetting on the voluntary carbon market .................................... 94
5.4 Selling peatland rewetting on the compliance carbon market ................................ 99
5.5 Voluntary emission reduction projects – how to start in Belarus .......................... 105

## 6 Land use options for rewetted peatlands .............................................................. 107
6.1 Overview on land use options after rewetting....................................................... 107
6.2 Biomass use for food and fodder .......................................................................... 110
6.3 Biomass use for raw material ................................................................................ 113
6.4 Biomass use for energy......................................................................................... 115
6.5 Benefits from land use on rewetted peatlands ..................................................... 128

## 7 The BMU-ICI project ................................................................................................. 133
7.1 Summary of the project ......................................................................................... 133
7.2 Site selection and rewetting actions ..................................................................... 137
7.3 Climate actions...................................................................................................... 141
7.4 Biodiversity actions................................................................................................ 145
7.5 Policy actions......................................................................................................... 147
7.6 Communication and awareness raising................................................................ 149
7.7 Capacity building ................................................................................................... 152

7.8 Lessons learnt ................................................................................................ 154
7.9 The BMU-ICI twin project in Ukraine ............................................................ 165

**8 Practical rewetting examples** ........................................................................ 169
8.1 Introduction .................................................................................................. 169
8.2 Dalbeniski .................................................................................................... 170
8.3 Zada ............................................................................................................ 172
8.4 Hrycyna-Starobinskaje ................................................................................. 174
8.5 Scarbinski Moch .......................................................................................... 177
8.6 Dakudauskaje .............................................................................................. 181
8.7 Jelnia ........................................................................................................... 184

**9 Recommended research and monitoring activities in rewetted peatlands** ........................................................................................ 189
9.1 Recommended research activities ............................................................... 189
9.2 Recommended monitoring activities ............................................................ 193

**10 Acknowledgements** ..................................................................................... 197

**References** ........................................................................................................ 199

**List of contributors** ........................................................................................... 217

**Index** ................................................................................................................. 219

# 1 Introduction

Peatland rewetting; carbon emission reductions; establishment of a peatland carbon standard; development of a methodology; facilitation of emission trading; habitat restoration; stimulating biodiversity; dialogue enforcement; capacity building; and promotion of paludiculture. Visions of a decade long programme?

This publication presents all of the above, as an outcome of a three year project: 'Restoring peatlands and applying concepts for sustainable management in Belarus – climate change mitigation with economic and biodiversity benefits', conducted in the framework of the International Climate Initiative (ICI) of the German Federal Ministry for the Environment, Nature Conservation and Nuclear Safety (BMU).

For the project team, mires and peatlands are an extremely valuable habitat to work in. None of the team members ever had the common negative image of an inaccessible place full of myth, danger, and mosquitoes. In recent years, a positive impression of peatlands as fascinating ecosystems has spread among a wider audience. The importance of their functions, including groundwater recharge, cooling, storage of carbon, as habitat for a wide diversity of species, as water reservoir (to name only a few) are now largely understood. Particularly the role of peatlands in the global carbon cycle has become apparent. Peatlands cover only about 3% of the world's land surface, but store c. 550 Gt of carbon. Peatland drainage leads to fast mineralization of the carbon and nitrogen stocks in the peat, which transforms the peatland from a sink to a potentially very strong carbon and nitrogen source. Rewetting of peatlands is a very promising option to combat climate change.

In Europe, over the last centuries (culminating in the 1970s), the perceived mystical danger of peatlands was challenged. Peatlands were drained, and land was claimed for agriculture and forestry and for the extraction of peat as a heating material. As a consequence, the drained peatlands of temperate Europe (especially in Germany, Poland, Belarus, Ukraine, and European Russia) now constitute an important source of greenhouse gas emissions and are (after Southeast Asia) the second most important global hotspot in this respect (chapter 3.2). Belarus, the country with the highest proportion of peatlands in Europe (almost 15% of the country's land area), has a global responsibility for the protection of mires and for reducing peatland related emissions. At present, c. 1.5 million ha (7% of the country's land area) have been drained for agriculture, forestry, and peat extraction (chapter 2.2).

The aim of the project was to demonstrate reduction of greenhouse gas emissions and enhancement of biodiversity values through the restoration and sustainable management of large areas of currently degraded peatland in Belarus and to develop a scheme for the sale of carbon credits to secure further peatland rewetting activities over the longer-term.

In 2008, the BMU-ICI project was developed by the initiative of the Michael Succow Foundation (MSF), Germany, in partnership with the Royal Society for the Protection of Birds (RSPB), UK and with APB-BirdLife Belarus ('Achova ptusak Backauscyny'). Funding was granted by BMU through Kreditanstalt für Wiederaufbau (KfW) Entwicklungsbank and the project received support from a number of partners and related institutions:

- United Nations Development Programme (UNDP) in Belarus;
- The Ministry of Natural Resources and Environmental Protection and the Ministry of Forestry of the Republic of Belarus;
- Scientific organisations such as the University of Greifswald and the Leibniz-Centre for Agricultural Landscape Research (ZALF) Müncheberg (both Germany) and the National Academy of Sciences of Belarus;
- The German Centre for International Migration and Development (CIM);
- Companies such as TerraCarbon (USA) and Silvestrum (Netherlands), and Climate Focus (Netherlands).

This publication will not only pick up all the issues mentioned above and present the results and lessons learnt of very ambitious project but it will also present state of the art knowledge on peatland rewetting and conservation. After an introduction into peatlands in Belarus (chapter 2), the importance of peatlands with respect to climate and particularly greenhouse gas emissions is explained (chapter 3), followed by a chapter on biodiversity aspects (chapter 4). The driving forces and funding options for peatland restoration are presented in chapter 5 and a particular focus is given to sustainable land use options (chapter 6). The approach and outcomes of the BMU-ICI project are presented in detail in chapter 7, enabling the reader to gain an insight into the project, its results and the lessons learnt. Chapter 8 provides information on practical rewetting examples in Belarus, accompanied by detailed maps, site descriptions, and pictures of rewetting measures. At the end of the book, an outlook on recommended research and monitoring in rewetted peatlands (chapter 9) is provided.

The project partners are very grateful to have the opportunity to release this best practice guidebook within the BMU-ICI project and to make it available to a wide audience, from stakeholders involved in peatland rewetting (engineers, scientists, land owners etc.), to individuals and companies interested in carbon offsetting, and last but not least to universities and research institutions.

Viktar Fenchuk
(chief executive of APB-BirdLife Belarus),
Mike Clarke (chief executive of the RSPB) &
Sebastian Schmidt (chief executive of the MSF)

# 2 Peatlands in Belarus

Belarus is one of Europe's key peatland countries. Its mires used to cover about 15% of the country's area. No other country holds such a high-quality collection of peatland types representative for temperate lowland Europe. Extent, distribution, and the impressive variety of peatlands in Belarus are explained in chapter 2.1. Here, also important definitions and concepts with regard to peat and peatlands are presented. Chapter 2.2 details the history of peatland investigation and drainage. The use of peatlands and peat in Belarus is described in chapter 2.3.

Peatland drainage leads to fast mineralization of the carbon (C) and nitrogen (N) stocks in the peat, which transforms the peatland from a C and N sink to a potentially very strong C and N source. Since the beginning of the 1990s, socio-economic changes and soil degradation have led to a declining use of drained peatlands in Central Europe. Thanks to the strong dedication of the Belarusian government and the successful preparatory work of several European non-governmental organisations, extensive peatland rewetting projects have been carried out (chapter 2.4). The two largest being: a project with core funding by the United Nations Development Programme (UNDP) and the Global Environmental Facility (GEF), realised on 28,000 ha in 2006–2010; and the BMU-ICI project (see chapter 7).

## 2.1 Extent and types of mires and peatlands in Belarus

Nina Tanovitskaya

### 2.1.1 Extent and distribution of peatlands

Belarus (207,600 km$^2$) is located in the geographic centre of Europe. Before drainage and peat extraction started, peatlands (see box 1) covered 2,939,000 ha which equals 14.2% of the total land area (Bambalov et al. 1992). Despite its small size, the country comprises a wide variety of peatland types depending on factors important for peat formation such as climate, bedrock, relief, and hydrological network. Due to large variation in these factors, amount and types of peatlands are not equally distributed within Belarus. Five peatland districts and three peatland regions have been described (Fig. 1 Colour plates I; Pidoplichko 1961, Tanovitskiy 1980, Bambalov 2005).

### 2.1.2 Conditions of mire formation

Mire formation in Belarus has been closely related to the development of the climate throughout the Holocene with peat formation intensifying when the climate became wetter. When climate conditions are favourable for paludification and peat formation, other factors, above all relief and hydrogeological conditions, exert large influence. This influence is illustrated by the differences in mean peat thickness of 2.5 m in Viciebsk region in the north and only 1.4 m in Brest region in the south of Belarus. In Viciebsk region and the north of Minsk region, initial peat formation was related to overgrowth and terrestrialization of lakes, i.e. of deep basins (Tanovitskaya & Ratnikova 2010a), whereas the flat or slightly undulating relief in Paliessie favoured extensive peatland development by paludification. Furthermore the Belarusian peatland zonation reflects the geomorphological and hydrogeological diversity of the country (Pidoplichko 1961, Bambalov & Dubovets 1990). The overall area of fens in Belarus is 2,103,800 ha (82% of all peatland area), that of transitional peatlands 106,200 ha (3%), and that of bogs 333,700 ha (15%; Bambalov 2005).

### 2.1.3 Peatland districts and regions

Based on physical-geographical conditions of peat formation and geobotanical and geomorphological characteristics, Belarusian peatlands can be allocated to five peatland districts (Fig. 1, Table 1; Pidoplichko 1961):
- District I is the northernmost and comprises bogs with thick peat layers (frequently accumulated over lake sediments; cf. Fig. 2 Colour plates I) and fens with rather steep slopes. Geomorphologically, this district is characterised

## Box 1

### What are mires and peatlands?

Hans Joosten

Mires are wetland ecosystems that are characterized by the accumulation of dead organic matter (peat), which is derived from dead and decaying plant material under conditions of permanent water saturation. In most natural ecosystems the production of plant material is counterbalanced by its decomposition by bacteria and fungi. In those wetlands where the water level is stable and near the surface, the dead plant remains do not fully decay but accumulate as peat. A wetland in which peat is actively accumulating is called a mire. Every mire is a peatland, i.e. a land with peat, but not every peatland is necessarily a mire. When a mire has been drained and does not accumulate peat anymore, it is still a peatland, but not a mire anymore.

The major characteristics of natural mires are permanent water logging, development of specific vegetation, the consequent formation and storage of peat and the continuous (upward) growth of the surface. Mire distribution and peat formation and storage are primarily a function of climate, which determines water conditions, vegetation productivity and the decomposition rate of dead organic material. Mires are found in almost every country, but occur primarily in the boreal, subarctic and tropical zones as well as in appropriate zones in mountains. Many peatlands are not recognised as such but are classified as marshes, meadows, or forests.

As a result of different climatic and biogeographic conditions, a large diversity of mire types exists. However, because of similar ecohydrological processes, they share many ecological features and functions. A major distinction is between bogs (which are fed only by precipitation and are nutrient-poor) and fens (which are fed by surface or ground water as well as precipitation and tend to be more nutrient rich) (Fig. 3). Peatlands may be naturally forested or naturally open and vegetated with mosses, sedges or shrubs. The complex relationship between plants, water, and peat makes peatlands vulnerable to a wide range of human interference.

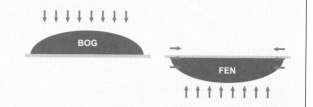

Fig. 3: The classical difference between 'bog' and 'fen' peatlands. Shaded = peat; arrow = water flow. From Parish et al. (2008).

by ridges, hills, and other formations of terminal moraines as well as by depressions with sandy sediments and occasionally plains with ground moraine or glacio-limnic sediments;

- District II comprises fens with thick peat layers in western Belarus. Terminal moraines predominate and only few lakes have been formed;
- District III includes large bogs and fens of the slightly undulating ablation plain in central Belarus;
- District IV is located in eastern Belarus but consists of small bogs and fens formed predominantly on loess formations (cf. Fig. 4 Colour plates I);
- District V corresponds to Belarusian Paliessie, where under special hydrological conditions, vast fens with a rather thin peat layer have been formed (cf. Fig. 5 Colour plates I). Paludification of this region has mainly been induced by the small hydraulic gradient of the rivers, the low elevation above sea level, and the position of aquifers close to the soil surface. This region is characterised by a large low-lying, slightly dissected alluvial outwash plain.

Table 1: The characteristics of the Belarusian peatland districts (after Pidoplichko 1961).

| Peatland district | I | II | III | IV | V |
|---|---|---|---|---|---|
| Area covered by peatlands (%) | 11 | 8 | 16 | 6 | 18 |
| Mean thickness of the peat layer (m) | 2.0 | 2.0 | 1.9 | 1.6 | 1.6 |
| Fen peat<br>Transitional peat<br>Bog peat<br>(% of overall peat stock) | 55<br>7<br>38 | 94<br>4<br>2 | 70<br>6<br>24 | 85<br>4<br>11 | 87<br>7<br>6 |

In addition to the five peatland districts three natural peatland regions that coincide with the geobotanical zonation of Belarus have been described (Fig. 1 Colour plates I; Tanovitskiy 1980):
- The northern region with >10% of the area covered by peatlands, mainly bogs with thick peat layers and some fens located on rather steep slopes;
- The central region with 7% (eastern and western parts) to 16% (central part) fens with thick peat layers and rather small transitional peatlands and bogs;
- The southern region with >18% peatlands, mainly large fens.

### 2.1.4 Typology of mires and peatlands

In Belarus, mires ('bolota') are defined as permanently waterlogged areas covered by hygrophilous vegetation where peat is being formed. Mires are classified according to the chemical characteristics of their water supply and to the living plant cover. Three main types are distinguished: (eutrophic) fens, (mesotrophic) transitional mires, and (oligotrophic) bogs. Fens are formed under rich supply of ground or river water next to atmospheric precipitation and have 60 to 400 mg/l of dissolved mineral salts in their water (Bambalov & Rakovich 2005). Transitional mires receive less ground water and relatively more precipitation and surface-discharge (40 to 80 mg/l of dissolved mineral salts). Bogs are mainly/solely fed by atmospheric precipitation (less than 50 mg/l of dissolved mineral salts). The Belarusian distinction between fens and bogs is similar to that in common west European typologies (e.g. Sjörs 1950), whereas Belarusian 'transitional peatlands' would commonly be named 'poor fens'.

Each type can be further divided into three subtypes on the basis of vegetation structure and humidity in the root layer: forested (low humidity), half-open (medium humidity), and open (high humidity). The half-open subtype is further divided into peatlands dominated by forest and grasses or by forest and mosses, whereas the open subtype is subdivided into grasses, grasses and mosses, and mosses, respectively. The lowest taxonomical unit of peatland vegetation is the phytocoenosis characterised by a specific floristic composition and particular environmental conditions. For example, alder phytocoenoses grow where the root layer is well aerated and ground waters are nutrient rich, while *Sphagnum fuscum* phytocoenoses grow under wet and nutrient-poor conditions. In Belarus, approximately 40 peatland phytocoenoses have been described.

Parallel to the three main mire types, three main peat types are distinguished: (eutrophic) fen peat, (mesotrophic) transitional peat, and (oligotrophic) bog peat. They constitute four types of peat deposits: fen deposits, transitional deposits, bog deposits, and mixed deposits (Tiuremnov 1976, Tiuremnov et al. 1977). Mixed deposits are formed in situations where during the formation of the peatland changes in water and nutrient supply occurred (e.g. a fen developing into a bog). This typology is officially acknowledged and recommended for the assessment of peat resources in all countries of the Commonwealth of Independent States (CIS). Since the uppermost 50 cm of a peatland are not considered in the classification of peat deposits, the current vegetation does not necessarily correspond to the type of peat deposit. So a peatland that has developed from a eutrophic into a meso- or oligotrophic type may still be classified as eutrophic although the current vegetation reflects nutrient-poor conditions. Neglecting the evolution of peatlands and the current vegetation is seen as a major shortcoming of this typology. Therefore, based on stratigraphical research of Belarusian peat deposits, seven genetic types of peat deposits (instead of only four) have been described: fen, fen-transitional, transitional, fen-bog, fen-transitional-bog, transitional-bog, and bog (Bambalov et al. 1981, Dubovets 1981). This typology includes the uppermost peat layer regardless of its thickness.

## 2.2 Investigation and drainage of peatlands

Nina Tanovitskaya

### 2.2.1 History of peatland investigation and drainage

Peat and peatlands are well documented in the works of Russian scientists in the 18th century and this focus probably goes back much further in time. The first drainage activities on the territory of contemporary Belarus were carried out in the 16th century. The first rural reform in 1557, which bequested land to peasants, enhanced drainage of peat soils. In the second half of the 18th century, Mateusz Butrimovich initiated the first large-scale drainage activities in the Paliessie region. In 1773 the first manors were established on drained land. 100 years later, in 1873, the geodesist and general-lieutenant Josif Ippolitovich Zhilinsky (1834–1916) elaborated a general plan for the drainage of Paliessie. In the same year he started the 'Western Paliessie expedition' that was to cover c. 9 million ha of land. After 25 years of work, 2.5 million ha had been drained and 4,657 km of channels, 549 bridges and 30 sluices built. 100,000 ha land were drained by the 'Northern expedition' in 1877–1899. At the same time, a study was made of which crops, grasses and vegetables could be cultivated on drained soils. Arable cultivation (of sugar beet) on drained peat soils was first mentioned for the territory of contemporary Belarus, i.e. Dubreyka village (today Mahilou region) in 1851. From 1883 onwards, sand cover cultivation on drained fen soils is reported. As a result of large-scale drainage mainly in the 1960s, another 1.5 million ha were drained for agriculture in Paliessie.

Utilization and study of peat as a fuel was reported for the territory of contemporary Belarus for the first time in 1793 when peat sods were dug by hand in Hrodna and Minsk regions (Radzevich 1991). Scientific investigations (mainly descriptions of peatland vegetation and distribution of peatland species) of Belarusian peatlands started in the middle of the 19th century (Bambalov et al. 1992). G.I. Tanfilev studied extensively the vegetation of peatlands and developed its first typology (Tanfilev 1895). Further important research has been carried out by Stanislaw Kulczynski who described physiognomic groups of Paliessie mires (Kulczynski 1949) and by Vladimir Semyenovich Dokturovski. The process of peat formation was thoroughly studied by V.V. Dokuchaev, who linked peat formation to conditions of relief, bedrock, climate, hydrological regime etc. (Dokuchaev 1949). After the October Revolution, the peat industry grew rapidly. Local sources of energy were extremely important in the Soviet Union, where fossil fuels were mainly located in the east and energy demands expanded in the west (including Belarus) and where the transfer of energy over large distances was not yet technically solved. An inventory of Belarusian peatlands started in 1922, and systematic research in 1928. In-depth search for and description of peatlands took place in 1935–1938 and again in 1948–1953. The results were summarised in the peatland cadastre of the Belarusian Socialist Soviet Republic (founded in 1940). After the Second World War, a period of intensive investigation and large-scale utilization of Belarusian peat resources for energy and agricultural use started. The handbook 'Peat resources of the Belarusian Socialist Soviet Republic' issued in 1953 comprises data from 5,945 peatlands with a total area of 1,467,550 ha peatland regarded as industrial peat stocks. The 1979 update of the handbook contains 7,055 peatlands with an area of 2,543,780 ha. Peat stocks have been investigated in detail on 35% of the overall peatland area, preliminarily on 18%, and surficially on 47% (Tanovitskaya & Bambalov 2009).

Until very recently, Belarusian peatlands were investigated in order to facilitate drainage for agriculture and peat extraction. The perception of peatlands has widened gradually since 1990: Among peatland scientists and governmental administration, peatlands are now not only regarded as objects of human utilization, but also as areas of nature conservation (see also box 9 in chapter 4.1). A milestone formed the criteria for the formation of 'peat funds' (see chapter 2.3) and the 'scheme for the wise use and protection of peat resources of Belarus until 2010' developed by I.G. Tanovitsky and colleagues (Academy of Sciences of Belarus) and endorsed by the government in 1991. The scheme comprises 9,192 peatlands with an overall area of 2,393,000 ha and justifies the enlargement of the nature conservation fund from 13% to 30% of the overall peatland area in 2010.

### 2.2.2 Methods of peatland drainage

Belarusian peatlands have been drained for agriculture (Fig. 6 Colour plates I) and forestry (Fig. 7 Colour plates I), for peat extraction (Fig. 8 Colour plates I), and for constructing roads and other infrastructure. Each land use type requires a particular method of peatland drainage. In selecting the most appropriate type of drainage, it is important to take the quality (especially the permeability) of the peat layer and the underlying sediment into account.

Drainage for agriculture must guarantee discharge of water in periods of excess precipitation and spring floods and retention of water in periods of draught, especially in the root layer. More intensive drainage is required for the industrial use of peat. In Belarus, peat extraction is mainly done by milling, i.e. by extracting one thin surface layer after another. To allow for milled peat extraction, drainage must be sufficient to provide peat soil carrying capacity for the milling equipment. Drainage for milled peat extraction must also eliminate the moistening of the uppermost peat layer by groundwater. According to Smelovskiy (1988), the water content of the uppermost (milled) peat layer decreases by 2% with every 0.3–0.4 m of additional drainage depth.

The most common drainage method for milled peat extraction in Belarus is discharging ground and surface water to the receiving watercourse using open main and auxiliary channels and gravity drainage or pumping stations. Main channels are built in parallel with up to 500 m distance between each other. The distance between auxiliary channels varies between 20 and 40 m, depending on the characteristics of the peat layer. The drainage network may have a central channel (in case the receiving water course is far away) or not. In the latter case, main channels discharge directly into the receiving watercourse. A peculiarity of such networks is that the main and auxiliary channels must be designed perpendicularly to allow milled peat extraction. This requirement has to be taken into account when designing the drainage network.

An intensification of drainage in milled peat extraction fields was pursued in Russia and Belarus by various means. Until the 1950s networks of drainage tubes in combination with a network of open ditches was common. Using drainage tubes the water level in extraction fields of peat sods was lowered by 0.1–0.2 m (in exceptional cases up to 0.35 m). In milled peat extraction, it increases the yield of an extraction cycle by 10%. Its main shortcoming is its lack of permanency (Smelovskiy 1988). In the 1960s research was carried out to intensify drainage of milling fields by a network of deep channels that go down to the mineral soil. Such deep channels allowed for better drainage, but were not widely adopted in the peat extraction industry. A combined and very efficient method for the deep drainage of bogs includes a network of open auxiliary channels with a 20 m interspace with additionally one or two 2.5 m deep drainage tubes per section parallel to the channels. In fen drainage, the construction of plastic drainage tubes in the mineral soil under the peat layer enhances the potential of agricultural use of the peatland after the end of peat extraction.

In 1959 the Peat Institute of the Belarusian Academy of Sciences carried out research to compare the efficiency of subsurface networks of drainage tubes with those of open auxiliary channels. Drainage tubes at 1.3–1.9 m depth on a distance of 15 m proved to result in a 10% higher seasonal yield of milled peat, in a more space-efficient way of peat extraction, and in three times lower costs compared to drainage with open channels. In 1962–1972 research was continued as large-scale trials on peat extraction fields of companies of the Ministry of Fuel Industry of the Belarusian Socialist Soviet Republic. Also these trials showed that subsurface drainage of peat fields on fens is more efficient and profitable than open auxiliary channels. Drainage networks of clay or polyethylene tubes, either in combination with open channels or as stand-alone drainage, were identified as the most promising way of drainage for milled peat extraction (Tanovitskaya & Smelovskiy 1972). Generally, subsurface drainage works more reliably than drainage with open auxiliary channels. Positioned at a certain depth, subsurface drainage maintains drainage of the entire peat layer down to this depth throughout its milling. In contrast, open channels are prone to collapse during milled peat extraction.

In peatland drainage for agricultural use, subsurface drainage networks have been widely used since the 1960s and preferred over open drainage networks since the 1980s. Subsurface drainage does not divide the soil surface into compartments and allows for quicker and more reliable regulation of the water level and of soil aeration and temperature (e.g. Skoropanov 1961). Trials with sparse channels in 300–400 m distance with a depth of 3 m proved highly efficient in peatlands with a peat layer of only 1.5–2.5 m and an underlying sediment of high permeability. Additional drainage with tubes may be needed at the beginning, but can be removed after some time (Smelovskiy 1988).

## 2.3 Use of peatlands and peat

Nina Tanovitskaya

### 2.3.1 Land use of drained peatlands

As a result of large-scale drainage, more than half of the peatland area of Belarus has been drained (Tanovitskaya & Bambalov 2009). Today the overall area of drained peatland in Belarus is 1,505,000 ha of which 1,085,200 ha (72%) are drained for agriculture, 383,000 ha (26%) for forestry, and 36,800 ha (2%) are currently used for industrial peat extraction.

Drained peatland in agricultural use includes 122,200 ha of cutover peatland with a c. 0.5 m thick layer of remaining peat. 96% of the agriculturally used drained peatlands have fen soils, whereas only 4% (situated in large fen complexes) have bog or transitional peatland soils.

Drained peatland in forestry use includes 103,000 ha of cutover peatland and 24,000 ha of (not effectively drained) soils with low forest productivity. Drained treed peatlands, in contrast to peatlands drained for agricultural use or for peat extraction, are regarded as 'suppressed peatlands' (Bambalov 2005) since the peatland flora and fauna has not been erased completely, but has partly survived in a 'suppressed' way and gradually revives by overgrowth of drainage channels and subsequent natural rewetting.

### 2.3.2 Land ownership and designation

All peatlands in Belarus are owned by the state and the government regulates their use and protection within the 'scheme for the wise use and protection of peat resources of Belarus until 2010' (developed in 1991; see also chapter 2.2). According to this scheme, all peatlands are allocated to 'peat funds' (Table 2; Tanovitskiy 1980, Tanovitskiy & Obukhovsky 1988):

- The nature protection fund (11% of all Belarusian peatlands) includes peatlands in nature conservation, scientific and recreational use;

- The fund of extracted peat deposits (9%) includes all peatlands where peat has been extracted for fuel or fertiliser. Until 1990 these peatlands were mainly used for agriculture. This proved to be inefficient. Their future use is addressed by two new normative documents that were elaborated by the Global Environment Facility (GEF) project (see chapter 2.4) and will facilitate rewetting. Currently only some 10% are rewetted but it is expected that this proportion will increase thanks to active and 'natural' (overgrowth of channels) rewetting;

- The reserve fund (1%) includes deposits of slightly decomposed *Sphagnum* peat that can be used for medicinal mudpacks or chemical/microbiological processing. Prior to extraction these peatlands have a nature protection function;

- The land fund (42%) includes all peatlands drained for agriculture and forestry, 34% of them have a peat layer of more than 1 m thickness. The proportion of strongly degraded soils is expected to increase with continuing agricultural use;

- The undetermined fund (18%) includes mainly undrained peatlands that have not yet been sufficiently investigated and currently have a nature protection function;

- Peatlands currently under peat extraction (1%).

The total peat fund currently includes 2,415,200 ha of peatland with 4.37 billion t of peat. Another 523,800 ha of small peatlands with shallow peat layers are not included in the peat fund since they are not regarded as industrially exploitable peat deposits (Bambalov 2005).

### 2.3.3 Peat extraction

As a result of industrial peat extraction (Fig. 9 Colour plates I and Fig. 10 Colour plates I), peat stocks in Belarus have decreased substantially. Whereas the first peat cadastre (1953) estimated the total size of the peat stock to be 5.7 billion t, including 1.5 billion t that can be extracted, this estimate changed to 4.9 billion t and 1.0 billion t, respectively, in 1979, and to 4.4 billion t and 320 million t, respectively, in 1988. Since the beginning of peat extraction, the peat stocks of Belarus shrunk by 1.4 billion t (Radzevich 1991). The natural accumulation of peat in the same period was c. 60–70 million t (Bambalov 2005). Today, the overall peat stock of Belarus is estimated to be 4.0 billion t, including peat in small and shallow

Table 2: Allocation of mires and peatlands in Belarus according to 'peat funds' in 1988 and 2010 (from Tanovitskaya & Ratnikova 2010b).

| Name of fund | 1988 (ha) | 2010 (ha) |
|---|---|---|
| Nature protection fund | 312,600 | 326,500 |
| Fund of extracted peat deposits, including: | 183,400 | 255,600 |
| in agricultural use | | 122,200 |
| in forestry use | | 103,000 |
| 'recultivated' (rewetted) | | 28,200 |
| Reserve fund | 31,100 | 30,800 |
| Land fund | | 1,468,200 |
| Agricultural land, including: | 963,100 | 1,085,200 |
| strongly degraded peat soils | | 250,520 |
| mineral soils after peat extraction | | 31,100 |
| drained peatland of the forestry fund | | 383,000 |
| of which ineffectively drained | – | 24,000 |
| Undetermined fund | 797,500 | 522,500 |
| Peatlands currently under peat extraction | 109,000 | 36,800 |
| Small/shallow peatlands not included in the 'peat fund' | 542,300 | 523,800 |
| Total | 2,939,000 | 2,939,000 |

## 2.4 Rewetting of peatland

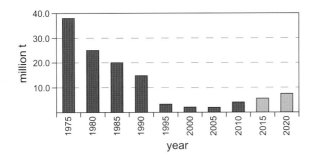

Fig. 11: Annual peat extraction volume in Belarus since 1975 and forecast until 2020.

peatlands that have an ecological but no industrial importance.

In 1959, the proportion of peat in overall fuel consumption of the country was 48%. In the same year, the production of peat briquettes started. Their production soon exceeded 2 million t per year. After 1961, peat extraction for fertiliser production grew rapidly. In 1975, peat extraction reached 39 million t (Fig. 11) but decreased thereafter. About one third of the annual yield was used to fuel power stations and to produce peat briquettes, and about two thirds were used in agriculture (Radzevich 1991). Already in 1980 major large fen areas were extracted and the remaining smaller peat deposits were not sufficiently large to keep up this level of peat extraction. In 1991, an analysis of the peat funds proved that there are no peatlands in Belarus that justified the construction of new peat factories (Bambalov et al. 1992, Radzevich 1991). In the early 1990s, Belarus gained access to other cheap energy resources that resulted in a reduction in peat extraction for fuel. However, the recent deficit in energy resources gave cause for a state programme for the modernisation of the national energy production sector (2007; including the promotion of domestic energy sources), a conception of national energy security (2007), and the programme 'peat' for the period until 2010 (2008). These governmental programmes demanded an increase in the proportion of domestic, renewable and alternative sources of energy in national energy production to 25% in 2010 and 30% in 2020. Therefore, the annual yield should be increased more than twofold until 2020. Since peat and wood are regarded as the safest sources of energy, the amount of peat used for energy production should increase to 4.4 million t by 2020 according to the conception of national energy. According to the programme 'peat', the amount of extracted peat shall increase until 2020 to 7.5 million t, including 5.1 million t for energy production and 2.4 million t for agricultural use (Fig. 11). Today 25 peat extraction companies are active on 40 Belarusian peatlands.

### 2.3.4 Agricultural use of peatlands

A great portion of the peatlands belonging either to the fund of extracted peat deposits or to the land fund are in agricultural use. The following agricultural land use types are common on peatlands (in order of magnitude):
- Arable land;
- Grassland for mowing;
- Meadow;
- Abandoned land.

The peatlands are mainly used as grasslands for fodder production for dairy farms. These farms function very differently and produce 2,000 to 8,000 L milk per cow per year. The average productivity is 3,500 L. Highly productive farms produce most of their milk from arable land; those with lower performance in milk production produce most of their milk from peatland grasslands (B. Schuster pers. comm., Wichtmann & Tanneberger 2009).

## 2.4 Rewetting of peatland

Alexander Kozulin

### 2.4.1 Key initiatives and projects

The first initiative for the rewetting of degraded peatlands goes back to N.N. Bambalov (Academy of Sciences Belarus) in the 1970s/80s. At that time, such ideas focused on the restoration of peatlands after peat extraction. But the idea was generally rejected by the authorities and until the end of the 20th century, all rewetting activities in Belarus were small-scale and usually implemented by peat extraction companies to prevent or, more often, to extinguish peat fires.

After the draught in 2002, when the majority of the large depleted peatlands burnt (Fig. 12 Colour plates II), the opinion of the authorities changed completely. As a first direct result, an expert group financed by the Royal Society for the Protection of Birds (RSPB), UK, and the United Nations Development Programme (UNDP) carried out a survey that showed that peat fires occurred mainly on drained and abandoned peatlands, which are extremely fire prone in dry years. Based on this analysis, the Ministry of Natural Resources and Environmental Protection approved the preparation of a United Nations Development Programme (UNDP) – Global Environment Facility (GEF) project on the restoration of peatlands. It took another four years to elaborate application details and agree them with various stakeholders, including ministries, before the project was approved as a four-year project in 2006. The project 'Renaturalization and Sustainable Management of Peatlands in Belarus to Combat Land Degradation, Ensure Conservation of Globally Valuable Biodiversity and Mitigate Climate Change' (www.peatlands.by) of UNDP and the Belarusian Ministry of Forestry was mainly funded by the Global Environment Facility (GEF) and financially supported by the RSPB.

In addition, a number of rewetting projects have been prepared and implemented by APB BirdLife Belarus with funding of the GEF Small Grants Programme, including activities on famous peatlands like Jelnia (see chapter 8.6) and Dakudauskaje. Also in key fen mires such as Sporava and Zvaniec, rewetting activities have been implemented during the last decade based on management plans funded by the UK Darwin Initiative and implemented by APB. The need for action was identified and hydrological measures were implemented with funding of the UK Darwin Initiative and other sources. The BMU-ICI project is now the second large-scale rewetting project in Belarus.

The Government of Belarus, specifically the Ministry of Finance, Ministry of Interior, Ministry of Forestry, and Ministry of Natural Resources and Environmental Protection supports peatland restoration projects. An area of 260,000 ha of degraded peatland, mainly peatland after peat extraction and degraded peatland ineffectively used for forestry, has a high potential for rewetting. Another 250,000 ha of degraded peatland ineffectively used for agriculture are feasible for rewetting in the future (N. Tanovitskaya pers. comm.).

### 2.4.2 The legal foundation for rewetting

Until 2006, there was only one normative document in Belarus that addressed peatland restoration. A directive on the recultivation of land that had been degraded as a result of peat extraction stated that nature protection and rewetting is one possibility for land use of depleted peatlands (see also chapter 5.1).

In the framework of the GEF project two Technical Codes of Established Practices according to Belarusian law were developed and have been implemented since January 1, 2009. 'Rules and procedures for definition and change of directions for use of depleted peat deposits and other damaged mires' and 'Order and rules for implementation of work on environmental rehabilitation of depleted peat deposits and other damaged mires and on prevention of damage of the hydrological regime of natural ecosystems during meliorative activities'. Additionally, practical guidance on peatland rewetting was developed with the 'methodological recommendations' published in 2010 (Kozulin et al. 2010a, b; see box 2). The developed normative documents are legally binding and form a sound basis for the implementation of further rewetting activities in Belarus.

### 2.4.3 Practical rewetting work

Within the GEF project, a total area of 28,208 ha of degraded peatland located at 15 project sites all over Belarus has been rewetted. Ten sites (21,115 ha) were rewetted using GEF funds and five sites (7,093 ha) were restored using funds provided within the BMU-ICI project. Among the 15 rewetted peatlands, there are seven fens with a total area of c. 13,280 ha.

When implementing peatland rewetting in Belarus, the following procedure has to be performed:
- Development of the 'scientific justification';
- Approval of the site selection process;
- Elaboration of an engineering plan;
- Formulation of an ecological (if necessary) and/or state expertise for the engineering plan;
- Implementation of the works.

The 'scientific justification' provides a detailed assessment of the current condition of the peatland and provides data on water regime and quality, peat properties, current biotope types, composition of flora and fauna, and land use characteristics. Based on this, it is determined which peatland type should be restored (target conditions) and which water levels and regime should be achieved through rewetting (rewetting strategy). The 'justification' also provides an assessment of the short and long term environmental and social impact of the measures and recommendations for the management of the rewetted site.

The engineering plans for the peatlands rewetted within the GEF project were prepared by two state organisations: 'Belgiprovodkhoz' and 'Belgiproles'. These organisations possess most

## 2.4 Rewetting of peatland

> **Box 2**
>
> **The new hydrotechnical recommendations for rewetting of drained peatlands in Belarus**
>
> Frank Edom & Franziska Tanneberger
>
> Like many other regions and countries, Belarus lacked, until recently, practical recommendations on methods, approaches, devices and technologies for constructing hydro-technical facilities to restore degraded peatlands and to prevent disturbances to the hydrological regime of natural mires. Thanks to the GEF project (see main text), this gap has been filled and recommendations have been agreed with all relevant authorities and published in Russian (400 copies) and English (200 copies) (Kozulin et al. 2010a, b). In Russian language this is currently the most detailed and comprehensive hydro-technical guidebook on rewetting.
>
> The recommendations are applicable to rewetting of depleted peat deposits, to improving the hydrological regime in damaged peatlands, and to designing and operating drainage systems in areas adjacent to natural mires. The guidebook contains a general description of the terminology, of the hydrology of natural and damaged peatlands as well as of technical approaches and methods currently applied in international practice, and in the GEF project to improve peatland hydrology. In addition, the guidebook summarises the key outcomes of the GEF project regarding vegetation succession and its prediction on rewetted peatland (see also chapter 3.5).
>
> Rewetting is largely a process of trial and error and site-adapted solutions. The Belarusian publication presents some new and special methods and technologies developed especially for large rewetting areas. As the guidebook provides, for each method, various pictures, a detailed construction scheme with work schedule, and a list of equipment and building materials required. It is especially useful for hydro-technical engineers both during planning and construction in the field. A key advantage of the publication is that it also describes the errors most frequently made during the building process. This reflects the large Belarusian experience gained from a continuous learning process (which is missing in a lot of western literature) and illustrates the very practical approach of the publication. The extensive practical part may have been prepared at the expense of details on peatland hydrogeology, hydromorphology, and processes in peatland development, but such background is missing in many rewetting guidebooks. For further peatland rewetting in Belarus, and also for exchanging the experiences with other countries, the recommendations are a real milestone. They should be widely used in Belarus and abroad both in practical rewetting projects and in the education of hydro- and landscape engineers.

relevant documents such as maps, levelling data etc., and experienced (Kozulin et al. 2010a). The building works were carried out by local building organisations, usually district companies for melioration systems and building companies.

At some sites (e.g. Marocna and Dakudauskaje) new methods to prevent damage to natural peatlands from adjacent peat extraction fields were tested. To prevent discharge from the part not used for peat extraction, an anti-leakage dam of c. 1 km length was built along the border between this area and the peat extraction field (Fig. 13 Colour plates I).

In most rewetting sites, the main objective was to raise the water level over all the rewetting site, therefore, water-regulating facilities had to be built in cascades following the macrorelief of the sites. To ensure effectiveness and reliability, dams were built with a maximum height difference of the water level on both sides of the dam of 0.3–0.4 m. To close channels with little water flow, mainly dams of solid earth (often peat; Fig. 14 Colour plates I) or combinations of earth and poles (Fig. 15 Colour plates I) were used. The main purpose of the dams is to fully stop the draining effect of the channels and to raise the water level to the peatland surface. The water in the channel rises at the dam to the height of the surrounding surface and bypasses the dam in a wide front along the lowest areas. Wooden dams last some 40 years and during this period, thanks to the lack of permanent water flow, the channels become overgrown by peatland plants such as *Sphagnum* mosses. To close channels with high water flow or straightened riverbeds, water-discharge structures were used (Fig. 16 Colour plates I and Fig. 17 Colour plates I). Such structures allow excess water to flow out and retain water during dry periods. In Belarus these structures usually consist of piling walls made of boards or rock-filled dams with piling walls. In other countries plastic piling walls are more widespread. In some cases, especially in main canals, standard water regulating facilities or their modifications are used to regulate the water level. Most widespread are crossing pipes with a plunging portal and a stoplog gate, regulator-pipes with a bucker gate, and crossing pipes with a metal portal with stoplog gate and a plastic pipe (Fig. 18 Colour plates I and Fig. 19 Colour plates II).

# 3 Peatlands and climate

Climate is the most important determinant of the distribution and character of peatlands. It determines the location and biodiversity of peatlands throughout the world. Past climate changes have led to expansion and contraction of peatland. The strong relationship between climate and peatland distribution suggests that future climate change (causing e.g. rising temperatures or changes in the amount, intensity, and seasonal distribution of rainfall) will exert a strong influence on peatlands.

Peatlands, in return, affect climate via a series of feedback mechanisms including the sequestration and release of carbon dioxide and methane, changes in albedo and alteration of the micro- and mesoclimate. Peatlands are some of the most important carbon stores in the world and influence the global balance of three main greenhouse gases (GHG) – carbon dioxide, methane, and nitrous oxide ($CO_2$, $CH_4$, and $N_2O$).

This section introduces basic information on the relationship of peatlands and GHGs (chapter 3.1) and presents key figures on the extent and distribution of global $CO_2$ emissions from peatlands (chapter 3.2). It further establishes how GHG fluxes in peatlands can be measured (chapter 3.3). In rewetting projects aiming at generating 'carbon credits', it is particularly important to assess GHG emissions from peatlands over large, diverse areas and throughout a project period. Chapter 3.4 describes why and how vegetation can be used as a proxy for assessing GHG emissions. Finally, chapter 3.5 offers guidance on how to predict vegetation development in peatlands in response to changes in water level after rewetting.

## 3.1 Peatlands and greenhouse gases

Jürgen Augustin, John Couwenberg & Merten Minke

### 3.1.1 Relevant gases, source-sink function, and climate impact

Numerous gaseous compounds are involved in the cycling of matter in peatlands:
- C-cycle: $CO_2$ (carbon dioxide), CO (carbon monoxide), $O_2$ (oxygen), $CH_4$ (methane), non-methane-hydrocarbons (NMHCs, like isoprene, terpene, alcohols, ethers and esters);
- N-cycle: $N_2O$ (nitrous oxide), $N_2$ (nitrogen), NO and $NO_2$ (nitric and nitrogen oxide), $NH_3$ (ammonia);
- S-cycle: inorganic sulphides like $H_2S$ (hydrogen sulphide) and COS (carbonyl sulphide), organic sulphides like $(CH_3)_2S$ (dimethyl sulphide) or $(CH_3)_2SO$ (DMSO or dimethyl sulphoxide), thiols like $CH_3SH$ (methanethiol), disulphides like $CS_2$ (carbon disulphide);
- P-cycle: $PH_3$ (phosphine)

In addition, molecular hydrogen ($H_2$) has been demonstrated to occur in the peat body (Shotyk 1989, Chanton & Whiting 1995, Reddy & Delaune 2008).

In principle, the exchange of gases between a peatland and the atmosphere can occur in both directions. When gas flux into the peatland dominates over a prolonged period of time, e.g. one year, the peatland is a sink. In the opposite case, the peatland is a source. Depending on whether the peatland or the atmosphere is chosen as reference, gas fluxes are expressed using a plus or a minus sign. By convention, climate science takes the atmosphere as reference and expresses flux 'as the atmosphere sees it'. Gas fluxes into the atmosphere are expressed as positive, gas fluxes into the peatland as negative fluxes (cf. chapter 3.3).

With the exception of nitrogen and oxygen, the two most important components of the Earth's atmosphere (78% and 21% by volume, respectively), all other gases involved in matter cycling in peatlands belong to the so-called trace gases. In ambient air the total fraction of trace gases rarely exceeds 0.04%. After entering the atmosphere, the majority of trace gases are immediately involved in chemical conversion (e.g. CO, NMHC, NO, COS, DMSO) or taken up by plants or absorbed by the soil (e.g. $NH_3$, NO) (Bliefert 1995). Only relatively inert trace gases with high transfer rates like $CO_2$, $CH_4$ and $N_2O$ are exchanged in notable volumes on a permanent basis between peat soils and the higher atmosphere (Fig. 20).

$CO_2$, $CH_4$ and $N_2O$ are among the most important climate relevant trace gases (greenhouse gases, GHG) that affect the radiative forcing of the atmosphere and contribute to the 'greenhouse effect' (Table 3). Since the middle of the 18th century, atmospheric concentrations of these trace gases have increased strongly. Until the present, atmospheric carbon dioxide ($CO_2$) concentration has increased from 280 to 388 ppm, atmospheric methane ($CH_4$) concentration from 715 to 1800 ppb and nitrous oxide ($N_2O$) from 270 to 323 ppb (IPCC 2007, NOAA 2010; cf. Table 3). This fast rise in atmospheric concentration caused absorption of heat radiated from the Earth surface and in all likelihood contributed significantly to the 0.6°C increase of mean global temperatures (IPCC 2007, Table 3). The increase in atmospheric concentration of GHGs is the main cause of anthropogenic climate change. The contribution of $CO_2$ to the anthropogenic greenhouse effect is 63%, that of $CH_4$ 18%, and that of $N_2O$ 6%. The climate effect of the latter two gas species is mainly because of their much higher global warming potential (GWP) compared to $CO_2$ (IPCC 2007; Table 3).

To assure comparability over longer periods of time, the national inventory reports on greenhouse gases to the United Nations Framework Convention on Climate Change (UNFCCC) use the GWP values published by IPCC in 1996 (Houghton et al. 1996). These values depart slightly from the more recent values listed in Table 3, which are used here to convert flux values of $CH_4$ and $N_2O$ to $CO_2$ equivalents in order to calculate the total GWP of a peatland site (see chapter 3.3 and 3.4). If the total GWP is negative, the peatland acts as a sink that has a 'cooling effect' on the climate. If the total GWP is positive, the peatland acts as a source that has a 'warming effect'. The GWP of trace gases changes with time. This change makes it essential that a fixed reference period is chosen to compare the climate effect of different gas species. As a rule, the choice of international conventions like the Kyoto

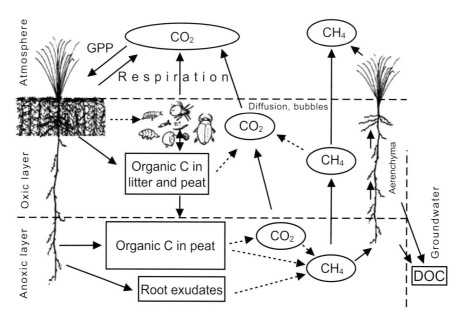

Fig: 20: Simplified carbon cycling between the atmosphere and a peatland with an oxic upper part and an anoxic layer beneath. Encircled symbols represent gases and dashed arrows show microbial processes. GPP is gross primary production; DOC is dissolved organic carbon leaching out from both the oxic and anoxic layers of the peatland via the groundwater (After Rydin & Jeglum 2006 and Parish et al. 2008).

# 3.1 Peatlands and greenhouse gases

Table 3: Characteristics of GHGs relevant to peatland ecosystems.

| Parameter | $CO_2$ | $CH_4$ | $N_2O$ |
|---|---|---|---|
| Current atmospheric concentration | 388 ppm (v) | 1.80 ppm (v) | 323 ppb (v) |
| Annual increase in concentration | 1.5 ppm (v) | 0.01 ppm (v) | 0.8 ppb (v) |
| Relative global warming potential* | 1 | 25 | 298 |
| Contribution to anthropogenic climate change | 63 % | 18 % | 6 % |

\* Global warming potential is a measure of radiative climate forcing integrated over time (here: 100 y) and put in relation to the effect of $CO_2$ (IPCC 2007, gas concentrations corrected after NOAA Annual Index 2010).

Protocol, is followed and a reference period of 100 years is used (Lashof 2000), but accepted GWP values are also available for time horizons of 20 and 500 years. The GWP of methane, for example, is 72 over a 20-year horizon, but only 7.6 over a 500-year horizon (IPCC 2007). The choice of time horizon thus strongly affects the resulting GWP, particularly in the case of peatland ecosystems (Whiting and Chanton 2001, Frolking et al. 2006).

## 3.1.2 Gas formation, gas exchange, and peat C budgets

Like in all other ecosystems, the carbon cycle of peatlands is driven by photosynthesis. During photosynthesis plants capture $CO_2$ from the atmosphere and fix it in reduced carbon compounds like carbohydrates. The process is referred to as gross primary production (GPP). The resulting products are distributed over the entire plant and sequestered in various places. Part of the fixed carbon reaches the soil through the plant's roots, so-called rhizodeposition. In parallel, a large fraction of fresh photosynthetic products are oxidised in respiration by above and below ground plant parts and the resulting $CO_2$ flows back into the atmosphere. This plant respiration is also referred to as autotrophic respiration. The difference between gross photosynthesis and plant respiration is called net primary production (NPP).

Particularly in the oxic part of the peat soil microorganisms through respiration continuously decompose the rhizodeposition, other plant remains and the peat to $CO_2$. This microbial activity is also referred to as heterotrophic respiration. In the upper, oxic peat layers relatively rapid rates of decomposition occur. In the lower, water saturated, anoxic peat layers the rate of decomposition is extremely slow. The sum of autotrophic and heterotrophic respiration is the so-called ecosystem respiration ($R_{eco}$). $R_{eco}$ represents the total amount of $CO_2$ that flows into the atmosphere out of the peatland ecosystem with its plants and soil, in opposite direction to the total $CO_2$ uptake represented by GPP (Fig. 21).

Gas fluxes from the peat to the atmosphere strongly differ between water saturated and drained soils. Drained peat has a large fraction of permanently air filled pores that allow for fast transfer of gases. Diffusion is the main mode of gas flux and responsible for more than 90% of total gas fluxes in drained peat soil. The rate of diffusion is determined by differences in concentration in the soil or between the soil and the atmosphere as well as by the gas species and geometry of the soil pores. A further mode of gas exchange in drained fen peatlands is by convective mass flow, caused by pressure differences between soil and atmosphere and temperature gradients in the soil. In contrast, in water saturated peat soils neither convection, nor diffusion play an important role in gas exchange, which is mainly because the diffusion rate in water saturated soil is 10,000 ($CO_2$) to 300,000 times ($O_2$) smaller than in dry soil (Step-

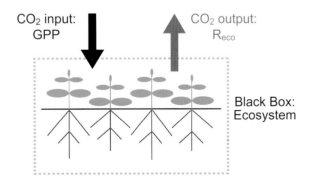

Fig. 21: The net balance of $CO_2$ fluxes in a peatland is expressed as gross $CO_2$ uptake (gross primary production, GPP) minus gross $CO_2$ emission (ecosystem respiration, $R_{eco}$).

niewski & Glinski 1988). Especially in absence of vegetation cover a substantial part of soil gases is transported by bubbles to the water surface (ebullition). Whereas their oxygen content is very low, these bubbles often contain very high concentrations of $CO_2$ (1–10%) and methane (10–90%) (Chanton & Whiting 1995).

In the anoxic part of the peat soil $CH_4$ is formed by a group of microorganisms called methanogens, which phylogenetically belong to *Archaea*. Methane is a highly reduced compound produced as the end product of anaerobic decomposition. Methanogenesis is preceded by rapid decomposition of plant litter, rhizodeposition and peat to low molecular compounds by other microorganisms. The strictly anaerobic *Archaea* can utilize only a limited set of substrates with $H_2/CO_2$ and acetate being the most important (Reddy & DeLaune 2008). Unlike $CO_2$ and $N_2O$, $CH_4$ is transported from the peat to the atmosphere through the interior of vascular marsh and water plants. The coarse aerenchymous conduit tissue of these plants provides a shortcut or shunt between the rhizosphere and the atmosphere; plant species that provide aerenchymous pathways for $CH_4$ efflux are also referred to as 'shunt-species'. The main pathway of plant mediated $CH_4$ efflux is active transport driven by overpressure in the plant tissue. In return the aerenchymous tissue allows for a fast transfer of oxygen to the roots and the surrounding anoxic peat (rhizosphere), essential for metabolism (Brix 1993, Chanton & Whiting 1995, Lai 2009). Systems may be open to the atmosphere in both directions (gas influx through young leaves, flux through the rhizome, efflux through old leaves; e.g. *Phragmites, Schoenoplectus, Typha, Alnus glutinosa*) or they are directed from the leaves to the submerged plant parts (e.g. *Sagittaria, Carex, Oryza sativa*). The presence of oxygen in the rhizosphere enables a group of microorganisms called methanotrophs to oxidise part of the methane produced before it can be transported to the atmosphere. The same process occurs in the oxic, upper peat layer as soon as methane from lower peat layers moves there. As a result of these oxidation processes only part of the methane produced in the peat reaches the atmosphere (Fig. 20, Megonigal et al. 2004, Reddy & DeLaune 2008, Lai 2009). If the peat becomes oxic over a larger depth as a result of drainage the dominance of methane oxidation can turn peatland sites into weak sinks of atmospheric methane (Augustin et al. 1996, Flessa et al. 1998, Byrne et al. 2004).

Conventionally, the production of $N_2O$ is linked to the microbial soil processes of nitrification and denitrification. During nitrification ammonia is oxidised to nitrate with $N_2O$ as a stable side-product. During denitrification nitrate is initially also turned into $N_2O$. The major part of the $N_2O$ formed is further reduced to $N_2$. Whereas nitrification is a strict aerobic process, denitrification is bound to anoxic conditions. Recent research has shown that processes involved in $N_2O$ evolution are much more complex. The number of organisms able to form $N_2O$ apparently is much larger than previously assumed, which also applies to the relevant processes. $N_2O$ is also formed in the course of so-called aerobic denitrification, of co-denitrification, of methanotroph-dependent denitrification, and not in the least in the course of dissimilatory nitrate reduction to ammonium. $N_2O$ is furthermore formed during the purely chemical process of chemodenitrification (Robertson & Groffmann 2007, Baggs 2008, Kool et al. 2011). In general, formation of $N_2O$ only occurs when inorganic N is made available as ammonium or nitrate through the mineralization of peat, through fertilizer application or through N deposition. In deficiency of N, undisturbed peatlands will even act as a sink of $N_2O$, as microorganisms are obviously able to use $N_2O$ to a limited extent for $N_2$ formation during denitrification (Roobroeck et al. 2010).

The processes involved in C gas exchange ($CO_2$, $CH_4$) decide whether peat accumulates or whether peat decomposition prevails. Carbon balances based on gas flux measurements provide a fast and precise estimate as to whether a peatland is a carbon sink or a carbon source. Different approaches exist to establish a carbon balance. The simplest budget is drawn up by the difference between the simultaneous, but opposed $CO_2$ fluxes of gross primary production (GPP) and ecosystem respiration ($R_{eco}$) (Fig. 21). As already explained, GPP and $R_{eco}$ are integrated fluxes in and out of the ecosystem and their difference is referred to as net ecosystem exchange (NEE) of $CO_2$ (Eq. 1) or net ecosystem production (NEP). NEE (or NEP) is the net $CO_2$ flux between ecosystem and the atmosphere; it is the balance of all $CO_2$ entering and all $CO_2$ leaving the ecosystem during a certain time period, typically one year.

$$NEE = GPP - R_{eco} \quad (1)$$

By convention NEE is opposite in sign to NEP. While NEE is defined from the viewpoint of the atmosphere, NEP is defined from the viewpoint of the ecosystem; a negative NEE (positive NEP) indicates a net carbon sink (Chapin et al. 2006, Smith et al. 2010).

More complex and therefore more precise ecosystem carbon budgets include next to $CO_2$ exchange also fluxes of other gases like $CH_4$ as well as of solid and liquid carbon compounds between the ecosystem and its environment. The net ecosystem carbon budget (NECB) refers to the total rate of organic carbon accumulation (or loss) from ecosystems (Chapin et al., 2006, Eq. 2). Analogous to NEP, NECB takes the ecosystem as reference point and a positive NECB indicates a net carbon sink or peat accumulation. Extrapolation of NECB to larger spatial scales has been termed 'net biome productivity' (NBP, Chapin 2006 et al.).

$$NECB = NEP + F_{CO} + F_{CH4} + F_{VOC} + F_{DIC} + F_{DOC} + F_{PC} \quad (2)$$

NEP: net ecosystem production or net $CO_2$ uptake from the atmosphere (or net $CO_2$ efflux to the atmosphere [negative sign] = NEE);

$F_{CO}$: net carbon monoxide (CO) absorption (or efflux [negative sign]);

$F_{CH4}$: net methane ($CH_4$) consumption (or efflux [negative sign]);

$F_{VOC}$: net volatile organic C (VOC) absorption (or efflux [negative sign]);

$F_{DIC}$: net dissolved inorganic C (DIC) input to the ecosystem (or net DIC leaching loss [negative sign]);

$F_{DOC}$: net dissolved organic C (DOC) input (or net DOC leaching loss [negative sign]);

$F_{PC}$: net lateral transfer of particulate (non-dissolved, non-gaseous) C into the ecosystem (or out of the ecosystem [negative sign]) by processes such as animal movement, soot emission during fires, water and wind deposition and erosion, and anthropogenic transport or harvest (Chapin et al. 2006).

### 3.1.3 Factors controlling gas formation and exchange

The most relevant factors controlling annual gas exchange rates and the climate impact of peatland sites are the mean water table and the vegetation type (Augustin et al. 1996, Jungkunst & Fiedler 2007, Dias et al. 2010, Couwenberg et al. 2011; see also chapter 3.4).

The lowering of water tables by drainage allows oxygen to penetrate the peat causing a rapid mineralization of the accumulated organic material. While oxygenation of the peat leads to cessation of methane efflux, it primarily causes a drastic increase in net $CO_2$ emissions and DOC losses (Gorham 1991, Kalbitz & Geyer 2002, Holden et al. 2004). Deeply drained, agriculturally used (fen) peatlands show high emissions over

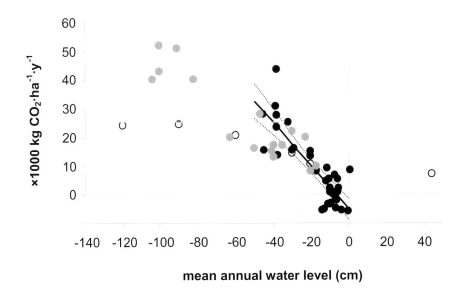

Fig. 22: Net annual $CO_2$ fluxes (×1,000 kg $CO_2$ ha$^{-1}$ year$^{-1}$) from peat soils in relation to the mean annual water level. Open dots refer to controlled lysimeter experiments and one flooded site; grey dots to long term subsidence measurements; the linear regression applies to the filled black dots, representing NEE measurements corrected for harvest export (from Couwenberg et al. 2011).

20 t $CO_2$ ha$^{-1}$ yr$^{-1}$. Only when mean water levels are above -50 cm there is a marked decrease in emissions reaching near zero (or net uptake) at mean water levels close to the surface (Fig. 22). In contrast to $CH_4$, the distribution of $CO_2$ flux values is not or hardly skewed. Most of the available $CO_2$ flux data from fen peatlands lack indication of pH. In case of bogs, emissions are highest at a pH value of ~4 (not shown). Higher carbon to nitrogen (C:N) ratios are associated with low emissions and vice versa, but a clear trend is lacking (not shown). Net $CO_2$ efflux rates are on average 10 times higher than the rate of $CO_2$ uptake in undrained sites (Augustin et al. 1996, Byrne et al. 2004, Drösler et al. 2008, Parish et al. 2008, Couwenberg et al. 2011). This strong negative climate effect of drained peatlands is frequently increased by the emission of the more effective greenhouse gas $N_2O$ (cf. Fig. 22, 24).

Annual methane fluxes from temperate European peatlands show a clear relationship with mean annual water level. Emissions are negligible (<2 kg ha$^{-1}$ yr$^{-1}$) at low mean water levels (<-20 cm), while values rise steeply with mean water levels above -20 cm (Fig. 23). Mean water levels above the surface are associated with lower $CH_4$ emissions than those at or just below the surface. The variation in measured fluxes is high, with very low values still occurring at higher water levels. Extremely high methane emissions (ranging from 1,300 to 3,800 kg ha$^{-1}$ yr$^{-1}$; not included in Fig. 23) were observed from ditches (Schrier-

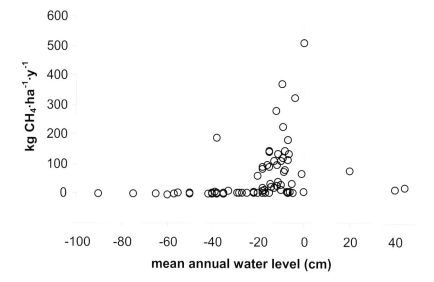

Fig. 23: Annual methane emissions (kg $CH_4$ ha$^{-1}$ year$^{-1}$) from peat soils in relation to the mean annual water level (from Couwenberg et al. 2011).

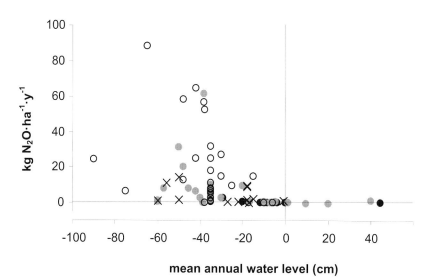

Fig. 24: Annual nitrous oxide fluxes (kg $N_2O$ ha$^{-1}$ year$^{-1}$) from peat soils in relation to the mean annual water level. Filled black dots denote bog sites, filled grey dots fen sites without fertilizer application, open dots fen sites with fertilizer application; crosses denote treed sites (from Couwenberg et al. 2011).

## 3.1 Peatlands and greenhouse gases

Uyl et al. 2008), from compacted gyttja soils with water levels near or above the surface (Hendriks et al. 2007) and particularly also from flooded harvests (*Avena, Zea, Phalaris*, Augustin & Chojnicki 2008; Hahn-Schöfl et al. 2011). Closer inspection reveals that sites with high water levels but low emissions lack plant species with coarse aerenchyma (shunt species). Such cases include bare peat surfaces as well as rewetted cultivated grasslands still dominated by short grasses. There is a clear relationship between methane fluxes and the density of aerenchymous shunts (Drösler 2005). Methane emissions from acidic bog peatlands show high values only at pH values above 4 (not shown), which is in agreement with findings of Tomassen et al. (2004). Methane emissions from fen peatlands seem to have their potential for highest values at pH values of 5.5 (cf. Best & Jacobs 1997). In general, there is no clear relationship between pH and water level (Couwenberg et al. 2008).

While $CH_4$ emissions are associated with mean water levels above -20 cm, $N_2O$ emissions are restricted to mean water levels below -15 cm (Fig. 24). In bog peatlands $N_2O$ emissions are low (<3 kg $N_2O$ ha$^{-1}$ yr$^{-1}$) even at the lowest observed mean water levels (-29 cm). A lack of variation in the data from bogs makes assignment of a sensible relationship between $N_2O$ fluxes and C:N or pH difficult (not shown). All $N_2O$ flux data stem from (semi-)natural bogs. Emissions from deeply drained, fertilized bog grasslands may be expected to turn out higher. There is no clear distinction between $N_2O$ fluxes from fertilized, agriculturally used fens and fluxes from unfertilised fens. Fens were classified as unfertilised or fertilised based on details provided in the literature sources. Differences in atmospheric nitrogen deposition were not taken into account. Fens with mean annual water levels below -15 cm showed a wide range of $N_2O$ emission values. Whereas rewetted peatland sites will have negligibly low $N_2O$ fluxes, the amount of $N_2O$ emitted from drained sites is unpredictable. Thus, estimation of the decrease in $N_2O$ fluxes after rewetting is uncertain.

In general, $CH_4$ production is higher in peatlands with higher vascular plant cover and higher water tables, like rich fens (Vitt 2006). Higher nutrient availability promotes the growth of vascular plants, primarily of sedges. Poor fens, with lower vascular plant cover, generally have lower potentials for $CH_4$ production. Like in *Sphagnum*-dominated bogs, $CH_4$ production via the $H_2/CO_2$ pathway is more important in poor fens. Input of labile carbon substrates necessary for methanogenesis into the anaerobic peat layer is reduced by the dominance of mosses (without roots) and mycorrhizal vascular plants (without deep carbon-rich roots), and a lower abundance of sedges with well developed deep roots. The eventual efflux of methane to the atmosphere depends on factors that control methane production, its transport and oxidation during transport. A key control on transport and oxidation is the density, root distribution and phenology of gas-conducting vegetation (shunt species) that can transport both oxygen from the atmosphere to the rhizosphere (enhancing $CH_4$ oxidation) and $CH_4$ from the rhizosphere to the atmosphere (enhancing $CH_4$ efflux) (Byrne et al. 2004, Lai 2009, Fritz et al. 2011).

Besides the water level, the type and intensity of land use play a major role in determining GHG fluxes from drained peatlands (Byrne et al. 2004, Drösler et al. 2008, Maljanen et al. 2010). Additional factors like peat type and trophic conditions seem less important (Parish et al. 2008, Couwenberg et al. 2008, Couwenberg et al. 2011). Activities to raise the water level in peatland ecosystems, expecting to re-establish their C-sink function on the short run, have started worldwide (Van Diggelen et al. 2006, Schumann & Joosten 2008). Elevating the mean annual water level indeed slows decay (Richert et al. 2000, Van der Peijl and Verhoeven 2000) and associated $CO_2$ efflux (Fig. 22). With respect to the C-sink function and total gas exchange and climate impact the detailed effects of rewetting are still unclear (but see chapter 3.4). Rewetting reduces $CO_2$ emissions, but often leads to a sharp increase in $CH_4$ release, as confirmed by field studies in fen (Chojnicki et al. 2007, Wilson et al. 2008, Höper et al. 2008) and bog ecosystems (Tuittila et al. 2000; Drösler 2005, Höper et al. 2008). It took several decades until disturbed bogs behaved as pristine bogs again in terms of carbon balance and climate effect (Yli-Petäys et. al. 2007, Samaritani et. al. 2010). Similar measurements are still lacking for fens, but the time required to re-establish typical vegetation may be decisive here as well. Typical fen vegetation dominated by reed or sedges unites the qualities to accomplish the desired C-sink function with reduced climate impact: high net primary production, supply of slowly biodegradable litter and promotion of methane oxidation (Couwenberg et al. 2008, Lai 2009, Hahn-Schöfl et al. 2011).

## 3.2 The global peatland $CO_2$ picture

Hans Joosten

### 3.2.1 Introduction

Conventional land use on peat soils commonly involves drainage, which results in substantial emissions of carbon dioxide and nitrous oxide to the atmosphere. This chapter presents a global overview of drainage related $CO_2$ emissions from peatlands. It identifies the emission hotspots where peatland rewetting and restoration activities are most urgently needed.

### 3.2.2 Methods

For global comparison, the International Mire Conservation Group (IMCG) Global Peatland Database was used (www.imcg.net/gpd/gpd.htm, cf. Joosten 2009d; Fig. 25 Colour plates II). This database adjusts the variety of regional data to uniform global standards, using clear and unambiguous definitions (Joosten & Clarke 2002; see also box 1 in chapter 2.1):
- Peat is sedentarily accumulated material consisting of at least 30% (dry mass) of dead organic material;
- A peatland is an area with a naturally accumulated peat layer at the surface. To provide a uniform standard, the data concern peatlands with a minimum peat depth of 30 cm.

Many different concepts and approaches exist on what constitutes a peatland between and even within countries and disciplines. Peatlands are often not recognized and feature in national inventories under other land categories. Ecosystems that might be peatlands, but are generally overlooked as such include mangroves, salt marshes, paddy rice fields, paludified forests, cloud forests and elfin woodlands, paramos, dambos, and cryosols.

Emissions (Table 4) were calculated using the default $CO_2$ emission factors of Couwenberg (2009a, 2011) or are based on interpolations and educated estimates (Joosten 2009d). Only emissions from microbial peat oxidation in drained peatlands have been included, emissions from fires and peat extraction are excluded. Following the United Nations Framework Convention on Climate Change (UNFCCC), only anthropogenic emissions were considered, $CO_2$ fluxes in pristine peatlands were not addressed.

### 3.2.3 Results

On a global scale some 500,000 km² of degraded peatlands emit 1.3 Gt of $CO_2$ through microbial peat oxidation. This figure does not include the emissions from peatland fires, nor the ex situ emissions from extracted peat. Including these sources, total emissions from peatlands are estimated at around 2 Gt.

The country with the largest emissions is Indonesia (500 Mt from drained peatland excl. extracted peat and fires) followed by the EU 27 (174 Mt), Russia (161 Mt), China (77 Mt), and USA (67 Mt). Big emitters that until now were less apparent are China and Mongolia (Table 5, Fig. 26 Colour plates II).

The global $CO_2$ emissions from drained peatland have strongly increased from 1,058 Mt in 1990 to 1,298 Mt in 2008. This 240 Mt increase is equivalent to >20% of the 1990 emissions from drained peatland.

The large majority of emission increases have taken place in developing countries. A more than 50% increase in emissions occurred in Papua New Guinea, Malaysia, Burundi, Indonesia, Kenya, Gabon, Togo, Trinidad and Tobago, Dominican Republic, Colombia, Rwanda, China and Brunei. Included in this list of countries with >50% increase in their peatland emissions are Indonesia, China, Malaysia and Papua New Guinea, who feature in the global peatland emission top 20 (Table 5a).

The developed countries (UNFCCC Annex 1) emit c. 0.5 Gt $CO_2$ from c. 250,000 km² of drained peatland (excl. extracted peat and fires). These emissions seem to have decreased from 655 Mt in 1990 to 492 Mt in 2008, i.e. a decrease of c. 25% compared to 1990. Part of these reductions only emerges because peatlands abandoned since 1990 have been assigned to 'un-managed'

Table 4: Default values used for $CO_2$ emissions from drained peat soils (in t $CO_2$ ha⁻¹ year⁻¹). Bold: Figures derived from Couwenberg (2009a, 2011), italics: interpolated in Joosten (2009d).

|  | Forest land/Agro-forestry | Cropland | Grassland | Extraction sites |
|---|---|---|---|---|
| Tropical | **40** | **40** | **40** | **30** |
| Subtropical | *30* | *35* | *30* | *25* |
| Temperate | *20* | *25* | *20* | *15* |
| Boreal | **7** | **25** | **10** | **10** |

## Colour plates, Part I

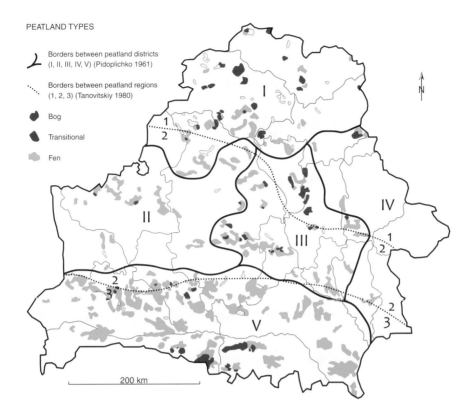

Fig. 1: Distribution of peatlands in Belarus (modified after Bambalov & Rakovich 2005; map: Stephan Busse).

Fig. 2: Jelnia – one of the country's most beautiful and least disturbed bogs in the northern peatland region (photo: Annett Thiele).

Fig. 4: Transitional mire in Zadzienauski Moch in the northern peatland region, with *Sphagnum* spec., *Vaccinium oxycoccos*, and Bogbean *Menyanthes trifoliata* (photo: Annett Thiele).

Fig. 5: Zvaniec mire is one of the largest and best preserved fens in the southern peatland region of Belarus (photo: Franziska Tanneberger).

Fig. 6: Drained peatland in agricultural use after peat extraction in Hrycyna-Starobinskaje peatland (photo: Nina Tanovitskaya).

Fig. 7: Drained peatland in forestry use in Sviatoje peatland (photo: Nina Tanovitskaya).

Fig. 8: Industrial peat extraction in Hrycyna-Starobinskaje peatland (photo: Nina Tanovitskaya).

Fig. 9: Industrial peat extraction in Hrycyna-Starobinskaje peatland (photo: Nina Tanovitskaya).

Fig. 10: Machinery for industrial peat extraction in Hrycyna-Starobinskaje peatland (photo: Nina Tanovitskaya).

Fig. 12: A peat fire in Dakudauskaje peatland (photo: Nina Tanovitskaya).

Fig. 13: An anti-leakage dam at the border of a peat extraction field in Marocna peatland (photo: Alexander Kozulin).

Fig. 14: Earth dam compacted with a bulldozer in Hrycyna-Starobinskaje peatland (photo: Alexander Kozulin).

Fig. 15: Solid earth dam with walls made of poles in the main channel of Haloje bog (photo: Alexander Kozulin).

Fig. 16: Piling dam made of wooden boards in a channel with little water flow in Barcianicha peatland (photo: Alexander Kozulin).

Fig. 17: Rock-filled dam with piling wall made of concrete slabs with crossing for vehicles at Zvaniec peatland (photo: Alexander Kozulin).

Fig. 18: The main sluice at Hrycyna-Starobinskaje peatland after installation of dam beams (blocks, 100x100 mm; photo: Alexander Kozulin).

## 3.2 The global peatland $CO_2$ picture

**Box 3**

**Where is Belarus?**

Hans Joosten

Belarus is one of the more important peatland countries in the world (see also chapter 2.1). With a total peatland area of 22,352 km² Belarus ranks 15th among all countries of the world, with respect to peatland proportion (% of the country) it ranks 20th and with respect to actual carbon stock 21st. Much higher is the score of Belarus with respect to peatland emissions; with 41 Mt $CO_2$ year$^{-1}$ Belarus is the 8th most important country in the world (see Table 5a, Fig. 26 Colour plates II). In terms of total emissions per unit land area, Belarus ends up third after Indonesia and Estonia with 1.99 t ha$^{-1}$ (Table 5b, Fig. 27 Colour plates II).

Table 5a: The countries/areas with the largest total emissions from degrading peatland in 2008. Emissions from fire and extracted peat are excluded.

| Country/area | Emissions from degrading peatland in 2008 (Mt $CO_2$ year$^{-1}$) |
|---|---|
| Indonesia | 500 |
| European Union | 174 |
| Russia – European part | 139 |
| China | 77 |
| USA (lower 48) | 67 |
| Finland | 50 |
| Malaysia | 48 |
| Mongolia | 45 |
| Belarus | 41 |
| Germany | 32 |
| Poland | 24 |
| Russia – Asian part | 22 |
| Uganda | 20 |
| Papua New Guinea | 20 |
| Iceland | 18 |
| Sweden | 15 |
| Brazil | 12 |
| United Kingdom | 10 |
| Estonia | 10 |
| Ireland | 8 |
| Lithuania | 6 |
| Netherlands | 6 |
| Norway | 6 |
| Vietnam | 5 |
| Ukraine | 5 |
| Zambia | 5 |
| Japan | 5 |
| Canada | 5 |
| Latvia | 4 |

Table 5b: Countries (with a land area > 1,000 km² and peatland $CO_2$ emissions >1 Mt) with the largest emissions from degrading peatland per unit national land area in 2008. Emissions from fire and extracted peat are excluded.

| Country/area | Peatland $CO_2$ emissions in 2008 (t ha$^{-1}$) |
|---|---|
| Indonesia | 2.63 |
| Estonia | 2.11 |
| Belarus | 1.99 |
| Iceland | 1.70 |
| Finland | 1.48 |
| Malaysia | 1.46 |
| Netherlands | 1.39 |
| Ireland | 1.17 |
| Brunei | 1.09 |
| Lithuania | 0.93 |
| Germany | 0.90 |
| Uganda | 0.83 |
| Poland | 0.75 |
| Latvia | 0.66 |
| Denmark | 0.64 |
| Papua New Guinea | 0.43 |
| Russia – European part | 0.40 |
| United Kingdom | 0.39 |
| Sweden | 0.32 |
| Mongolia | 0.29 |

land and emissions are no longer accounted for (Barthelmes et al. 2009). Without rewetting, drainage related emissions will remain as high as when these lands were still used as cropland or grazing land, however.

### 3.2.4 Conclusions

Total $CO_2$ emissions from the worldwide 500,000 $km^2$ of degraded peatland amount to 2 Gt. Even when taking into account that only part of this area is available for rewetting and that a part of the $CO_2$ emissions reduction may be annihilated by increased $CH_4$ emissions, peatland rewetting may globally reduce greenhouse gas emissions with several hundred Mt $CO_2$-eq. year$^{-1}$. The inventory highlights that the perspectives of better peatland management for climate change mitigation are global and not limited to a few selected countries. Large reduction opportunities exist for both Annex 1 (developed) and non-Annex 1 (developing) countries. The countries that require priority action (emissions hotspots) are presented in Table 5a (Fig. 26 Colour plates II; with respect to total peatland $CO_2$ emissions) and Table 5b (Fig. 27 Colour plates II; with respect to peatland $CO_2$ emission density).

## 3.3 Measuring GHG emissions from peatlands

Merten Minke, Hanna Chuvashova, Andrei Burlo, Tatsiana Yarmashuk & Jürgen Augustin

### 3.3.1 Introduction

Measuring the full GHG balance of peatlands requires techniques that allow for accurate and continuous monitoring of the net exchange of all relevant trace gases (carbon dioxide $CO_2$, methane $CH_4$, and nitrous oxide $N_2O$) over prolonged periods of time. To assess daily, seasonal and interannual variability observations over multiple years are necessary. Covering the net exchange of $CO_2$ is challenging, considering the net $CO_2$ exchange between ecosystem and atmosphere is the result of two opposite fluxes: $CO_2$ uptake by plants during photosynthesis and $CO_2$ release by respiration of plants, microbes, and animals (Chapin III et al. 2002, cf. chapter 3.1). Measurement techniques need to work under different weather conditions to provide a high temporal and spatial resolution and because of the remoteness of most peatlands, they need to be largely independent from power sources (Drösler et al. 2008). The most appropriate and widely used techniques are closed chambers and the ground based eddy covariance method (Byrne et al. 2004).

In the closed chamber approach gas exchange is calculated as a function of concentration changes over time in an enclosed air volume. Closed chambers cover a small area of soil (typically less than 1 $m^2$) and allow for assessing the spatial variability of GHG sources or for the study of different site types in parallel. The temporal resolution of closed chamber measurements is limited, however, particularly in case of manually operated chambers. The eddy covariance method is a direct micrometeorological technique that quantifies trace gas exchange rates by measuring fluxes within the lower atmospheric boundary layer; air is sampled as it flows past a sampling point determining its vertical wind speed and direction and its gas concentration (Lenschow 1995, Moncrieff et al. 1997). The eddy covariance method allows for continuous measurements but provides only a low spatial resolution as the different small scale sources are integrated over a larger area.

### 3.3.2 The closed chamber technique

Closed chambers are placed airtight on the soil and restrict the volume of air available for gas exchange (Fig. 28 Colour plates II). Across the covered surface any net emission or uptake of the enclosed gases (the headspace) can be measured as a concentration change (Livingston & Hutchinson 1995). Fluxes of $N_2O$ and $CH_4$ are usually measured with opaque, static closed chambers (Rochette & Eriksen-Hamel 2008). In static closed chambers a series of air samples from the headspace is taken with syringes or evacuated gas-tight flasks and their gas concentration is measured later in the laboratory using a gas chromatograph. Gas fluxes are derived from concentration changes in this series of samples. Sample series are taken during short enclosure times (typically 15 to 60 minutes) and repeated every two to three weeks, sometimes more often. In between measurements the chambers are removed from their permanent collars which are placed airtight on the soil.

Natural gas dispersion in response to temperature gradients or pressure fluctuations within the headspace of the chamber is mostly sufficient to ensure uniform headspace mixing (Livingston et al. 2006). No significant difference between measured fluxes was found when a fan was operated or not, both for $N_2O$ (Matthias et al. 1980) and $CH_4$ (M. Drösler pers. comm.). A test of static chambers against a reference $CH_4$ flux (Pihlatie et al. in prep.) produced mixed results: static closed chambers with continuously mixed headspace produced significantly higher fluxes compared to chambers without mixing on dry sand, but not on wet sand.

Net ecosystem $CO_2$ exchange (NEE, chapter 3.1) is always, at least during the growing season, measured with dynamic closed chambers, a system where air is circulated in a closed loop between the chamber headspace and an infrared gas analyser used to measure the $CO_2$ concentration in situ (cf. Whiting et al. 1992, Alm et al. 1997, Drösler 2005). Short enclosure times (1–3 min) and high frequency air sampling (every 1–5 s) are necessary to reproduce the $CO_2$ exchange rates accurately. Especially during sunny summer days the $CO_2$ concentration in a transparent chamber can decrease by over 100 ppm within one minute. Transparent chambers are necessary to allow photosynthetic uptake of $CO_2$ by the enclosed vegetation. Continuous headspace mixing by fans is important in closed dynamic $CO_2$ chambers, but care must be taken that the fan does not induce mass flow from the soil (Pumpanen et al. 2004).

Closed dynamic chambers are sometimes also used to measure $CH_4$ fluxes, for example in combination with the photo acoustic infrared gas spectrometer Innova 1412 (Sachs et al. 2010) or for $CH_4$ and $N_2O$ fluxes with Quantum cascade laser spectroscopy (Guimbaud et al. 2011) but these sensors require high technical and financial efforts.

In general, linear regression functions are applied to calculate gas exchange rates from the observed changes in concentration. However, chambers can slow down the gas exchange, for example when saturation of the headspace impedes further gas efflux or when decreasing $CO_2$ concentrations in transparent chambers reduce the $CO_2$ uptake for photosysnthesis, making non-linear regression functions more appropriate (cf. Kutzbach et al. 2007). Non-linear regression functions introduce additional uncertainties, however, and the 'damping' effect of the chamber can alternatively be reduced by a shorter enclosure time and larger headspace volume (c.f. Matthias et al. 1978, Livingston & Hutchinson 1995, Pedersen et al. 2010). The 'damping' effect is particularly pronounced in transparent chambers used for $CO_2$ flux measurements and it is recommended to check concentration changes carefully and use only the initial, linear change in gas concentration to calculate fluxes (Drösler 2005).

Disturbance of the microclimate and light conditions of the measurement site by the chambers needs to be minimized, especially when plants are involved in the trace gas exchange processes (Livingston & Hutchinson 1995). Short enclosure times and cooling systems effectively reduce heating of transparent chambers. Whiting et al. (1992) and Alm et al. (1997) use chilled water and a heat exchanger. The cooling system used by Drösler (2005) consists of frozen cooling packs fixed in a frame and placed within the chamber. The cooling packs do not only cool the air, but also cause water vapour to condense at their surface, thus reducing water vapour increase in the chamber as well as condensation at the chamber walls (which causes diffusion of light). White opaque chambers do not need a cooling system. Physical perturbation of the site, for example damage of vegetation or compaction of the soil must be avoided as it can alter the measured fluxes. Boardwalks are especially important in wet peatlands where treading may cause ebullition of gases.

Closed chambers are ideal for small scale studies, can be applied in nearly any terrain, offer low cost and portability, and allow for studying different site types in parallel. On the downside, closed chambers do not provide continuous time records. To arrive at total annual flux estimates, models are applied to extrapolate the punctuated measurements. Gross primary production (GPP, equalling NEE + $R_{eco}$, chapter 3.1) and total ecosystem respiration ($R_{eco}$) are modelled using as input continuous records of site parameters like PAR (photosynthetic active radiation), soil and air temperature, vascular green area, and water level. Net annual $CO_2$ exchange rates are then calculated from the difference between GPP and $R_{eco}$ (Alm et al. 1997, Whiting et al. 1992, Drösler 2005, Wilson et al. 2007). Annual $CH_4$ emission can be modelled using soil temperature and/or water level as input parameters (Saarnio et al. 1997, Kettunen et al. 2000). If correlations with these site parameters are low, annual $CH_4$ emissions are calculated by linear interpolation between subsequent measurements (Drösler 2005, Hendriks et al. 2007, Hyvönen et al. 2009). Robust models to derive annual $N_2O$ fluxes from punctuated closed chamber measurements are not yet available for peatland sites. $N_2O$ fluxes can be very erratic in time and space (chapter 3.1) and when linear interpolation is applied to arrive at annual flux estimates, measurement schemes should take into account large flux events associated, for example, with fertilizer application and freeze-thaw cycles (Smith & Dobbie 2001).

**Box 4**

**How to calculate the global warming potential (GWP) of a peatland site?**

Merten Minke

This box explains how the GWP is calculated e.g. within the BMU-ICI project for various sites in Belarus (see also box 5).

To measure annual exchange rates of $CO_2$, $CH_4$, and $N_2O$ manually operated static and dynamic closed chambers are used (Fig. 28 Colour plates II). At every selected site type $CH_4$ and $N_2O$ fluxes are measured with opaque static chambers once every second week during the snow free period and monthly during winter. Four air samples are taken from the chamber headspace during a 15–30 min enclosure and subsequently analysed in the laboratory with a Chromatograph ('Khromatek – Kristall 5000 version 2 (5)'), using an electron capture detector (ECD) for analysing $N_2O$ and a flame ionization detector (FID) for $CH_4$ (Fig. 28 Colour plates II). $CO_2$ exchange is measured every third week with transparent and opaque dynamic closed chambers designed after Drösler (2005) (Fig. 28 Colour plates II). About ten transparent chamber measurements of 1–3 min per soil collar and measuring day are carried out from before sunrise until late evening, equally distributed over the daily range of PAR (photosynthetic active radiation) to determine the light response curves of GPP. Similarly, about ten opaque measurements per soil collar provide the data to calculate the temperature response of $R_{eco}$ for the measuring day.

To calculate the exchange rates of $CO_2$, $CH_4$ and $N_2O$ from the linear concentration changes in the chamber headspace during the enclosure time, the following equation is used:

$$F = \frac{1}{R} \times \frac{P \times V}{T \times A} \times \frac{dc}{dt} \qquad (3)$$

F = flux rate of $CO_2$, $CH_4$, or $N_2O$ [µmol m$^{-2}$s$^{-1}$]
P = atmospheric pressure in the chamber [Pa]
V = chamber volume [m³]
R = gas constant [8.3143 m³ Pa K$^{-1}$ mol$^{-1}$]
T = air temperature in the chamber [K]
A = surface area of the chamber [m²]
dc = concentration change within the chamber atmosphere [ppm]
dt = time over which concentration change was measured [s]

To convert µmol into µg, F is multiplied with the molar mass M of the molecules ($CO_2$, $CH_4$ and $N_2O$) or elements (C and N).

Annual fluxes of $CH_4$ and $N_2O$ are calculated by simple linear interpolation between measurements. For sites where $CH_4$ emissions are significantly correlated with temperature or water level, annual fluxes are calculated based on these relationships. In order to calculate annual NEE values – with NEE = GPP + $R_{eco}$ – the approach of Drösler (2005) is used, in which the light response of GPP and temperature response of $R_{eco}$ is determined for every measuring day. These relationships are then used to model GPP and $R_{eco}$ stepwise from one measuring day to the next. For the modelling only PAR and temperature are required while changes of water table and plant development are addressed by the repeated calibration of the model at least every third week. Other authors model GPP and $R_{eco}$ for the entire season or year, using next to PAR and temperature water table and/or plant cover or vascular green area as model inputs to address environmental changes within the season or year (cf. Alm et al. 1997, Wilson et al. 2007). The approach of Drösler (2005) has proven to give a very strong fit between modelled and measured NEE values. The first step is to calculate the parameters for the $R_{eco}$ model, based on the Lloyd & Taylor (1994) equation:

$$R_{eco} = R_{ref} * e^{E_0 * (1/(T_{ref} - T_0) - 1/(T - T_0))} \qquad (4)$$

$R_{eco}$ = respiration [$CO_2$-C mg m$^{-2}$ h$^{-1}$]
$R_{ref}$ = respiration at the reference temperature [$CO_2$-C mg m$^{-2}$ h$^{-1}$]
$E_0$ = activation energy [K]
$T_{ref}$ = reference temperature: 283.15 [K]
$T_0$ = temperature constant for the start of biological processes: 227.13 [K]
T = soil or air temperature during measurement of best fit with the dataset [K]

$R_{ref}$ and $E_0$ are calculated by fitting Equation 4 into a regression of measured $R_{eco}$ and temperature (Fig. 30). After interpolating both parameters between two measuring campaigns $R_{eco}$ is modelled on a half hour resolution from one campaign to the next, based on Equation 4, the calculated parameters and the high resolution temperature data from the meteorological station near the measuring site.

The next step is to calculate GPP values by subtracting the modelled $R_{eco}$ from the meas-

## 3.3 Measuring GHG emissions from peatlands

ured NEE values, and to explore the regression between GPP and PAR by fitting a rectangular hyperbola equation (Michaelis & Menten 1913):

$$GPP = \frac{\alpha \cdot PAR \cdot GP_{max}}{\alpha \cdot PAR + GP_{max}} \quad (5)$$

GPP = gross primary production [mg $CO_2$-C $m^{-2}$ $h^{-1}$]
$\alpha$ = initial slope of the curve; light use efficiency [mg $CO_2$-C $m^{-2}$ $h^{-1}$/µmol $m^{-2}$ $s^{-1}$])
PAR = photon flux density of the photosynthetic active radiation [µmol $m^{-2}$ $s^{-1}$]
$GP_{max}$ = maximum rate of carbon fixation at PAR infinite [mg $CO_2$-C $m^{-2}$ $h^{-1}$]

GPP is modelled based on Equation 5, the calculated parameters and the PAR measured at the meteorological station (Fig. 31). Annual NEE is then calculated as the difference between the modelled GPP and the modelled $R_{eco}$ (cf. Fig. 32).

The annual exchange rates of $CH_4$ and $N_2O$ are converted into $CO_2$ equivalents on the basis of their radiative forcing over a 100 year horizon (IPCC 2007; Table 6). The global warming potential (GWP) of the site is then calculated as:

$$GWP = NEE + F_{CH4} + F_{N2O} \quad (6)$$

with all variables expressed in [kg $CO_2$-eq $m^{-2}$ $year^{-1}$]. Exported harvest is usually expressed as instantaneous $CO_2$ emission and deducted from NEE (see chapter 3.1). GWP is expressed from the viewpoint of the atmosphere with positive values indicating emissions and negative values indicating sequestration by the ecosystem.

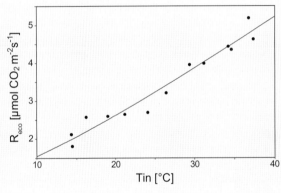

Fig. 30: Relationship of Ecosystem Respiration ($R_{eco}$) to air temperature measured 03.08.2010 at site B6-I (cf. Table 7).

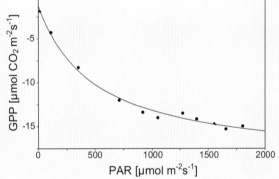

Fig. 31: Relationship of Gross Primary Production (GPP) to photosynthetic active radiation measured 03.08.2010 at site B6-I (cf. Table 7).

Table 6: Calculation of the global warming potential of site B6 (wet peat with *Phragmites australis*, Table 7) at the rewetted site Barcianicha for the period from 03.08.-20.09.2010. The roman numbers indicate the three soil collars each providing one repetition. Negative values denote uptake, positive values emissions. The negative GWP means the site has a net 'cooling effect' (cf. chapter 3.1).

| | flux [t $ha^{-1}$ $(49d)^{-1}$] | | | Conversion factor (IPCC 2007) | GWP flux [t $ha^{-1}$ $(49d)^{-1}$] | | | | |
|---|---|---|---|---|---|---|---|---|---|
| | B6-I | B6-II | B6-III | | B6-I | B6-II | B6-III | mean | SD |
| $CO_2$ | -3.621 | -8.628 | -2.597 | 1 | -3.621 | -8.628 | -2.597 | -4.948 | 3.227 |
| $CH_4$ | 0.109 | 0.129 | 0.143 | 25 | 2.717 | 3.234 | 3.573 | 3.175 | 0.431 |
| $N_2O$ | 0.00125 | 0.00157 | 0.00315 | 298 | 0.373 | 0.467 | 0.938 | 0.593 | 0.303 |
| | | | | total | -0.531 | -4.927 | 1.914 | **-1.181** | **3.456** |

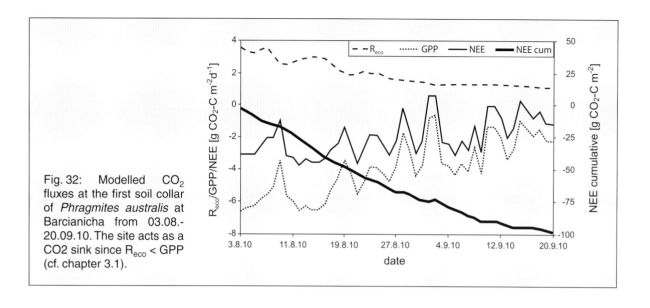

Fig. 32: Modelled $CO_2$ fluxes at the first soil collar of *Phragmites australis* at Barcianicha from 03.08.-20.09.10. The site acts as a CO2 sink since $R_{eco} <$ GPP (cf. chapter 3.1).

The closed chamber method is well suited for GHG flux measurements at most peatland sites. Closed chambers may be less suited to measure fluxes from open water surfaces, where ebullition (bubble transport) presents a significant pathway of methane emissions. Ebullition of methane from the sediment through the water column to the atmosphere is fast and nearly without oxidative losses. Methane ebullition is difficult to quantify using closed chambers because bubble fluxes are episodic and easily missed. When bubbles are 'caught' by the chamber, methane concentration increases abruptly, complicating the emission calculation (Chanton & Whiting 1995). Submerged bubble traps which are sampled periodically for volume and gas concentration of the trapped gases are recommended to account for ebullition (Baker-Blocker et al. 1977, Chanton & Whiting 1995). Methane efflux from vegetated wet peatland sites is dominated by transport through emergent aquatic plants, another 'shortcut' for methane (chapter 3.1).

### 3.3.3 The eddy covariance technique

The eddy covariance technique measures the turbulent air flux, composed of eddies of different sizes, in the lower atmospheric boundary layer (ABL). The ABL is closely coupled to the earth's surface by turbulent exchange processes and extends a few tens of metres over land at night to several kilometres on clear sunny days (Stull 1988, Lenschow 1995). Modern eddy covariance systems consist of a sonic anemometer for 3-D wind observation, mounted on a tower of suitable height (Fig. 29 Colour plates II), open-path or closed-path gas analysers to sample air from within a few decimetres of the sonic anemometer and a suite of meteorological sensors for continuous measurements of temperature, radiation components, atmospheric pressure, relative humidity and other variables of interest (Sachs 2009). Air velocity, temperature and gas concentration are measured with high frequency measurements (10 Hz or higher) over a period of time sufficient to capture all transporting eddy sizes. The upper limit of the sampling duration is defined by the requirement that the turbulent field needs to be stationary during the averaging period. Atmospheric turbulence is usually sampled and averaged over 15–60 min during daylight, but longer averaging times may be needed during the night when the air near the surface is stably stratified and turbulence is intermittent. The vertical GHG flux is calculated as the covariance between vertical air velocity and trace gas concentration (Stull 1988, Lenschow 1995, Baldocchi 2003).

While the modern eddy covariance studies of $CO_2$ exchange commenced already in the mid 1980s when sonic anemometers and sufficiently fast responding infrared gas analysers became available (Baldocchi 2003), implementation of eddy covariance measurements of methane and nitrous oxide fluxes became successful only in the late 1990s using closed-path systems with tuneable diode laser (TDL) spectrometers. Recently, quantum cascade laser (QCL) spectrometers have been tested successfully for the same purpose (Kroon et al. 2010). QSL has a higher stability of the laser frequency compared to TDL and no need for cryogenic cooling. These laser techniques are

## 3.3 Measuring GHG emissions from peatlands

expensive and complicated, however, and eddy covariance studies of methane and nitrous fluxes are much less common than those of carbon dioxide exchange. With their continued rapid development, reliable, cost and energy efficient eddy covariance techniques for methane and nitrous oxide may become available in the next years.

The major advantages of the eddy covariance technique are the minimal impact on the site conditions and the ability to provide continuous data over long periods. However, data gaps frequently occur, for example as a result of low turbulent conditions or technical problems and require appropriate gap filling procedures. According to Moffat et al. (2007) the typical coverage of the eddy covariance method is 80–90% during day and 35% during night time. Compared to closed chambers, eddy covariance measurements cover a much larger area. The measurement footprint is about 100–300 times larger than the sensor

---

**Box 5**

**GHG measurements in Belarus**

Merten Minke

Monitoring GHG emissions from peatlands by direct measurements with chambers and/or eddy covariance is very costly. To assess the greenhouse gas emission reduction from rewetting the Greenhouse Gas Emission Site type (GEST) approach has been developed, which uses vegetation as a proxy to estimate annual GHG fluxes (chapter 3.4). Chamber measurements are carried out only at selected site types to test, expand, and calibrate the GESTs for peatlands in Belarus. Measurements are done with closed chambers because these allow for measuring distinct vegetation types. Chamber design follows Drösler (2005). All studied site types are equipped with three soil collars allowing for three replicate measurements. Water level, soil and air temperatures, photosynthetic active radiation, atmospheric pressure, and precipitation are recorded every half hour at each site with an automated meteorological station and water level data loggers, but also manually at all site types during every chamber measurement. Soil and air temperatures, photosynthetic active radiation, and atmospheric pressure are necessary for the flux calculations. Gas flux measurements are being carried out for two consecutive years to allow assessment of inter annual variability of the GHG fluxes.

Before site selection started a list of the most important vegetation types of drained and rewetted fens and bogs was compiled taking into account the gaps in the GEST approach and the need for calibration of the existing set (chapter 3.4). Three measuring sites were selected from this list which are easily

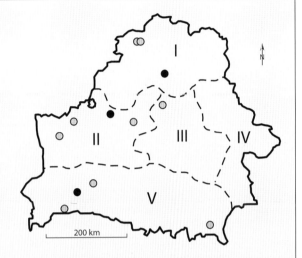

○ Sites for vegetation calibration
● Sites for GHG measurements and vegetation calibration

Fig. 33: Location of sites for GHG measurements and vegetation calibration studied within the BMU-ICI project in Belarus. Roman numbers refer to the five Belarusian peatland districts (Pidoplichko 1961; see chapter 2.1).

accessed by car and present as much of the desired site types as possible within small distances (Table 7, Fig. 33). The selected sites are:

- Sporauski zakaznik and surroundings with pristine, drained agricultural, and rewetted fen sites providing site types for gap filling, and some for GEST calibration;
- Barcianicha, an extracted and recently rewetted fen with some dry parts, also for gap filling;
- Raznianskaje balota a pristine bog in the Biarezinski zapaviednik providing site types for GEST calibration.

Table 7: Measured site types with water level (WL) class and reason of type selection (for GEST calibration or gap filling).

| Site | Site type | WL class | Aim |
|---|---|---|---|
| Barcianicha | | | |
| B3 | moist bare peat | 3+ | calibration |
| B2 | moist peat with sparse herbs | 3+ | gap filling |
| B1 | moist peat with small *Pinus sylvestris* | 3+ | gap filling |
| B7 | moist forbs with *Urtica dioica* | 3+ | gap filling |
| B6 | wet peat with *Phragmites australis* | 5+ | gap filling |
| B5 | wet peat with *Carex rostrata* | 5+ | gap filling |
| B4 | wet peat with *Eriophorum angustifolium* | 5+ | gap filling |
| B8 | recently rewetted meadow with *Agrostis stolonifera* and *Potentilla anserina* | 5+ | gap filling |
| B9 | recently flooded meadow, vegetation died | 6+ | gap filling |
| Sporauski zakaznik and surroundings | | | |
| T3 | grassland with *Phleum pratense* | 2+/2- | calibration |
| T1 | arable land with *Triticum aestivum* | 3+/2+ | gap filling |
| T2 | grassland with *Phleum pratense* | 3+/2+ | calibration |
| V1 | wet tall reeds with *Carex elata*, *Phragmites australis*, *Comarum palustre* | 5+ | gap filling |
| V2 | wet tall sedges with *Carex elata* and *Menyanthes trifoliata* | 5+ | gap filling |
| Z1 | wet sedge marsh with *Carex rostrata*, *C. nigra*, and *Drepanocladus aduncus*, not harvested | 5+ | gap filling |
| Z2 | wet sedge marsh with *Carex rostrata*, *C. nigra*, and *Drepanocladus aduncus*, biomass harvest in winter | 5+ | gap filling |
| M2 | flooded *Typha latifolia* | 6+ | gap filling |
| M3 | flooded *Carex* | 6+ | gap filling |
| M1 | flooded *Phragmites australis* | 6+ | gap filling |
| Raznianskaje balota | | | |
| R3 | hummocks with *Sphagnum magellanicum*, *Polytrichum strictum*, and *Chamaedaphne calyculata* | 3+ | calibration |
| R4 | wet peat moss lawn with *Sphagnum magellanicum* and *Andromeda polifolia* | 5+ | calibration |
| R2 | wet peat moss hollow with *Sphagnum cuspidatum* | 5+ | calibration |
| R1 | wet peat moss hollow with *Sphagnum cuspidatum* and *Scheuchzeria palustris* | 5+ | calibration |

height, and changes in response to wind velocity and direction, surface roughness, and atmospheric stability (Denmead 2008). The eddy covariance method is therefore ideal to get integrated flux values at the landscape scale, but does not allow for small scale studies. The eddy method does not work equally well for all landscapes. To meet the assumption of a constant vertical flow, the footprint area should be flat because sloping landscapes tend to modify air flow (Baldocchi 2003). While this criterion can usually be met by peatlands, the measurement of flooded areas introduces new difficulties. During the night, $CO_2$ from ecosystem respiration of the surroundings is transported to the warmer water surface where it affects the measured flux (Eugster et al. 2003).

## 3.4 Vegetation as a proxy for greenhouse gas fluxes – the GEST approach

John Couwenberg

### 3.4.1 The need for a proxy

Rewetting of drained peatlands may lead to considerable reductions in greenhouse gas (GHG) emissions to the atmosphere. Both the compliance market (Kyoto Protocol, www.unfccc.org; see chapter 5.4) and the voluntary market (e.g. the VCS, www.v-c-s.org; see chapter 5.3) require that these emission reductions are reliably quantified. An obvious approach would be to measure on site all GHG fluxes that occur before, during and after the project intervention. Indeed, adequate techniques exist to measure these fluxes in detail (see chapter 3.3). The chamber method, for example, enables measurements on the scale of a few $dm^2$ up to one $m^2$, whereas eddy-covariance allows assessing GHG fluxes over larger areas (typically 1,000 $m^2$–1 $km^2$). GHG fluxes are dependent on a wide spectrum of site parameters that vary strongly over the year and between years, including temperature, water level, plant growth and land use. Therefore, assessing annual GHG balances require highly frequent and prolonged observations to cover daily, seasonal and interannual variability. Moreover, a sufficiently dense net of observations is necessary for the chamber method to cover the often fine-scale spatial patterns that are typical for degraded and rewetted peatlands. Assessing the effect of peatland rewetting by comprehensive, direct flux measurements might currently cost in the order of 10,000 Euro per ha per year, which is prohibitively expensive. In practise, direct measurements are only feasible for selected pilot sites in order to develop, calibrate and verify models that allow assessment by proxy of GHG fluxes over much larger areas (Couwenberg et al. 2011).

### 3.4.2 Water levels vs. vegetation

In order to develop easily applicable proxies for GHG fluxes from peatlands, a meta-analysis of available flux data was carried out (Couwenberg et al. 2008, 2011, see chapter 3.1). Mean annual water level turns out to be the best single explanatory variable for annual GHG fluxes. With this relationship in mind, it would be possible to assess and monitor GHG fluxes from peatlands by measuring water levels. However, like direct flux measurements, using water level as a proxy for GHG fluxes would require a dense network with frequent measurements over a prolonged period of time. Remote sensing is not yet suited for direct monitoring of groundwater levels of drained peatlands. Alternatively, vegetation can be used as an indicator for water levels with the advantage that additional site characteristics that influence GHG fluxes may also be differentiated.

Plant species and vegetation have already for a long time been used to indicate site conditions (Trommer 1853, Ramenski et al. 1956, Scamoni 1960, Hundt & Succow 1984, Ellenberg et al. 1992). The approach followed here builds on the German 'vegetation form' concept (Koska et al. 2001, Koska 2007, box 6), which integrates floristic and environmental parameters to derive comprehensive vegetation based proxies. Vegetation is well suited for indicating GHG fluxes (Couwenberg et al. 2011) because:

- It is a good indicator of water level, which in turn strongly correlates with GHG fluxes;
- It is controlled by various other site factors that determine GHG emissions from peatlands, including nutrient availability, soil reaction (pH) and land use (history);
- It is itself directly and indirectly responsible for the predominant part of the GHG emissions by regulating $CO_2$ exchange, by supplying organic matter (incl. root exudates) for $CO_2$ and $CH_4$ formation, by reducing peat moisture and by providing possible bypasses for methane fluxes via aerenchymous 'shunts' (chapter 3.1);
- It reflects long-time water level conditions and thus provides indication of average GHG fluxes on an annual time scale;
- It allows fine-scaled mapping, e.g. on scales 1:2,500–1:10,000.

### 3.4.3 The GEST approach

Based on the site descriptions provided in the GHG literature, vegetation forms as described by Koska et al. (2001) and Koska (2007) were associated with each flux measurement site. For vegetation forms that are indistinct between moist and very moist sites, moist and very moist subtypes were derived based on indicator species or on shifts in dominance. Net $CO_2$ and $CH_4$ flux values were assigned to the vegetation forms using the following protocol (Couwenberg et al. 2011):
1. Compare the distinguished vegetation types with the vegetational and floristic character-

## Box 6

### The 'vegetation form' concept

John Couwenberg

The 'vegetation form' concept is a classification approach that integrates floristic and environmental parameters. It departs from the observation that in an environmental gradient (e.g. from dry to wet) some plant species occur together, whereas others exclude each other (see Fig. 34). The combined occurrence of specific species groups, as well as their mutual exclusion, provides a much sharper indication of site parameters than individual plant species (e.g. the well known Ellenberg indicator values). The amplitudes of various species groups allow the differentiation of factor classes (see Table 8).

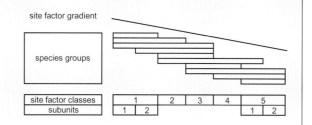

Fig. 34: Co-occurrence of species groups along a site factor gradient (modified after Koska et al. 2001).

Vegetation forms are named by the combination of names of characteristic plant species and a term referring to physiognomy and/or type of land use (e.g. *Caltha-Filipendula*-forbs). They can be rapidly identified in the field by checking the presence and/or absence of species groups.

Table 8: Major peatland site factor classes used in the vegetation form approach (modified after Koska et al. 2001). WLw: long-term median water level wet season; WLd: long term median water level dry season; WD: water supply deficiency; C/N: carbon/nitrogen-ratio in the topsoil measured with Kjeldahl; $pH_{kts}$: pH of topsoil measured in KCl-solution.

| Factor and description | Classes | | Characteristics |
|---|---|---|---|
| Water level/supply<br><br>water supply:<br>+: wetlands and aquatic habitats<br>-: non-hydric terrestrial habitats | 7+ | upper sublitoral | WLw/WLd: +250 to +140 cm |
| | 6+ | lower eulitoral | WLw: +150 to +10 cm; WLd: +140 to +0 cm |
| | 5+ | wet (upper eulitoral) | WLw: +10 to -5 cm; WLd: +0 to -10 cm |
| | 4+ | very moist | WLw: -5 to -15 cm; WLd: -10 to -20 cm |
| | 3+ | moist | WLw: -15 to -35 cm; WLd: -20 to -45 cm |
| | 2+ | moderately moist | WLw: -35 to -70 cm; WLd: -45 to -85 cm |
| | 2- | moderately dry | WD: <60 l/m² |
| | 3- | dry | WD: 60–100 l/m² |
| | 4- | very dry | WD: 100–140 l/m² |
| | 5- | extremely dry | WD: >140 l/m² |
| Seasonally alternating wetness is indicated by a combination of different water level classes, e.g. 5+/4+ refers to a WLw within 5+ range and a WLd within 4+ range. Strongly alternating wetness is indicated by a tilde-sign, e.g. 3~ refers to a WLw within 4+ range and a WLd within 2+ range. | | | |
| Nutrient availability<br><br>availability of main nutrients (especially N, P, K, resp. the limiting element), crucial for bioproductivity<br>Used proxy: soil C/N relation | o-vp | oligotrophic – very poor | C/N >40 |
| | o-p | oligotrophic – poor | C/N 33–40 |
| | m-lm | mesotrophic – rather poor | C/N 26–33 |
| | m-hm | mesotrophic – medium | C/N 20–26 |
| | e-mr | eutrophic – moderately rich | C/N 13–20 |
| | e-r | eutrophic – rich | C/N 10–13 |
| | p-vr | polytrophic – very rich | C/N <10 |
| Base richness (pH)<br>Used proxy: $pH_{KCl}$ of soil | ac | acid | $pH_{kts}$ <4.8 |
| | sub | subneutral | $pH_{kts}$ 4.8–6.4 |
| | alk | alkaline | $pH_{kts}$ >6.4 |

## 3.4 Vegetation as a proxy for greenhouse gas fluxes – the GEST approach

Table 9: Vegetation types in Astrauskoje and Vyhanascanskaje with associated flux measurements and their standard deviations and best estimates of global warming potential (GWP) (from Couwenberg et al. 2011).

| Vegetation type | $CO_2$ emissions ($CO_2$-eq. ha$^{-1}$ yr$^{-1}$) | $CH_4$ emissions ($CO_2$-eq. ha$^{-1}$ yr$^{-1}$) | GWP estimate ($CO_2$-eq. ha$^{-1}$ yr$^{-1}$) | Remarks |
|---|---|---|---|---|
| Astrauskoje | | | | |
| Bare peat | 7.0 (±2.6) for active extraction sites ($n$ = 12) / 7.4 (±0.9) for abandoned extraction sites ($n$ = 3); Maljanen et al. (2010) | 0.4 (±0.6) for active extraction sites ($n$ = 13) / 0.06 (±0.0) for abandoned extraction sites ($n$ = 2); Maljanen et al. (2010) | 7.5 | |
| *Calluna* | as 'moist bog heath' | | 12.5 | Drier than 'bare peat' |
| *Eriophorum* | 3.3 (±2.1) ($n$ = 8); Tuittila et al. (1999), Maljanen et al. (2010) | 0.3 (±0.1) ($n$ = 8); Tuittila et al. (2000), Maljanen et al. (2010) | 3.5 | Litter accumulation of *Eriophorum vaginatum* counteracts carbon losses from degrading peat |
| *Polytrichum* | as 'bare peat' | | 7.5 | Mosses lack roots (cf. [very] moist bog heath) |
| Dry grassland | as 'moderately moist forb meadows' | | 20 | Direct measurements from dry bogs are lacking, but water level fluctuations are expected to be similar to those in moderately moist forb meadows |
| Moist bog heath | 12.6 (±4.0) ($n$ = 3); Drösler (2005) | negligible; Drösler (2005) | 12.5 | With the same water levels emissions are higher than from bare peat because plant roots change the water regime, improve aeration and add labile organic compounds in the form of recently dead roots and root exudates that stimulate the decomposition of the more recalcitrant peat (Kuzyakov 2006). |
| Very moist bog heath | 9; Drösler (2005) | 0.7; Drösler (2005) | 10 | |
| Moderately wet *Sphagnum* hummocks | neglected | 0.7 (±0.2) ($n$ = 4); Bortoluzzi et al. (2006) | 0.5 | $CH_4$ emissions increase with higher water levels. $CH_4$ emissions from wet bog sites in boreal regions are much lower than the values cited here (Couwenberg 2009b). Shannon & White (1994) found similar $CH_4$ fluxes at comparable sites in temperate South-Michigan bogs. Whereas net emissions of $CO_2$ have been reported from rewetted bog sites (Drösler 2005), published measurements generally show uptake of $CO_2$ (Tuittila et al. 1999, 2004, Bortoluzzi et al. 2006, Maljanen et al. 2010). Carbon sequestration is overestimated when water-borne carbon losses are not taken into account (Roulet et al. 2007, Nilsson et al. 2008). Water borne carbon export is generally larger before rewetting (Holden et al. 2004) and this flux can thus conservatively be neglected. As a strictly conservative approach potential carbon sequestration is discarded and zero $CO_2$ flux is assumed for rewetted sites. |
| Wet *Sphagnum* lawn | neglected | 5.2 (±3.2) ($n$ = 5); Drösler (2005) | 5 | |
| Very wet *Sphagnum* hollows | neglected | 12.8; Drösler (2005) | 12.5 | |

| Vegetation type | $CO_2$ emissions ($CO_2$-eq. ha$^{-1}$ yr$^{-1}$) | $CH_4$ emissions ($CO_2$-eq. ha$^{-1}$ yr$^{-1}$) | GWP estimate ($CO_2$-eq. ha$^{-1}$ yr$^{-1}$) | Remarks |
|---|---|---|---|---|
| Vyhanascanskaje | | | | |
| Moderately moist (forb) meadows | data lacking; assumed somewhat lower than moderately moist cultivated lands | negligible; Fig. 23; Couwenberg (2009b), Maljanen et al. (2010) | 20 | Very similar to $CO_2$ fluxes derived for boreal peat grasslands (21.7 (±6.8) t $CO_2$-eq. ha$^{-1}$ yr$^{-1}$; n = 4; Lohila et al. 2007, Maljanen et al. 2010). |
| Moist forb meadows | data lacking; corresponding bog sites: 12.6 (±4.0) (n = 3); Drösler (2005) | negligible; Fig. 23; Couwenberg (2009b) | 12.5 | |
| Very moist reeds | data lacking; assumed near zero (Fig. 23) | 3.5 (±1.6) (n = 12); van den Pol-van Dasselaar et al. (1999), van Huissteden et al. (2006), Hendriks et al. (2007) | 3.5 | Assumed much lower than from very moist forbs and (cultivated) grasslands because of considerable (below ground) accumulation of biomass (see wet reeds and sedge fens below). |
| Wet reeds and sedge fens | -4.1 (±4.3) (n = 4); Bonneville et al. (2008), Drösler (2008) | 12.7 (±8.4) (n = 10); Augustin (2003, unpubl.), Drösler (2008) | 8.5 | Typically dominating aerenchymous plant species cause high $CH_4$ emissions. Net $CO_2$ uptake may be higher (cf. Whiting & Chanton 2001, Hendriks et al. 2007, Zhou et al. 2009). The GWP of very wet (flooded) sites is estimated to be similar. |
| Additional site types | | | | |
| Moderately moist cultivated lands | 24.1 (±8.2) (n = 14); Mundel (1976), Jacobs et al. (2003), Veenendaal et al. (2007), Drösler (2008), Augustin (unpubl.) | negligible; Fig. 23; Couwenberg (2009b), Maljanen et al. (2010) | 24 | Using site specific subsidence measurements van den Akker et al. (2008) and Verhagen et al. (2009) arrive at considerably higher fluxes from meadows and croplands at low water levels (Fig. 22). The value used here can be considered conservative. |
| Moist cultivated grasslands | 15.5 (n = 2) Veenendaal et al. (2007) | negligible; Fig. 23; Couwenberg (2009b) | 15 | |
| Very moist forbs and (cultivated) grasslands | conclusive data lacking; assumed in between 0 (Fig. 22) and ~15 (Jacobs et al. 2003) | negligible; Fig. 23; Couwenberg (2009b) | 7.5 | See very moist reeds |
| Wet cultivated grasslands | 1.4 (±3.5) (n = 4); Augustin (unpubl.) | 3.1 (±3.5) (n = 4); Augustin (unpubl.) | 5.5 | Data for wet fen peatlands dominated by cultivated short grassland species are lacking. The data of Augustin (unpubl.) are from an atypically wet year. |
| Drowned crop- and grasslands | | extremely high, up to 77; Augustin & Chojnicki (2008), Augustin (unpubl.) | | Upon flooding, large amounts of labile carbon become available for methanogenesis. At the flux measurement site Zarnekow studied by Augustin & Chojnicki (2008) easily degradable litter is washed in from the surrounding area. |

## 3.4 Vegetation as a proxy for greenhouse gas fluxes – the GEST approach

istics described in GHG literature. In case of identical presence and absence of species groups (Koska et al. 2001), the GHG flux values from literature were adopted.
2. In order to verify and specify the flux values, compare in a second step the water level data acquired from field observation, vegetation form indication (Koska et al. 2001) and Ellenberg values with regression models of GHG fluxes against mean annual water level (see chapter 3.1). In case the water level data do not provide conclusive results, apply expert judgement, taking into account similarities with well documented vegetation types.
3. In case the vegetation does not have sufficient similarity with literature, compare only the water level data and the presence of aerenchymous shunt species with the regression models to arrive at flux values.
4. In case these data do not provide conclusive results, apply expert judgement, taking into account the overall character of the site and its water level conditions and the flux values of related vegetation types.

Many publications on direct flux measurements provide insufficient information on the vegetation composition of their research plots. However, the available site parameters, vegetation descriptions and plant lists in most cases allow to derive 'vegetation forms' that serve as a comprehensive proxy for site conditions (Koska et al. 2001). In this way a water level class (integrating mean levels and fluctuations), trophic state (C:N ratio) and base richness (pH) as well as a vegetation type could be appointed to all but a few emission values. The enhanced dataset was then used to calibrate the vegetation – flux relationship for the group of vegetation forms representing the same or similar site conditions. Meta-analysis of available annual $CO_2$ fluxes data (chapter 3.1) denied a distinction between bog and fen sites, allowing the use of 'bog' measurements to specify $CO_2$ fluxes from 'fen' sites and vice versa.

The vegetation forms of Koska et al. (2001) and Koska (2007) provide a matrix system of all vegetation types that may occur in peatlands in Central Europe. This matrix allows extra- and interpolation of measured flux data along various axes of site characteristics (Couwenberg et al. 2008, 2011). The resulting Greenhouse gas Emission Site Types (GESTs, Table 9) are currently mainly based on water level class and presence of aerenchymous shunts, but also nutrient status, pH and land use were considered and may allow better differentiation once more flux data are added. Additional high-quality measurements are a necessity. Important gaps in the database still exist for $CO_2$ fluxes in general and for $CH_4$ fluxes in mosaics of bare peat with sparse vegetation, for $CH_4$ in relation to specific vegetation types of wet sites, particularly also in relation to shunt species abundance and generally for sites with tree cover (Couwenberg et al. 2011).

### 3.4.4 Applying GESTs in rewetting projects

The current set of GESTs provides a coarse, but good tool that can be improved and expanded (e.g. by regional calibration and gap-filling) when additional flux measurements become available. At present, the GEST approach already delivers more detailed assessments than the use of IPCC default values (IPCC 2006, Couwenberg 2011). Expansion of the tool may allow for sharp delineation of GHG flux classes along multiple axes, including water level (fluctuations), nutrient status, pH, and (former) land use and land use intensity. The use of GESTs allows for a rapid and relatively cheap estimate of baseline and project scenario emissions and thus of emission reductions from rewetting.

Available flux studies from treed sites are thus far inconclusive on the net effect of the multitude of processes associated with tree growth on drained peatlands. Simultaneous processes affecting net GHG fluxes include peat oxidation, above-ground biomass growth, below-ground biomass growth, above-ground litter deposition, below-ground litter deposition and waterborne losses of dissolved and particulate organic carbon (DOC, POC). On the one hand, drainage leads to oxidative peat losses resulting in release of $CO_2$; on the other hand, considerably more carbon is sequestered by the increasing tree biomass and litter than by the original mire vegetation (Crill et al. 2000; Minkkinen et al. 2008). To assess the net climate effect of forested peatlands the better studied fluxes from comparable unforested peat soils are combined with an estimate of carbon sequestration in woody biomass.

$CO_2$ fluxes from shallow peats where the entire peat layer is above the water table are determined more strongly by peat thickness than by other site characteristics (cf. Mundel 1976, Meyer 1999). Further investigation into shallow peat soils with deep water levels is needed to arrive at good emission estimates.

In order to estimate emission reductions, scenarios of vegetation development must be formulated for the situation with rewetting ('project scenario') and the situation without rewetting ('baseline scenario') (see chapter 3.5). Emission reductions can be conservatively estimated by applying low estimates for the baseline and by leaving out emissions from ditches and of $N_2O$, while applying high estimates for the project scenario. Particularly $CH_4$ emissions from ditches can be substantial (Van de Pol-van Dasselaar et al. 1999, Schrier-Uyl et al. 2009, Maljanen et al. 2010), but usually ditches cover only small areas. Moreover, ditches are expected to be overgrown after rewetting measures, which will substantially reduce emissions. Disregarding emissions from ditches thus means emission reductions are underestimated.

Whereas rewetted peatland sites will have negligibly low $N_2O$ fluxes, the amount of $N_2O$ emitted from drained sites is unpredictable (chapter 3.1). Although the processes involved in the soil $N_2O$ budget are fairly well understood, broadly applicable proxies for soil $N_2O$ fluxes are still lacking (Jungkunst et al. 2006, Lamers et al. 2007, Baggs 2008, chapter 3.3). $N_2O$ fluxes from peat soils can be substantial and identification of mapable proxies should be a priority for further research. As N2O emissions always decrease after rewetting, disregarding these fluxes will also result in conservative emission reduction estimates. The current set of GESTs is based on $CO_2$ and $CH_4$ fluxes only.

Flux measurements from Belarusian peatlands are not yet available, but a large research programme has recently started to fill this gap (Minke et al. 2009, Thiele et al. 2009 and chapters 3.3 and 7.2).

## 3.5 Prediction of vegetation development with and without rewetting

Annett Thiele, Frank Edom & Nazdeya Liashchynskaya

### 3.5.1 The necessity of predicting vegetation development

To be able to account for greenhouse gas (GHG) emission reductions from peatland rewetting, an emission baseline needs to be set. The GHG emissions after implementing a mitigation activity are compared against this baseline to determine the accountable GHG emission reductions or removals resulting from the activity. Baseline approaches differ for different mechanisms. Most (voluntary) 'activities' under art. 3.4 of the Kyoto Protocol (cropland management, grazing land management, re-vegetation) and the associated emission trading under article 17 of the Protocol use a static 'historical' baseline of 1990. In contrast, the Clean Development Mechanism (CDM) under art. 12 of the Protocol (see chapter 5.4) as well as the Verified Carbon Standard – Peatland Rewetting and Conservation (VCS-PRC, chapter 5.3) use a forward-looking baseline that describes emissions that would have occurred during the intended project duration in absence of the project. Emission reductions strongly depend on the baseline approach chosen (Table 10).

In the case of peatland rewetting projects, forward-looking baseline emissions could be quantified by the monitoring of a similar but not rewetted reference site. In practice, such a reference is difficult to achieve, however, because of a lack of suitable sites of sufficient similarity, because of the extra costs involved, and because the purposive management as a reference site prevents the site to follow regional developments fully sponta-

Table 10: Emission reductions at Vyhanascanskaje based on different baseline assumptions. Project emissions are conservatively estimated as in Table 15.

| Criterion | Baseline emissions (t $CO_2$-eq. ha$^{-1}$year$^{-1}$) | Project emissions (t $CO_2$-eq. ha$^{-1}$ year$^{-1}$) | Resulting emission reduction (t $CO_2$-eq. ha$^{-1}$year$^{-1}$) |
|---|---|---|---|
| Fixed baseline 1990 | 20.0 | 8.0 | 12.0 |
| Forward- looking baseline (2010–2039) – intensified agriculture | 16.0 | 8.0 | 8.0 |
| Forward- looking baseline (2010–2039) – abandonment | 8.2 | 8.0 | 0.2 |

## 3.5 Prediction of vegetation development with and without rewetting

Table 11: Changes in soil, relief and hydrology during vegetational succession of drained, degenerating and rewetted, regenerating peatlands.

|  | Soil | Relief | Hydrology | Vegetation |
|---|---|---|---|---|
| Degeneration | • Peat oxidation<br>• Destruction of pore-structure<br>• Development of degraded soil horizons<br>• Increased nutrient availability | • Shrinkage<br>• Increased unevenness | • Loss of water supply<br>• Lowered water table<br>• Increased water table fluctuations | • Change from peat-forming vegetation to non-peat-forming vegetation |
| Regeneration | • Rebuilding of pore-structure<br>• Decreased nutrient availability | • Evening out of relief through new peat formation | • Restoration of water supply<br>• Rising water table<br>• Decreased water table fluctuations | • Re-establishment of peat-forming vegetation |

neously. Moreover, opting for the monitoring of baselines does not solve the issue of providing ex-ante estimates for total project GHG benefits, as required in both compliance and voluntary markets. The use of vegetation as a proxy for greenhouse gas (GHG) fluxes (chapter 3.4) then raises the need for sound prediction of vegetation development both in the project and baseline scenarios.

### 3.5.2 Succession on peatlands – principles

The term succession refers to a sequence of changes in the species composition of a community in reaction to a changing environment. In peatlands changes in soil, relief, hydrology and vegetation interact closely during successional development (Table 11). Whereas in a baseline scenario of continued land use succession is suppressed, abandonment and particularly rewetting invoke successional changes in vegetation. A brief summary of succession observations from (the extensive) literature and own experience is presented below. It provides the basis for the description of succession schemes that differ in their starting position and water management. Associated GHG emissions (chapter 3.4, Fig. 35, Fig. 36) are illustrated for 30 year project scenarios for two example sites in Belarus.

### 3.5.3 Succession on peatlands without rewetting ('baseline scenarios')

Peat extraction is a drastic disturbance that leaves behind vegetation free areas of bare peat. Bare peat has very small pores, meaning that water level fluctuations are large and will remain so until new loose peat has been formed. In the 'black deserts' of abandoned milled peatlands extreme fluctuations in temperature and humidity occur in a nutrient poor situation and strongly obstruct the establishment of plant species. Very dry sites can remain bare for more than 15 years (Edom et al. 2009b). Under these conditions the indicator value of species and vegetation is indistinct and should be interpreted with critical awareness; additional water level observations will be needed to assess GHG fluxes (Couwenberg et al. 2011). On effectively drained abandoned bare bog sites pioneer species like *Calluna vulgaris, Polytrichum strictum, Eriophorum vaginatum, E. angustifolium, Drosera rotundifolia, Molinia caerulea, Deschampsia caespitosa*, and *Betula pendula* occur as first ground cover (Salonen 1990, 1992, Lavoie et al. 2005b, Edom et al. 2009b, Kozulin et al. 2010a,b, Wendel 2010, Table 12a and 13). Mono-dominant dwarf shrub communities with *Vaccinium myrtillus, V. vitis-idaea*, and *Calluna vulgaris* have also been observed on dry bog sites in Southeast-Germany and Belarus (Edom et al. 2009b, Kozulin et al. 2010a, own observations). In wet depressions, communities with *Eriophorum vaginatum, E. angustifolium, Vaccinium oxycoccos*, and *Sphagnum* spec. can develop. Autonomous re-colonization with peat-forming *Sphagnum* species was not observed eight years after abandonment of milled peat extraction sites in Canada (Lavoie et al. 2005b), and 30 to 50 years after abandonment peat-forming *Sphagnum* species covered ≤10% of two other Canadian bog sites (Price & Whitehead 2001, Girard et al. 2002). In southern Germany various mono-dominant non-peat forming vegetation stands were found 20 years after abandonment on a milled peat extraction site and had hardly changed 20 years later (Poschlod et

Table 12: Range of ground water level (m) at which peatland vegetation communities occur in Belarus (after Kozulin et al. 2010a).

| 12a) Vegetation on peat extraction sites with oligotrophic/oligomesotrophic conditions | | | | | | | | | | | |
|---|---|---|---|---|---|---|---|---|---|---|---|
| Vegetation community | -1.0 | -0.5 | -0.3 | -0.2 | -0.1 | 0.0 | 0.1 | 0.2 | 0.3 | 0.5 | 1.0 |
| Weeping Weeping birch-dwarf shrub-moss | | | | | | | | | | | |
| Heather-moss | | | | | | | | | | | |
| *Eriophorum vaginatum* | | | | | | | | | | | |
| White birch White birch-dwarf shrub moss | | | | | | | | | | | |
| *Eriophorum vaginatum*-moss | | | | | | | | | | | |
| *Eriophorum vaginatum-Sphagnum* | | | | | | | | | | | |
| Slender sedge | | | | | | | | | | | |
| Common Reed | | | | | | | | | | | |
| 12b) Vegetation on peat extraction sites with mesotrophic/eutrophic conditions | | | | | | | | | | | |
| Vegetation community | -1.0 | -0.5 | -0.3 | -0.2 | -0.1 | 0.0 | 0.1 | 0.2 | 0.3 | 0.5 | 1.0 |
| Ruderal | | | | | | | | | | | |
| Weeping birch | | | | | | | | | | | |
| Meadow forb | | | | | | | | | | | |
| Bush grass-French willow | | | | | | | | | | | |
| Grey willow | | | | | | | | | | | |
| Swamp forb | | | | | | | | | | | |
| Cotton weed | | | | | | | | | | | |
| Beaked sedge | | | | | | | | | | | |
| Common Reed | | | | | | | | | | | |
| Aquatic vegetation | | | | | | | | | | | |
| 12c) Vegetation on peatlands drained for forestry with oligotrophic/mesotrophic conditions | | | | | | | | | | | |
| Vegetation community | -1.0 | -0.5 | -0.3 | -0.2 | -0.1 | 0.0 | 0.1 | 0.2 | 0.3 | 0.5 | 1.0 |
| Pine Pine-dwarf shrub | | | | | | | | | | | |
| Weeping birch Weeping birch-dwarf shrub-moss (burnt spot) | | | | | | | | | | | |
| Pine Pine-dwarf shrub-*Sphagnum* | | | | | | | | | | | |
| Blueberry-moss-*Sphagnum* (burnt spot) | | | | | | | | | | | |
| Marsh tea-*Sphagnum* | | | | | | | | | | | |
| Slender sedge-*Sphagnum* White beak sedge-*Sphagnum* | | | | | | | | | | | |

al. 2007). Colonization of abandoned bare bog peat by tree species depends on factors like water level, soil temperature fluctuations, remaining seed bank, shading, the presence of safe-havens for germination, and distance from already developed stands (Briemle 1978, 1980, 1990, Efremov 1987, Salonen 1990, 1992, Rosenthal 2010). The most important pioneer trees on dry bog peat are *Betula pendula* and *Pinus sylvestris* (Efremov 1987). Trees in acid peatlands can help by rebuilding the soil structure with their roots (Edom et al. 2009a), but they can also increase peat oxidation. The changing microclimate of tree stands (shading, increased air moisture and weak wind) allows for a faster re-colonization with mosses and grasses.

## 3.5 Prediction of vegetation development with and without rewetting

Table 13: Simplified succession after rewetting of temperate peatlands (bogs and fens) in relation to water level and nutrient availability. Low water level corresponds to water level class 4+ (very moist); Medium water level to 5+ (wet); High water level to 6+ (lower eulitoral) (see chapter 3.4). See text for more details.

| Peatland type | Water level | Water quality | Time after rewetting | | |
|---|---|---|---|---|---|
| | | | 1–5 years | 5–20 years | >20 years |
| Bog | Low | Ombrotrophic | Bare peat with some *Molinia/Eriophorum/Polytrichum* | *Eriophorum* and/or *Molinia* lawn | *Eriophorum/Calluna/Betula* wooded heath |
| | | Minerotrophic | *Molinia* and/or *Eriophorum* lawn | *Molinia/Betula* dry open birch forest | *Betula* birch woodland |
| | Medium | Ombrotrophic | *Eriophorum* lawn | *Eriophorum/Carex/Betula* Wet Forest | *Betula* wet birch forest |
| | | Minerotrophic | *Carex/Phragmites* | | *Salix/Sphagnum* willow carr |
| | High | Ombrotrophic | Open water | Open water | *Sphagnum/Eriophorum* mire |
| | | Minerotrophic | *Phragmites/Carex/Typha* beds | *Carex/Phragmites/Typha* reed | *Carex/Sphagnum* mire/fen or *Phragmites/Salix* willow carr |
| Fen | Low | Polytrophic | Dry meadows *Calamagrostis/Epilobium/Elymus* | *Phalaris/Cirsium/Urtica* weeds | *Phalaris/Cirsium/Urtica* weeds/*Betula* |
| | | Eutrophic | Dry meadows *Sanguisorbia/Molinia* | *Schoenoplectus/Phragmites* reeds | *Schoenoplectus/Phragmites* reeds |
| | Medium | Polytrophic | *Solanum/Bidens/Phragmites* | *Solanum/Bidens/Phragmites* with forbs | *Solanum/Phragmites/Salix* with forbs/shrubs |
| | | Eutrophic | *Phalaris/Juncus conglomeratus/Lysimachia vulgaris* | *Typha/Carex/Phragmites* | *Typha/Carex/Phragmites* |
| | High | Polytrophic | wet meadows/shrubs | *Typha/Carex/Phragmites* | *Typha/Carex/Phragmites* |
| | | Eutrophic | *Phalaris/Phragmites*/willow shrubs | *Typha/Carex/Phragmites* | *Typha/Carex/Phragmites/Sphagnum* |

On abandoned bare fen sites ruderal plants like *Calamagrostis epigeios*, *Epilobium angustifolium*, *Urtica dioica*, and *Elymus repens* establish on polytrophic and very dry sites; *Phalaris arundinacea* and *Lysimachia vulgaris* are typical species on eutrophic sites with medium water levels (Kozulin et al. 2010a, own observations, Table 12b and 13).

On abandoned fen sites drained for agriculture usually birch and/or willow trees establish (Koska et al. 2001, Succow et al. 2001). On abandoned bog sites drained for forestry an increase in tree and dwarf shrub growth is observed. Such sites often show a limited cover of *Sphagnum* and collapsed hummocks with lichens (Table 12c), but may also turn out rather wet, supporting *Sphagnum* spec. or *Eriophorum vaginatum* (Kozulin et al. 2010a, Wendel 2010, own observations). Tree stand densities in bogs differ corresponding to water table and microrelief (Ivchenko 2009). Tree growth is limited by competition between root systems that build a dense, but porous subsurface net, with coarse pores that regulate water level fluctuations (Pjavchenko 1963, Edom 1991, Edom et al. 2009a, Anderson 2010). Spontaneous establishment of peat forming vegetation has been observed in wooded or forested peatlands by several authors (Kästner & Flößner 1933, Pjavchenko 1963, Schneebeli 1991, Edom 1991, 2001a, Wendel 1992, 2010, Edom & Succow 1998, Edom & Wendel 1998, Wagner 1994, 2006, Zinke & Edom 2006, Bretschneider 2010). When conditions are less favourable, pioneer or planted tree species (*Betula*, *Pinus*) are succeeded by species like spruce (*Picea abies*) on bog peat (Kästner & Flößner 1933, Edom et al. 2010, Wendel 2010) or oak (*Quercus robur*) on (moderately) dry fen peat stands (Gremer &

Edom 1994, Clausnitzer & Succow 2001, Succow et al. 2001).

### 3.5.4 Succession on peatlands with rewetting ('project scenarios')

After rewetting of cutover bog sites, rapid changes from terrestrial species to typical peat-forming (oligo- to mesotrophic) species like *Sphagnum* spec., *Empetrum nigrum*, *Eriophorum vaginatum*, *E. angustifolium*, *Carex vesicaria*, *Carex nigra*, *C. canescens* or *C. rostrata* have been observed (Robert et al. 1999, Price & Whitehead 2001, Jauhiainen et al. 2002, Edom et al. 2009b, Wendel 2010, own observations; Table 12a and 13). On milled bog sites the rate and patterns of early succession stages can differ substantially because of small scale differences in extraction depth and resulting peat quality (Salonen 1990, 1992). Vegetation development within the first 20 years after rewetting is largely determined by water level and quality (see Poschlod 1990, 1992, Sliva 1997, Poschlod et al. 2009 for detailed succession schemes; Table 13). In Belarus, Kozulin et al. (2010a) observed a complete colonization by *E. vaginatum* within five years after rewetting. When dwarf shrubs (*Calluna vulgaris*, *Ledum palustre*, *Vaccinium uliginosum*) cover the extracted sites, *Calluna* will die off, *Ledum* and *Vaccinium* will be oppressed after rewetting and *E. vaginatum* and *Sphagnum* spec. will gradually colonize the area starting in the ditches. In case of flooding (20–30 cm above soil surface) sparse *Phragmites australis* stands develop on the margins accompanied by *Sphagnum* spec. (Table 12). If the flooding is deeper than one meter open water is expected to be colonized by floating *Sphagnum* species after a vegetation free phase (Kozulin et al. 2010a).

On extracted fen sites in Belarus, with oligotrophic site conditions and water levels after rewetting close to the surface (-10 cm to +10 cm), a transient vegetation cover of a sparse *Phragmites australis* reed with *Eriophorum angustifolium*, *Carex vesicaria*, *Lythrum salicaria*, and *Trichophorum alpinum* developed within two years. Under meso- and eutrophic conditions and with mean water levels up to the surface *Salix* spec. and herbs like *Calamagrostis canescens*, *Lysimachia thyrsiflora*, and *L. vulgaris* establish together with *Carex* species like *C. vesicaria* and *C. elata*. With water levels above 20 cm a mono-dominant *Phragmites australis* reed will develop. When water levels remain above 30 cm, water plants like *Myriophyllum alternifolium* and *Calla palustris* will establish (Kozulin et al. 2010a, own observations; Table 12). Raising the water level close to the surface on sites covered with *Betula pendula* forests leads to establishment of *Polytrichum strictum* (under oligotrophic conditions) or other bryophytes like *Calliergonella cuspidata*, *Calliergon cordifolium* and herbs (meso- to eutrophic conditions). If such a forest is flooded up to 30 cm above the surface in spring and autumn, it dies off and is replaced by *Salix* species and *Phragmites australis* reeds (Kozulin et al. 2010a).

On drained and degraded but not extracted bog sites with and without forest, rewetting may be expected to result in the re-establishment of hummock-hollow-lawn complexes, as the necessary dwarf shrubs and *Sphagnum* species have remained on site (cf. Weber 1902, Grummo et al. 2010). Establishment of peat forming vegetation may be initiated from anthropogenic structures like channels and pools (the latter may also originate from root plates of collapsed trees) (Edom & Succow 1998, Wendel 2010). In Belarus a nearly total die-off of *Betula pendula* and *Populus tremula* and a lower productivity of *Pinus sylvestris* trees was observed after raising water levels close to the soil surface. In general, vegetation succession after rewetting is expected to be rather slow. *Calluna vulgaris* disappears in the first three to ten years and only small patches of dwarf shrubs remain on the hummock level. Communities with *Eriophorum vaginatum*, *Carex vesicara*, and *Sphagnum* spec. develop. The expansion of *Sphagnum* spec. depends on the amount of donor mosses in channels or the surroundings (Kozulin et al. 2010a).

On drained and degraded but not extracted fen sites the transition from grassland species to wetland species often starts with open water and single *Typha* spec. islands, accompanied by *Phragmites* reeds with *Glyceria fluitans* and *G. maxima* and eventually developing into *Carex* reeds (Table 12b and 13, Timmermann et al. 2006, Steffenhagen et al. 2010, own observations). Colonization by *Carex* species may take as long as twelve years (Holsten et al. 2001, Steffenhagen et al. 2008). Strongly eu- or polytrophic stands with non-peat forming plants like *Typha* spec. develop on inundated sites, while eu- or mesotrophic stands with e.g. *Phragmites australis*, *Phalaris arundinacea*, *Angelica sylvestris*, *Epilobium palustre* develop on sites rewetted up to or slightly below the soil surface (Schulz 2005, Wendel 2010, own observations).

## 3.5 Prediction of vegetation development with and without rewetting

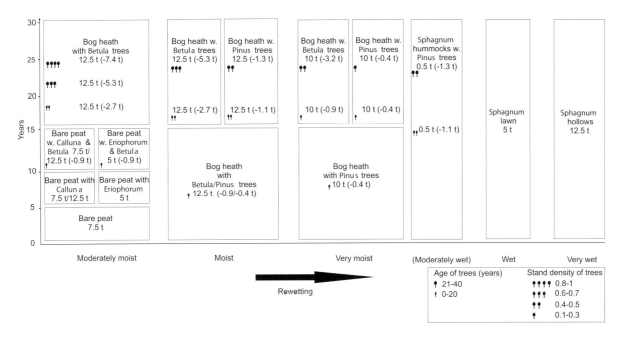

Fig. 35: Vegetation succession on bogs (such as Astrauskoje) in relation to site wetness, with values for GHG fluxes (in t $CO_2$-eq. ha$^{-1}$ yr$^{-1}$) from the peat soil and sequestration in tree biomass (in brackets). See chapter 3.4 for references on emission values. In absence of measured values, for 'Moderately moist bog heath' the same values were taken as for 'Moist bog heath'. Stand density is expressed as proportion of maximum stand density.

Once succession schemes are developed, the predicted vegetation is interpreted in terms of greenhouse gas emission site types (GESTs), which are vegetation types associated with GHG flux values (chapter 3.4). Comparison of the project scenario (with rewetting activities) and the baseline scenario (without rewetting activities) then provides an estimate of emission reductions achieved by the project.

### 3.5.5 Applying GESTs and vegetation prediction – the example of Astrauskoje

Astrauskoje (55°30'225"N, 27°42'631"E, Fig. 66) is a raised bog remnant located directly northwest of the largest bog in Belarus (Jelnia zakaznik, see chapter 8.6). The total peatland area is 701 ha. From the mid 1970s until 1994, the central 125 ha were subject to milled peat extraction (Kozulin 2010). As a consequence, the largely bare central area currently lies up to 1.2 m below the surrounding peatland (Fig. 37 Colour plates II). Drainage for forestry started before 1970 and has led to increased tree growth, especially along the drainage canals. Summer water levels in the extraction area are currently c. 0.6–1 m below the surface. The long-term (1980–2009) mean annual precipitation and air temperature are 640 mm and 6.1 °C (based on data from nearby climate stations, Lesnichiy et al. 2011).

The current (2009) vegetation of Astrauskoje consists of nine vegetation types (Fig. 38A Colour plates II; Couwenberg et al. 2011). *Calluna vulgaris, Eriophorum vaginatum*, and *Polytrichum strictum* occur in all types with high constancy. The GESTs (see chapter 3.4 for explanation of emission values) 'bare peat with *Calluna*' and 'bare peat with *Eriophorum*' are species-poor post-extraction re-vegetation stages without tree cover, the latter being slightly wetter. 'Bare peat with *Polytrichum*' has established after a peat fire in 2002 and is characterised by pioneer species such as *Funaria hygrometrica* and *Drosera rotundifolia*. The global warming potential (GWP) of these sparsely vegetated sites is calculated proportional to the respective cover of bare peat and vegetated patches. A GWP value of 7.5 t $CO_2$-eq. ha$^{-1}$year$^{-1}$ is assigned to 'bare peat' and 'bare peat with Polytrichum'. For the (drier) *Calluna* covered patches, the GWP value of 'moist bog heath' is used (12.5 t $CO_2$-eq. ha$^{-1}$ year$^{-1}$). The area effectively covered by *Eriophorum* is assigned a GWP value of 3.5 t $CO_2$-eq. ha$^{-1}$year$^{-1}$. 'Dry grassland' (20 t $CO_2$-eq. ha$^{-1}$year$^{-1}$), occur-

Table 14: Emissions from Astrauskoje in the current (2009) situation and in the baseline and project scenarios for 2039. See chapter 3.4 for references on emission factors and Fig. 38 Colour plates II for spatial distribution of vegetation types. From Couwenberg et al. (2011), see there for method to calculate the carbon sequestration of trees.

| Vegetation type | Emission factor [t $CO_2$-eq. ha$^{-1}$yr$^{-1}$] | 2009 Current | | 2039 baseline scenario (without active rewetting) | | 2039 project scenario (with rewetting) | |
|---|---|---|---|---|---|---|---|
| | | Area [ha] | Total emissions [t $CO_2$-eq. yr$^{-1}$] | Area [ha] | Total emissions [t $CO_2$-eq. yr$^{-1}$] | Area [ha] | Total emissions [t $CO_2$-eq. yr$^{-1}$] |
| Bare peat | 7.5 | 31.4 | 236 | – | – | – | |
| Bare peat with *Calluna* | 7.5/12.5 | 10.7/17.4 | 298 | – | – | – | |
| Bare peat with *Eriophorum* | 7.5/3.5 | 21.5/18.4 | 226 | – | – | – | |
| Bare peat with *Calluna* and *Eriophorum* | 7.5/12.5/3.5 | 2.9/2.9/3.0 | 69 | – | – | – | |
| Bare peat with *Polytrichum* | 7.5 | 12.9 | 97 | – | – | – | |
| Dry grassland | 20.0 | 11.3 | 226 | 11.3 | 226 | – | – |
| Moist bog heath | 12.5 | 209.3 | 2,616 | 330.4 | 4,130 | 28.0 | 350 |
| Very moist bog heath | 10.0 | 356.4 | 3,564 | 356.4 | 3,564 | 11.8 | 118 |
| Moderately wet *Sphagnum* hummocks | 0.5 | – | – | – | – | 110.9 | 56 |
| Wet *Sphagnum* lawn | 5.0 | 2.6 | 13 | 2.6 | 13 | 482.8 | 2,414 |
| Very wet *Sphagnum* hollows | 12.5 | – | – | – | – | 67.3 | 841 |
| Total emissions [t $CO_2$-eq. yr$^{-1}$] | | | 7,343 | | 7,933 | | 3,779 |
| Sequestration of trees [t $CO_2$-eq.yr$^{-1}$] | | | 1,886 | | 2,406 | | 1,376 |
| Total net flux [t $CO_2$-eq. yr$^{-1}$] | | | 5458 | | 5,527 | | 2,403 |
| Total net flux per ha [t $CO_2$-eq.ha$^{-1}$yr$^{-1}$] | | | 7.8 | | 7.9 | | 3.4 |

ring along drainage channels on excavated substrate, is the driest site type. It is characterised by *Calamagrostis epigeios*, *Dryopteris filix-mas*, and the moss *Dicranum scoparium*. Bog heaths occur on drained but not extracted sites. 'Moist bog heath' is characterised by dwarf shrubs (*Calluna vulgaris*, *Ledum palustre*, *Vaccinium vitis-idaea*), mosses (*Polytrichum strictum*, *Dicranum scoparium*, *Pleurozium schreberi*), and the absence of *Sphagnum*, which all reflect rather dry conditions. 'Very moist bog heath' comprises a vegetation of dwarf shrubs (*Calluna vulgaris*, *Ledum palustre*, *Chamaedaphne calyculata*) with some *Sphagnum magellanicum*, *S. angustifolium*, and *Aulacomnium palustre*. *Pleurozium schreberi* and *Cladonia* spec. indicate slight desiccation. Emission values of 12.5 t $CO_2$-eq ha$^{-1}$year$^{-1}$ for 'moist bog heath' sites and 10 t $CO_2$-eq. ha$^{-1}$year$^{-1}$ for 'very moist bog heath' are used. 'Wet *Sphagnum* lawn' (5 t $CO_2$-eq. ha$^{-1}$ year$^{-1}$) dominated by *Sphagnum angustifolium* and *S. cuspidatum* occurs occasionally in the west of the site and is influenced by mineral soil water (as is reflected by the occurrence of *Juncus filiformis*, *Carex canescens*, *C. nigra*, and *C. vesicaria*).

A possible forward-looking baseline scenario (2039; Fig. 38B Colour plates II; Couwenberg et al. 2011) assumes that, without economic interest in the site, drainage infrastructure in the former extraction site will not be actively maintained. The effectiveness of drainage will decrease because of collapsing ditch banks, vegetation partly filling in the ditches, beaver activity (Mitchell & William 1993), and ongoing peat subsidence (Stephens et al. 1984). As a result, relative (and absolute) water levels will slightly rise but not sufficiently to

## 3.5 Prediction of vegetation development with and without rewetting

stop peat oxidation over large areas during the 30 year project period. Birch trees will expand on the bare peat, favoured by the presence of *Eriophorum vaginatum* tussocks that provide 'safe sites' for germination (Salonen 1990). In the not extracted part, stand density will increase in recently burnt areas. In 2039, the area will be mainly covered by forested bog heath.

The rewetting scenario (2039; Fig. 38C Colour plates II; Couwenberg et al. 2011) assumes that on the area without peat extraction, very wet hollow and wet lawn *Sphagnum*-communities will expand at the expense of the 'very moist bog heath' and that the growth of dwarf shrubs and trees will be impaired leading to a lower quality class of trees. It is not expected that the rise in water level will cause a substantial die-off of the trees, except for some birch trees adjacent to the channels. Therefore, an eventual proportion of 10% 'very wet hollows', 70% 'wet lawn', and 20% 'moderately wet *Sphagnum* hummocks' is assumed. In the peat extraction area, the mean annual water level will be raised to around the surface. This will allow *Sphagnum* communities to develop from the diaspores still available in the ditches. As both the macro- (over the total extraction area) and meso-relief (between ditches) have height differences of a few decimetres up to almost a meter, this will lead to an alternation of (a) permanently inundated sites with open water and *Sphagnum* ('very wet *Sphagnum* hollows'), (b) 'wet *Sphagnum* lawn' and (c) sites where the water level will be more frequently below the surface and the vegetation will be characterised by moist *Eriophorum vaginatum*, dwarf shrubs and some trees ('very moist bog heath'). Based on initial field observation, a proportional distribution of these types of 10%, 80%, and 10%, respectively, is assumed. Upon rewetting, most trees that had established on the peat extraction site after abandonment are likely to die (except on ridges along the drainage channels).

Based on the mapped GESTs, the GWP ($CO_2$ and $CH_4$) for 2009 amounts to 5,486.6 t $CO_2$-eq.

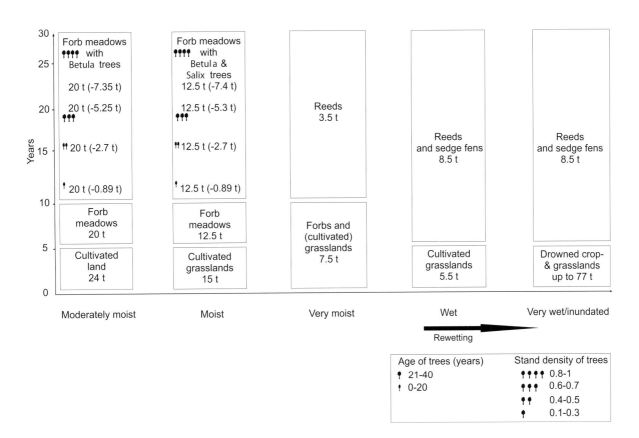

Fig. 36: Vegetation succession on abandoned fens (such as Vyhanascanskaje) in relation to site wetness, with values for GHG fluxes (in t $CO_2$-eq. ha$^{-1}$ yr$^{-1}$) from the peat soil and sequestration in tree biomass (in brackets). See chapter 3.4 for references on emission values. For sequestration in *Salix* trees, sequestration rates of *Betula pendula* yield class IV are used. Stand density is expressed as proportion of maximum stand density.

year$^{-1}$ (in average 7.8 t $CO_2$-eq. ha$^{-1}$ year$^{-1}$; Table 14). Compared to the 5,527 t $CO_2$-eq. year$^{-1}$ (7.9 t $CO_2$-eq. ha$^{-1}$ year$^{-1}$) in the baseline scenario, the project scenario (2,403 t $CO_2$-eq.year$^{-1}$ = 3.4 t $CO_2$-eq. ha$^{-1}$ year$^{-1}$) leads to a conservatively estimated (see chapter 3.4) emission reduction of 3,124 t $CO_2$-eq. year$^{-1}$ (4.5 t $CO_2$-eq. ha$^{-1}$ year$^{-1}$) in 2039. The outcome would be considerably higher in the case of a baseline scenario with continued peat extraction; this option has recently been raised by the Belarusian government.

### 3.5.6 Applying GESTs and vegetation prediction – the example of Vyhanascanskaje

Vyhanascanskaje (52°37'279"N, 25°50'441"E, Fig. 66) is a fen peatland in southwestern Belarus. The focal area for rewetting (1035 ha) is a polder system intensively drained by pumping, used for agriculture and partially abandoned (Fig. 39 Colour plates II).

The current (2009) vegetation consists of three types (Fig. 40A Colour plates II; Couwenberg et al. 2011; see chapter 3.4 for emission values): 'Moderately moist forb meadows' (20 t $CO_2$-eq. ha$^{-1}$ year$^{-1}$) are dominated by *Urtica dioica, Elymus repens,* and *Bidens frondosa*; 'Moist forb meadows' (12.5 t $CO_2$-eq. ha$^{-1}$ year$^{-1}$) developed after abandonment and next to relict grassland species such as *Agrostis gigantea, Festuca rubra,* and *Ranunculus repens,* vegetation is characterized by species such as *Phalaris arundinacea* and *Juncus effusus* that indicate strong water level fluctuations; 'Very moist reeds' (15.5 t $CO_2$-eq. ha$^{-1}$ year$^{-1}$) are characterised by *Carex lasiocarpa, Juncus articulatus, Calamagrostis canescens, Phragmites australis, Carex serotina,* and *Molinia caerulea* and constitute the wettest type on this site.

Forward-looking baseline scenarios (2039; Couwenberg et al. 2011) assume e.g. either that the entire area will be abandoned and drainage infrastructure will not be actively maintained (Fig. 40B Colour plates II) or that the current agricultural use and drainage will continue as in 2009 (Fig. 40C Colour plates II) or even increase (not displayed). In case of abandonment, relative and

Table 15: GHG fluxes from Vyhanascanskaje in the current (2009) situation and in the baseline and project scenarios for 2039. See chapter 3.4 for references on emission factors and Fig. 40 Colour plates II for spatial distribution of vegetation types. From Couwenberg et al. (2011), see there for method to calculate the carbon sequestration of trees.

| | | 2009 | | 2039 | | | | | | | |
|---|---|---|---|---|---|---|---|---|---|---|---|
| | | current | | baseline scenario (without rewetting) | | | | | | project scenario (rewetting) | |
| | | | | abandonment | | ongoing agriculture | | intensified agriculture | | | |
| Vegetation type | Emission factor (t $CO_2$-eq. ha$^{-1}$yr$^{-1}$) | Area (ha) | Total emissions (t $CO_2$-eq. yr$^{-1}$) | Area (ha) | Total emissions (t $CO_2$-eq. yr$^{-1}$) | Area (ha) | Total emissions (t $CO_2$-eq. yr$^{-1}$) | Area (ha) | Total emissions (t $CO_2$-eq. yr$^{-1}$) | Area (ha) | Total emissions (t $CO_2$-eq. yr$^{-1}$) |
| Moderately moist forb meadows | 20.0 | 421.2 | 8,424 | 421.2 | 8,424 | 421.2 | 8,424 | 421.2 | 8,424 | – | – |
| Moist forb meadows | 12.5 | 549.6 | 6,869 | 549.6 | 6,869 | 549.6 | 6,869 | 617.5 | 7,718 | – | – |
| Very moist reeds | 3.5 | 67.9 | 238 | 67.9 | 238 | 67.9 | 238 | – | – | 103.9 | 364 |
| Wet reeds and sedge fens | 8.5 | – | – | – | – | – | – | – | – | 934.8 | 7,946 |
| Total emissions [t $CO^2$-eq. yr$^{-1}$] | | | 15,531 | | 15,531 | | 15,531 | | 16,142 | | 8,310 |
| Sequestration of trees [t $CO_2$-eq.yr$^{-1}$] | | | 1,476 | | 6,980 | | 3,169 | | – | | – |
| Total net flux [t $CO_2$-eq. yr$^{-1}$] | | | 14,055 | | 8,551 | | 12,361 | | 16,142 | | 8,310 |
| Total net flux per ha [t $CO_2$-eq. ha$^{-1}$yr$^{-1}$] | | | 14 | | 8.2 | | 12 | | 16 | | 8.0 |

## 3.5 Prediction of vegetation development with and without rewetting

> **Box 7**
>
> **Qualitative hydromorphological analysis of Astrauskoje**
>
> Frank Edom & Andrei Shunko
>
> To plan and predict peatland rewetting measures, detailed hydro-ecological modelling depends on the following input data on the peatland and its surroundings:
> - A detailed height model to create streamlines (Fig. 41 Colour plates II) as a basis for optimal engineering of rewetting measures;
> - Hydrometeorological data to calculate the atmospheric water balance;
> - Prognosis of regional climate change;
> - Hydrologic data of rivers, lakes and groundwater in the surroundings to predict flooding and calculate water exchange;
> - Hydrochemical data on surface- and groundwater that could influence the vegetation development after rewetting;
> - Stratigraphical knowledge of the peatland basin and its mineral subsoil to calculate geo-hydrological interactions;
> - Understanding of the surface and groundwater catchments and of the location and structure of aquifers;
> - Knowledge on anthropogenic changes of aquifers and groundwater use in the surrounding area;
> - Data on the soil structure of the near surface peat.
>
> In September 2010, a hydro-ecological seminar was organised in the framework of the BMU-ICI project (see also chapter 7.6). The aim of this seminar was to improve the rewetting plan and prediction of vegetation succession of the Astrauskoje project site. Geodetic data from BELGIPROLES (Belarusian governmental agency for design and exploratory works) were complemented by the Geodetic and Geologic Planning Agency GEOPLAN Maladziecna with detailed height measurements of areas where substantial subsidence was expected. Contour lines and streamlines were drawn as a basis for hydromorphologic analysis after Ivanov (1975) and Edom et al. (2007). The streamlines show the direction of water flow with convergent and divergent situations and water divides (Fig. 41 Colour plates II).
>
> In the northeast of the project area convergent streamlines suggest that after rewetting a bog-stream (German: 'Flachrülle', Kästner & Flössner 1933, Edom 2001b, Bogdanovskaya-Gienef 1953, 1969) is likely to develop. Other parts in the north areas show very small slopes and can quickly develop 'very moist bog heath' vegetation. Height differences in the extracted area suggest that two areas in the south and the southeast will be flooded after rewetting. Whereas these flooded areas will have a positive effect on the higher elevated parts, large parts of the extracted area may still fall dry in summer, which will enhance tree growth and peat oxidation. Land use in the surroundings of the project site (gravel extraction and agriculture) is likely to lower the regional ground water level, increasing infiltration losses from Astrauskoje. The raised bog Jelnia (see chapter 8.6) is hydrologically connected to a large groundwater layer (Grummo et al. 2010) with likely geo-hydraulic connections to the gravel extraction site and to aquifers under the Astrauskoje peat body. Rewetting of Astrauskoje may be expected to contribute to the hydrological stabilization of the Jelnia nature reserve. In contrast, continuation of peat and gravel extraction in Astrauskoje will have a negative hydrologic effect on Jelnia.
>
> Without analysing the local hydrogeology and peat stratigraphy, possible water inputs and outputs cannot be understood. Based on geodetic data a fully quantitative hydromorphologic analysis can be carried out (Edom et al. 2007) using the approach of Edom & Golubcov (1996) to predict the development of sites.

absolute water levels will slightly rise during a 30 year project period. Trees and shrubs (*Betula pendula*, *Salix* spec., *Populus tremula*) will likely colonize the area, but it is expected that the established herb vegetation will prevent substantial colonization during the project period.

The project scenario (2039; Fig. 40D Colour plates II, Couwenberg et al. 2011) assumes the development of 'wet reeds or sedge fens' (with a conservatively estimated GWP of 8.5 t $CO_2$-eq. ha$^{-1}$ year$^{-1}$). Continued succession will allow *Carex* reeds to develop. Wet fen sites are typically dominated by aerenchymous plant species and $CH_4$ emissions are correspondingly high (chapter 3.1). Upon rewetting, trees that have established after abandonment are likely to die (Fig. 40D Colour plates II). It is assumed that 10% of the area will remain slightly drier ('very moist reeds'; 15.5 t $CO_2$-eq. ha$^{-1}$ year$^{-1}$).

The resulting GHG emissions for 2009 amount to 14 t $CO_2$-eq. ha$^{-1}$year$^{-1}$ (Table 15). In 2039, the baseline scenarios without rewetting yield substantially less in case of abandonment (8.2 t $CO_2$-eq. ha$^1$ year$^{-1}$), slightly less (12 t $CO_2$-eq. ha$^{-1}$ year$^{-1}$) with agriculture ongoing as in 2009, and slightly more (16 t $CO_2$-eq. ha$^{-1}$ year$^{-1}$) with intensified agriculture, i.e. with additional drainage. The project scenario (with rewetting: 8.0 t $CO_1$-eq. ha$^{-1}$ year$^{-1}$) thus avoids, depending on the choice of the baseline scenario, emissions of 0.2, 4.0, or 8.0 t $CO_2$-eq. ha$^{-1}$ year$^{-1}$, respectively.

### 3.5.7 Conclusions and way forward

The assignment of vegetation types to literature based emission factors (see chapter 3.4, Couwenberg et al. 2011) bears some uncertainties and limitations (Joosten & Couwenberg 2009):
- Differences in competitive interactions may lead to changes in indication values of species (Kotowski et al. 1998; Koska et al. 2001), which necessitates calibrating the relationship between vegetation, water level and GHG fluxes for different climatic and phytogeographical regions (cf. Schroeder 1998; Frey & Lösch 1998).
- Vegetation reacts slowly to environmental changes. It may take several years before changes in site conditions following rewetting are sufficiently reflected in the vegetation composition. Some direct and immediate indication is given by the die-off of species, but the establishment of new species may take more time. The time period over which vegetation adequately reflects changing environmental conditions generally does not exceed the five years of standard verification frequency of carbon projects (Koska et al. 2001).

Predicting vegetation development without detailed knowledge about future hydrology and soil structures is strongly simplified. For example, the assumption of optimal rewetting at Astrauskoje will not hold as changes in mesorelief and soil structure (as a result of drainage and extraction) will prevent optimal rewetting on every single spot. To make more robust predictions in terms of rewettability of severely degraded peatlands and subsequent vegetation succession, additional hydrological data are needed (box 7).

Available data and expertise on vegetation succession in peatlands already allows for realistic prediction of baseline and project scenarios. Combined with the GEST tool, these scenarios provide a solid basis for ex-ante assessment of emission reductions. Developments will be monitored during the verification process of carbon projects.

Fig. 19: Regulator pipe for active regulation of the water level at Barcianicha peatland (photo: Olga Chabrouskaya).

Fig. 28: Closed chamber measurements in Raznianskaje balota. In the foreground: measuring net ecosystem $CO_2$ exchange with transparent dynamic closed chambers that allow for photosynthesis of the enclosed vegetation to continue. Note the blue cooling packs and the air fans. A sensor for measuring photosynthetic active radiation is mounted on a pole nearby. In the background: measuring $CH_4$ and $N_2O$ exchange with opaque static closed chambers. Note the glass flasks used for taking air samples from the chamber atmosphere (photo: Merten Minke).

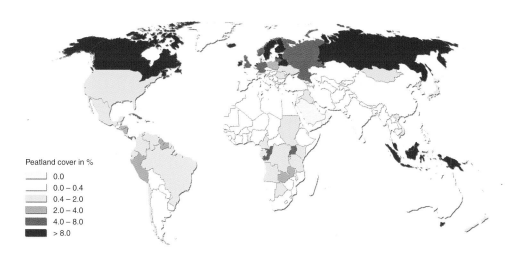

Fig. 25: Peatland cover (% of national land area) per country (based on data from the IMCG Global Peatland Database, Joosten 2009d) (map: Stephan Busse).

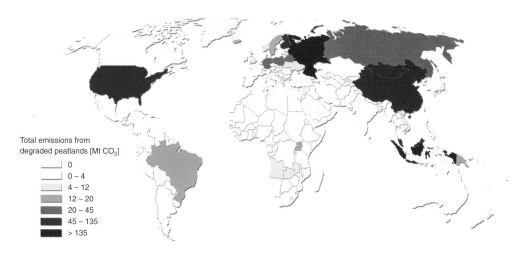

Fig. 26: Emissions from peatlands per country (in Mt Mt $CO_2$ yr$^{-1}$; based on data from the IMCG Global Peatland Database, Joosten 2009d) (map: Stephan Busse).

Fig. 27: Emissions from peatlands per unit national land area per country (in t ha$^{-1}$; based on data from the IMCG Global Peatland Database, Joosten 2009d) (map: Stephan Busse).

Fig. 29: Eddy covariance tower in Peene Valley near Anklam, Germany (photo: Michael Giebels).

Fig. 37: Border of the extracted and non-extracted part, separated by a channel, at Astrauskoje (photo: Annett Thiele).

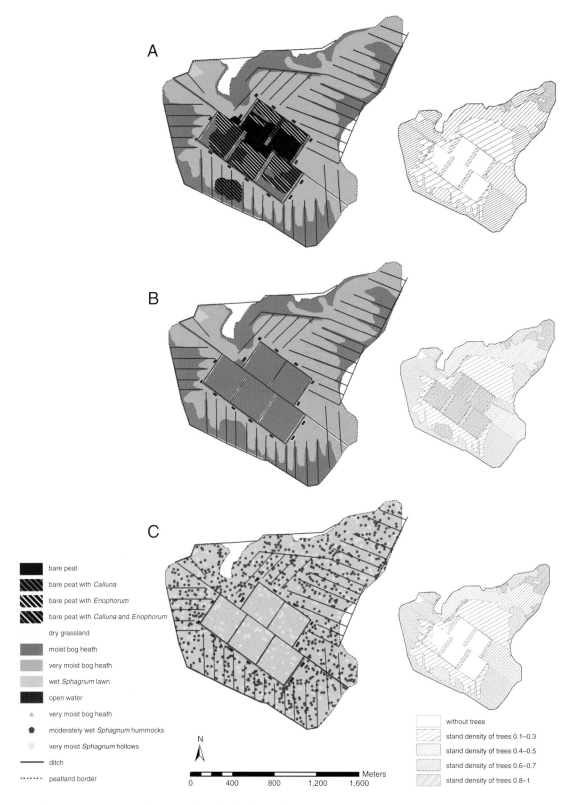

Fig. 38: Current and projected vegetation distribution of Astrauskoje. A In 2009 ('bare peat with *Polytrichum*' is located at the southern margin of the peatland and bears the same symbol as 'bare peat with *Calluna*'); B in 2039 without rewetting (baseline scenario); and C in 2039 with rewetting (project scenario; the distribution of hollows, hummocks, and very moist bog heath is spatially not explicit). The black and white maps at the right hand side show associated tree stand densities. From Couwenberg et al. (2011).

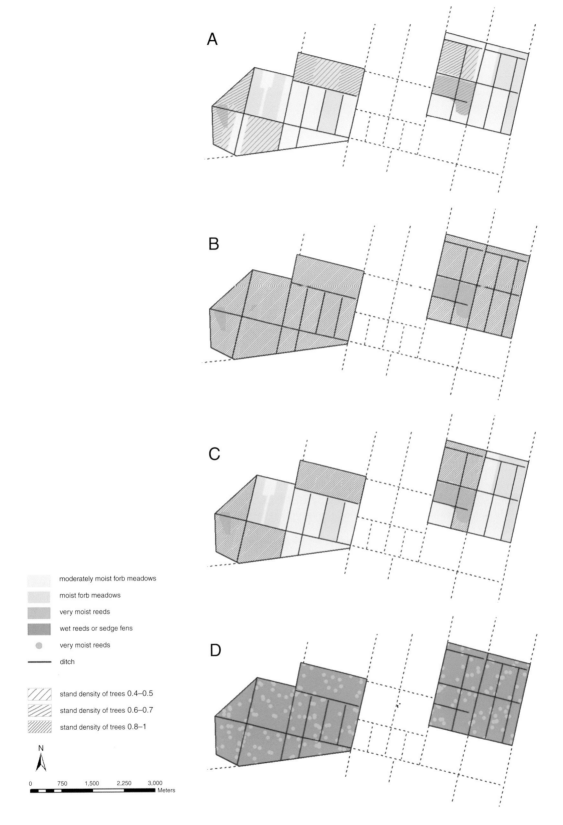

Fig. 40: Current and projected vegetation distribution of Vyhanascanskaje. A In 2009; B in 2039 without rewetting after abandonment; C in 2039 without rewetting, with ongoing agriculture; and D in 2039 with rewetting. Dotted lines depict ditches outside the project area. From Couwenberg et al. (2011).

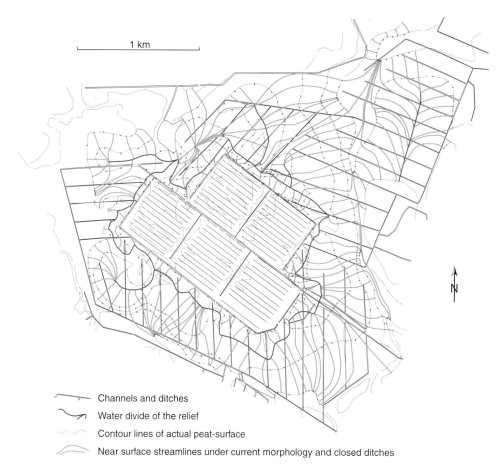

- Channels and ditches
- Water divide of the relief
- Contour lines of actual peat-surface
- Near surface streamlines under current morphology and closed ditches

Fig 41: Hydromorphological characterization (net of contour lines and streamlines) of Astrauskoje. Ideal rewetting is assumed to make all ditches ineffective leading to water flowing through near surface layers following the peatland relief as shown by streamlines. Geodetic survey by GEOPLAN Ltd. (Maladziecna) with additional data from BEL-GIPROLES (Minsk), streamline and watershed construction by F. Edom, digitized by GEOPLAN.

Fig. 44: Aquatic Warbler *Acrocephalus paludicola* (photo: Zydrunas Sinkevicius).

Fig. 45: Greater Spotted Eagle *Aquila clanga* (photo: Valery Dombrovsky).

Fig. 39: The poldered fen peatland Vyhanascanskaje (photo: Annett Thiele).

Fig. 46: Marsh Cinquefoil *Potentilla palustris*, a common fen plant (photo: Franziska Tanneberger).

Fig. 49: Rewetted peat extraction site in Poplau Mokh (May 2011; photo: Susanne Bärisch).

Fig. 50: Moor Frog *Rana arvalis* at Poplau Moch (June 2011; photo: Rob Field).

# 4 Peatlands and biodiversity

Peatlands exhibit highly characteristic ecological traits and are unique, complex ecosystems. They are of global importance for biodiversity conservation at genetic, species, and ecosystem levels.

This section introduces biodiversity values of Belarusian peatlands, including those of famous protected areas such as Zvaniec and Biarezinski zapaviednik and the 'national bird species' the Aquatic Warbler *Acrocephalus paludicola*, a species characteristic for fen mires (chapter 4.1). Chapter 4.2 further establishes the relationship between peatland condition and biodiversity values, discusses restoration of Aquatic Warbler habitats, and presents data from peatlands rewetted after the Chernobyl accident as well as in a recent rewetting project in Belarus. Considerations for the selection of target and indicator species are provided in chapter 4.3. The long-term management of rewetted sites is described in chapter 4.4, which is accompanied by a case study in Belarus.

## 4.1 Biodiversity values of Belarusian peatlands

Viktar Fenchuk & Norbert Schäffer

### 4.1.1 Introduction

Peatlands are of key importance for biodiversity and comprise a wide spectrum of rare, threatened and endangered habitats, plants, and animals (see also chapter 4.2). Due to the narrow range of environmental conditions, many species are peatland specialists and therefore have restricted distribution. The avifauna of bogs and fens, for example, has a high conservation value at the European level and is especially sensitive to changing environmental conditions (Anderson et al. 2010).

Although more than half of all Belarusian peatlands have been drained during the large-scale drainage campaigns in the middle of the 20th century, and a substantial part has suffered from adjacent drainage, more than 800,000 ha are still in near-natural condition (Kozulin & Tanovitskaya 2010). This large area of not, or slightly, disturbed wetlands is home to a variety of habitats and species that have a high European conservation status.

This chapter details the importance of peatlands for birds in Belarus, provides background information on the national Aquatic Warbler breeding population and describes the avifauna of two near-natural sites – Zvaniec zakaznik and Biarezinski zapaviednik.

### 4.1.2 Birds

In order to identify sites of key importance for bird conservation, APB-BirdLife Belarus has been implementing a programme of designation of Important Bird Areas (IBAs), as part of the international IBA programme initiated by BirdLife International. In early 2011, there were 49 IBAs in Belarus. These are located mostly in the northern Poozerie and southern Paliessie regions of Belarus, which are characterised by substantial areas of bogs and fens, respectively. Of the 49 IBAs, 84% are wetlands and include natural or slightly disturbed peatlands. The majority of peatland IBAs have been designated according to criterion A, i.e. they are of global importance.

Five bird species globally recognized as 'vulnerable' (VU) according to the IUCN Red List occur in Belarus. Two are breeding birds – the Aquatic Warbler *Acrocephalus paludicola* (see box 8 and Fig. 44 Colour plates II) and the Greater Spotted Eagle *Aquila clanga* (Fig. 45 Colour plates II) – and peatland specialists. The Aquatic Warbler is an indicator species of open fen mires and the Greater Spotted Eagle is confined almost exclusively to large undisturbed mire and forest complexes. Recent research shows a sharp decline of Greater Spotted Eagle at sites, affected by drainage (Dombrovsky 2010).

Another ten species are classified as 'near threatened (NT)', including five breeding species, such as peatland specialists Great Snipe *Gallinago media*, Black-tailed Godwit *Limosa limosa*, and Eurasian Curlew *Numenius arquata*.

The Belarusian populations of 13 species of high European conservation concern (SPEC 1–3, all with nationally unfavourable conservation status) represent more than 5% of their European population (see Table 16). The majority of Belarusian bird species with unfavourable conservation status are to some extent related to peatlands.

Protecting a substantial area of mires despite of large-scale drainage, Belarus has been always focused on protecting wetland habitats and species. Today, the occurrence of wetland birds (and all wetland fauna and flora) depends heavily on a network of protected areas. A large proportion of the national populations of several habitat restricted bird species occurs in protected areas, e.g. 95% of the Aquatic Warblers, c. 80% of the Black-tailed Godwits, 80% of the Eurasian Curlews, 95% of Eurasian Golden Plovers *Pluvialis apricaria*, 80% of Greater Spotted Eagles, and 80% of Great Snipes (A. Kozulin pers. comm.).

### 4.1.3 The Aquatic Warbler in Belarus

This species has been (re-)discovered in Belarus as recently as 1995. Currently, there are 23 regularly occupied breeding sites known in Belarus, home to between 3,000 and 5,500 singing males (latest available data collected in 2010, Fig. 42). With c. 40% of the world population (as of 2010), Belarus is therefore (together with Ukraine) the most important breeding country for this species. The most important site in the country (and the world) is the Zvaniec mire, contributing as much as 30% of the world population and c. 75% of the national population. Together with two other important sites (Sporava and Dzikoje) it houses 95% of the national population, while the other 20 sites contribute the remaining 5%. The total area occupied by Aquatic Warblers is estimated at 9,016 ha, while the total area of available, apparently suitable, habitat at all 23 sites is c. 14,500 ha. Potentially suitable habitat at occupied sites (areas that could

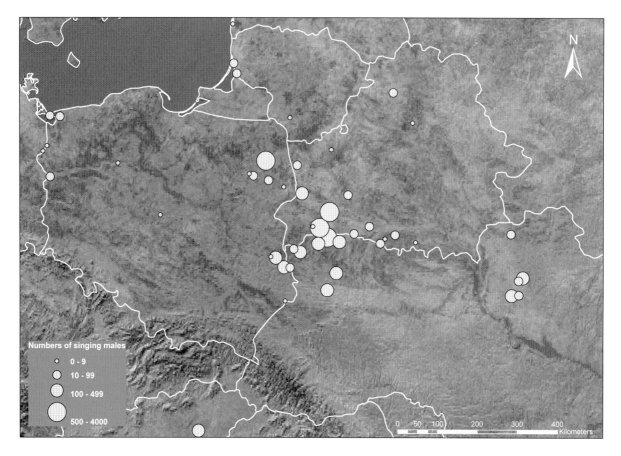

Fig. 42: The extent of all occupied Aquatic Warbler breeding sites in Europe for the year 2009. Not included is one small and apparently highly nomadic population of Aquatic Warblers which possibly still exists in western Siberia. The map highlights the extremely fragmented breeding distribution of the species (map: Uladzimir Malashevich).

# 4.1 Biodiversity values of Belarusian peatlands

**Box 8**

**Portrait of the Aquatic Warbler**

Franziska Tanneberger

The Aquatic Warbler (*Acrocephalus paludicola* Vieillot 1817) is a small (c. 12 g) song bird, which breeds in the fen mires in Europe and Siberia and winters in similar wetlands in West Africa. The main field characteristics of the species are the badger-like head pattern (a cream-coloured crown stripe separated by two clear dark stripes from the again cream-coloured over-eye stripes) and a back pattern with strongly contrasting cream and black longitudinal stripes (Schulze-Hagen 1991). In contrast to all other *Acrocephalus* warblers, the Aquatic Warbler has a breeding system characterised by uniparental care by the female. Males hold feeding territories, but stable pairs are not formed. Males and females mate with several partners, and this leads to intense sperm competition, and around 75% of broods are fathered by two or more males (Dyrcz at al. 2002). A rich food supply is essential because the female must feed her (usually four to five) nestlings alone (Leisler & Catchpole 1992). The males' home ranges cover up to 8 ha and overlap widely, according to a study from Biebrza Valley/Poland (Schulze-Hagen et al. 1999, Schaefer et al. 2000).

The Aquatic Warbler was once widespread and numerous throughout Europe (Schulze-Hagen 1991, Cramp 1992). Historical evidence from northern Germany and western Poland describes the species as ' … it occurs all over the vast fen mires … at times in particularly large aggregations' (Hesse 1910) and as 'exceptionally numerous' (Hübner 1908). Today,

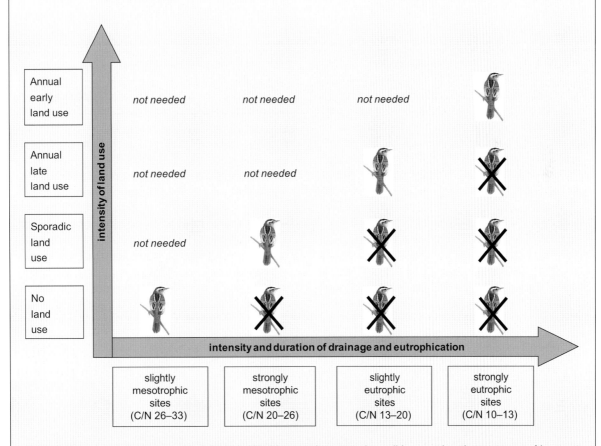

Fig. 43: Model of the occurrence of Aquatic Warblers under natural conditions and at three stages of increasingly intensive and prolonged drainage and eutrophication in relation to the intensity of land use (mire typology according to Succow & Joosten 2001, from Tanneberger 2008).

the Aquatic Warbler is the rarest and only globally threatened song bird of mainland Europe. It is classified as globally 'vulnerable' (BirdLife International 2009) based on the recent catastrophic population decline and its small current range of less than 1,500 km$^2$ (probably only about 400 km$^2$). It is also included in Annex I of the EU Birds Directive (79/409/EEC), in Appendix II of the Bern Convention and in Appendix I of the Bonn Convention. Under the latter convention, an 'International Memorandum of Understanding Concerning Conservation Measures for the Aquatic Warbler' was signed since 2003 by 14 out of 22 CMS-recognised range states, including all of the most important countries from the breeding, migration and wintering range.

Aquatic Warblers currently breed in seven countries, six of them in central and eastern Europe, and a small and possibly extinct population in the western Siberian region of Russia. The global population of the species is currently estimated between 11,500 and 16,400 singing males (data from 2009), with the biggest national populations in Belarus (4,000–7,600), Ukraine (4,000–4,700) and Poland (c. 3,200), together making up around 95% of the world population (Fig. 42). 68–78% of the total population is concentrated at only four key sites: Zvaniec (2,000–6,000) and Sporova (500–1,000) in Belarus, Biebrza Valley (c. 2,600) in Poland, and the Ukrainian Upper Prypiat (2,700–3,200) (data from 2009). The remaining part is concentrated at twelve other very important sites with 100–500 singing males each: Nemunas Delta, Curonian Lagoon (Lithuania), Poleski National Park, Chelm Marshes (Poland), Dzikoje (Belarus), Middle Styr, Chornoguzka, Lake Bile/Pisochne, Lower Stokhid, Udai, Supoy (Ukraine), and Hortobágy (Hungary). A further 20 sites exist with regular occurrence of more than ten singing males. Altogether, the European breeding population is currently restricted to about 61 regularly occupied breeding sites, of which only 35 sites holding more than ten males each (data from Aquatic Warbler Conservation Team database, pers. comm. U. Malashevich, M. Flade, and L. Lachmann).

Originally, Aquatic Warblers found their natural habitat probably in low productivity fen mires. With increased drainage and eutrophication, Aquatic Warbler habitats became increasingly dependent on human land use (Fig. 43). The most important and cost-effective management measure is therefore to reduce the intensity of drainage as much as possible. In mesotrophic sites (according to the mire typology of Succow & Joosten 2001), sporadic mowing, grazing or burning are sufficient to maintain Aquatic Warbler habitats (AWCT 1999, Kozulin & Flade 1999), whereas slightly eutrophic sites rely on annual late-summer mowing (Kloskowski & Krogulec 1999), and the more strongly eutrophic sites on annual early summer mowing (Tanneberger et al. 2008, 2010a). Early summer mowing creates conflicts of management, since it creates suitable habitat, but destroys nests. Protection of nesting areas, or the establishment of late and early mown habitat blocks is a viable solution to this conflict.

In addition to management to maintain current breeding sites ('maintenance management'), it may be necessary to implement management to restore former breeding sites that had become unsuitable through the lack of management in the past ('restoration management'). In such a situation, it will be necessary to temporarily go beyond the normal management practice for a site of a specific nutrient status. This may include the removal of bushes but also a more intensive mowing regime to restore unsuitable sites to suitable conditions. During the implementation of restoration management (e.g. annual early mowing) a site might be unsuitable for breeding Aquatic Warblers, so it should only be implemented on sites that are currently not occupied by the species and should be stopped and replaced by normal maintenance management as soon as the species appears on the site.

be restored to good habitat condition), is estimated at c. 35,000 ha for the seven most important sites, but may be as large as 70,000 ha across all Belarusian sites (U. Malashevich pers. comm.).

The national population trend is uncertain, since detailed monitoring has only recently been established. However, between 1996 and 2005, monitoring has concentrated on core population

## 4.1 Biodiversity values of Belarusian peatlands

Table 16: Species of European Conservation Concern (SPEC) breeding in Belarus with proportion of European population in % (from BirdLife International 2004, 2009).

| Species | | SPEC | Proportion of European population in Belarus (in %) |
|---|---|---|---|
| English name | Scientific name | | |
| Aquatic Warbler | Acrocephalus paludicola | 1 | 47.9 |
| Greater Spotted Eagle | Aquila clanga | 1 | 18.3 |
| Lesser Spotted Eagle | Aquila pomarina | 2 | |
| Black-tailed Godwit | Limosa limosa | 2 | 21.2 |
| Redshank | Tringa totanus | 2 | 6.1 |
| Lapwing | Vanellus vanellus | 2 | 12.4 |
| Black Stork | Ciconia nigra | 2 | 5.8 |
| Snipe | Gallinago gallinago | 3 | 11.4 |
| Black Tern | Chlidonias niger | 3 | |
| Garganey | Anas querquedula | 3 | 5.7 |

at the three key sites in the country. This has indicated strong fluctuations of the local populations due to flooding, drought and fire, and was instrumental in the development and implementation of conservation measures that has led to a stabilisation of local breeding conditions. Since 2006, representative sampling has been used at the largest site (Zvaniec), complemented by nearly full counts at the smaller sites. This method has delivered more accurate national population figures, but has required a downward correction of the national population estimate. The decrease of suitable open habitat at several sites, together with abandonment of small peripheral sites, may indicate a possible decrease of the overall national population. This, however, can at the moment not be tested using the available national monitoring data, although this should be possible in the future using data from the recently improved monitoring scheme (L. Lachmann pers. comm.).

Table 17: Average population size and density on monitoring plots of selected bird species in Zvaniec (from Shimova et al. 2010). + indicates presence on the peatland, but not on the plots.

| Species | | | Density in 2000–2006 on two monitoring plots per 1 ha/100 km² | | | | | | |
|---|---|---|---|---|---|---|---|---|---|
| English name | Scientific name | Average population size 1995–2001 (pairs) | 2000 | 2001 | 2002 | 2003 | 2004 | 2005 | 2006 |
| Bittern | Botaurus stellaris | 50–300 | 3.0 | 0 | 1.0 | 0 | 0 | 1.2 | 0 |
| Lesser Spotted Eagle | Aquila pomarina | 3–4 | no data | | | | | | |
| Greater Spotted Eagle | Aquila clanga | 2 | no data | | | | | | |
| Montagu's Harrier | Circus pygargus | 10–20 | no data | | | | | | |
| Spotted Crake | Porzana porzana | 1000–4000 | + | + | + | 0 | 0 | 4.4 | 0 |
| Corncrake | Crex crex | 50–100 | 0 | 0 | 0 | 2.2 | 0 | 1.1 | 0 |
| Common Snipe | Gallinago gallinago | 500–1000 | 0 | 6.6 | 16 | 6.6 | 10.5 | 13.4 | 8.7 |
| Great Snipe | Gallinago media | no data | 0 | 0 | 0 | 0 | 20.7 | 10 | 0 |
| Eurasian Curlew | Numenius arquata | 50 | 2 | 3 | 1 | + | + | + | 0 |
| Aquatic Warbler | Acrocephalus paludicola | 3000–6000 males | 96 | 0 | 103 | 81.1 | 34.2 | 40.3 | 102.2 |

## 4 Peatlands and biodiversity

**Box 9**

**How Belarusians value the biodiversity of their peatlands – a case study**

Sviataslau Valasiuk, Marek Giergiczny & Tomasz Zylicz

Unfavourable habitat changes triggered by massive drainage and almost complete cessation of traditional low intensity use of fen mires, grazing and hay making, have occurred at Zvaniec (see main text) during the last decades. This has negatively affected both the Belarusian and the world populations of the globally threatened Aquatic Warbler (see box 8). A biomass management programme could prevent or mitigate these undesirable consequences.

A biodiversity valuation study was conducted to estimate the willingness of Belarusians to pay for a complex conservation programme. 300 ordinary Belarusians (from Minsk, Brest, Hrodna, and Viciebsk, district towns and in rural areas) were proposed to participate in a choice experiment (CE) study in order to declare their preferences for different site conservation programmes. These programmes varied in their costs and the degree to which they meet the habitat requirements of the Aquatic Warbler and wider conservation needs (Table 19).

Each respondent was asked to rank their choices from most to least favoured. The results suggested that Belarusians are not only willing to pay for the protection of Zvaniec, but that they prefer scything (the best management for habitat conservation) over other cheaper management options. Furthermore, these results agree with other studies from Central and Eastern Europe that indicate that active conservation is an economically efficient spatial management option, where total benefits (including 'willingness to pay') exceed total project costs within a wide range of costs levels (for instance, see Markowska & Zylicz 1999).

Table 19: Attributes and levels used in the choice experiment. BAU: Business-As-Usual.

| Attribute | Description | Levels |
|---|---|---|
| Method of removing shrubs | Four different methods contemplated by reserve management team and biologists were accounted for. Respondents were informed about pros and cons of each technique. BAU*= none | 1) Manual (hand scything)<br>2) Mechanical mowing by site-adapted machinery (e.g. 'ratrac')<br>3) Controlled burning in winter<br>4) Using herbicides |
| Area | Annual area over which the shrubs would be removed (ha/year). BAU = 0 | 1) 1,000  3) 3,000<br>2) 2,000  4) 4,000 |
| Reserve | Enlarging protection level of the reserve from the current 10,000 ha. BAU = 0 | 1) 0         3) +4,000 ha<br>2) +2,000 ha  4) +6,000 ha |
| Cost | Annual cost per person (2010 prices). BAU = 0 | 1) 30,000 BYR   3) 110,000 BYR<br>2) 70,000 BYR   4) 150,000 BYR |

Table 18: Number of birds per km$^2$ of selected bird species in different peatland types in Biarezinski zapaviednik (from Byshnew et al. 1996).

| Species | | Birds/km$^2$ | Biotope type |
|---|---|---|---|
| English name | Scientific name | | |
| Mallard | *Anas platyrhynchos* | 1.8–2.6 | open fens |
| Teal | *Anas crecca* | 2.5 | open bogs and transitional mires, pine forests |
| Black Grouse | *Tetrao tetrix* | 2–4 | open bogs |
| Black Grouse | *Tetrao tetrix* | 2–8 | marshy pine forests |

## 4.1 Biodiversity values of Belarusian peatlands

| Species | | Birds/km² | Biotope type |
|---|---|---|---|
| English name | Scientific name | | |
| Black Grouse | *Tetrao tetrix* | 12 | transitional mires |
| Capercaillie | *Tetrao urogallus* | 1.6–5 | marshy pine forests |
| Crane | *Grus grus* | 0.77–0.85 | *Carex-Sphagnum* bogs |
| Crane | *Grus grus* | 0.11–0.13 | other bogs |
| Crane | *Grus grus* | 0.11–0.4 | fens and transitional mires |
| Lapwing | *Vanellus vanellus* | 0.8–15.3 | open mires |
| Wood Sandpiper | *Tringa glareola* | 3.4 | open bogs and transitional mires |
| Common Snipe | *Gallinago gallinago* | 23–38 | open eutrophic fens and low sparse birch forests |
| Common Snipe | *Gallinago gallinago* | 2.6–3.8 | open bogs and *Carex-Sphagnum* pine bogs |
| Nightjar | *Caprimulgus europaeus* | 4 | marshy pine forests |
| Great spotted woodpecker | *Dendrocopos major* | 4–12 | pine bogs |
| Tree Pipit | *Anthus trivialis* | 104–150 | pine bogs |
| Tree Pipit | *Anthus trivialis* | 13–78 | oligotrophic open bogs |
| Tree Pipit | *Anthus trivialis* | 12–24 | mesotrophic mires |
| Tree Pipit | *Anthus trivialis* | 6–18 | eutrophic mires |
| Meadow Pipit | *Anthus pratensis* | 50–70 | open oligotrophic and eutrophic mires |
| Meadow Pipit | *Anthus pratensis* | 8–20 | pine bogs |
| Yellow Wagtail | *Motacilla flava* | 5–95 | oligotrophic open bogs |
| Lesser Grey Shrike | *Lanius collurio* | 0–3 | bogs |
| Hooded crow | *Corvus cornix* | 6–8 | pine bogs |
| Raven | *Corvus corax* | 0.5–1.5 | pine bogs |
| Willow Warbler | *Phylloscopus trochilus* | 1.5–14 | oligotrophic open bogs |
| Willow Warbler | *Phylloscopus trochilus* | 13.2 | mesotrophic mires |
| Chiffchaff | *Phylloscopus collybita* | 1.6–6 | mesotrophic mires |
| Wood Warbler | *Phylloscopus sibilatrix* | 4–6.5 | pine bogs |
| Pied Flycatcher | *Ficedula hypoleuca* | 3.1 | pine bogs |
| Spotted Flycatcher | *Muscicapa striata* | 3.4–4 | pine bogs |
| Whinchat | *Saxicola rubetra* | 52 | oligotrophic open bogs |
| Whinchat | *Saxicola rubetra* | 3–19 | low pine stands on bogs |
| Whinchat | *Saxicola rubetra* | 16–26 | open parts of transitional bogs |
| Whinchat | *Saxicola rubetra* | 4–12 | open parts of fens |
| Chaffinch | *Fringilla coelebs* | 28–35 | pine bogs |
| Reed Bunting | *Emberiza schoeniclus* | 136 | open eutrophic mires |
| Reed Bunting | *Emberiza schoeniclus* | 7–96 | open transitional mires |

### 4.1.4 Zvaniec mire

Zvaniec is Europe's largest mesotrophic fen mire with numerous mineral islands. The total area included in the IBA is 15,873 ha and includes 10,460 ha protected as a national landscape reserve (zakaznik). A total of 110 bird species breed on the mire, of which 21 are listed in the National Red Data Book. The site is internationally important because it supports large populations of the globally threatened Aquatic Warbler, Corncrake *Crex crex*, and Greater Spotted Eagle. It also hosts more than 1% of the European populations of Bittern *Botaurus stellaris* and Water Rail *Rallus aquaticus*. The site is of national importance for the conservation of Common Crane *Grus grus*, Snipe *Gallinago gallinago*, and Eurasian Curlew (BirdLife International 2011).

In addition, 67 vascular plant species requiring various levels of protection are found in the IBA (Fig. 46 Color plates II), including 23 listed in the National Red Data Book. Most of these plants occur on the mineral islands and form a unique and outstandingly diverse flora. The mire hosts ten rare vegetation communities that were once widespread across Paliessie. Two mammals, 15 terrestrial and three water invertebrate species from the National Red Data Book occur on the IBA (BirdLife International 2011).

The biodiversity of Zvaniec in the period 1995–2006 has been summarised e.g. by Shimova et al. (2010). During this period, the hydrological regime has been slightly disturbed but the available data can be still seen as largely representative for undisturbed fen mires. Table 17 summarises information on population size and density of selected bird species.

### 4.1.5 Biarezinski zapaviednik

Another important reference site in Belarus is Biarezinski zapaviednik (85,000 ha), established in 1925. It harbours about 800 vascular plant species, 216 moss species, 238 lichen species, 317 water plants and 463 fungal species as well as 56 species of mammals, 230 species of birds, 5 reptiles, 11 amphibians, and 34 fish species (information from www.berezinsky.com). Table 18 summarises the number of individuals per km$^2$ of selected bird species in peatlands of Biarezinski zapaviednik.

## 4.2 Relationship between peatland condition and biodiversity values

Franziska Tanneberger

### 4.2.1 Introduction

Peatlands are unique, complex ecosystems of global importance for biodiversity conservation at genetic, species, and ecosystem levels. They contain many species found only or mainly in peatlands. These species are adapted to the special waterlogged conditions of peatlands. They are vulnerable to changes resulting from direct human intervention, changes in their water catchment, and climate change that may lead to loss of habitats, species and associated ecosystem services. The biodiversity values of peatlands demand special consideration in conservation strategies and land use planning.

Peatlands play a special role in maintaining biodiversity at the species and genetic level as a result of habitat isolation and at the ecosystem level as a result of their ability to self-organise and adapt to different physical conditions. The following should be considered:

- Although species diversity in peatlands may be lower, they have a higher proportion of characteristic species than dryland ecosystems in the same biogeographic zone;
- Peatlands may develop sophisticated self-regulation mechanisms over time, resulting in high within-habitat diversity expressed as conspicuous surface patterns;
- Peatlands are important for biodiversity far beyond their borders by maintaining hydrological and micro-climate features of adjacent areas and providing temporary habitats or refugia for dryland species;
- Peatlands are often the last remaining natural areas in degraded landscapes and thus mitigate landscape fragmentation. They also support adaptation by providing habitats for endangered species and those displaced by climate change;
- Peatlands are vulnerable to human activities both within the peatland habitats themselves and in their catchments. Impacts include habitat loss, species extinction and loss of associated ecosystem services;
- The importance of peatlands for maintaining global biodiversity is usually underestimated, both in local nature conservation planning and practices, as well as in international convention deliberations and decisions (after Parish at al. 2008).

## 4.2 Relationship between peatland condition and biodiversity values

**Box 10**

**Restoring Aquatic Warbler breeding sites**

Franziska Tanneberger &
Uladzimir Malashevich

In all countries with breeding Aquatic Warblers (see box 8 in chapter 4.1), the area occupied in the past was larger than the area currently occupied. In addition, several European countries have completely lost their breeding Aquatic Warblers and serve today only as stopover countries (e.g. Belgium, The Netherlands, France, Italy). This means that there is now a very large area of abandoned former breeding sites, which potentially could be restored.

The restoration of Aquatic Warbler breeding habitats can follow different strategies:
- Enlarge the suitable area within or adjacent to current breeding sites, especially in those with declining populations;
- Restore sites that became abandoned by Aquatic Warblers;
- Identify and manage sites that have a large potential as Aquatic Warbler habitat (but where no historical records exist).

The enlargement of current breeding sites is clearly most desirable. Site enlargement is usually done through the management of water levels and of vegetation (e.g. bush removal, mowing). Restoration of additional currently unoccupied sites is, however, also crucial for the conservation of Aquatic Warblers to achieve the long-term target of increasing its range by 50% by 2025 (BirdLife International 2009).

Conservation activities therefore need to focus on sites that either already provide suitable habitat or could do so after management intervention (e.g. degraded habitat formerly occupied or suitable). When choosing restoration sites, priority should be given to sites that can become 'stepping stones' to connect existing populations. Equally, they should not be too far from potential source populations to provide a reasonable chance of colonization. Such a management plan has been carried out for the Aquatic Warbler in Brandenburg/Germany (Tanneberger et al. 2010b) based on remote sensing and environmental data. Brandenburg is a federal state in northeast Germany of 29,478 km$^2$ bordering Poland. The state was formerly a stronghold of the species with at least 15 active breeding sites in the early 20th century. A total of 3,785 km$^2$ is covered with soil types consistent with those found at sites with historical Aquatic Warbler records. This figure, however, includes an unknown number of unsuitable sites (due to size, isolation, or vegetation). The management plan has selected a smaller set of potentially suitable sites. From this subset of sites, areas have been identified to which future surveys and planning should be targeted. Six priority areas for habitat restoration (covering 4,748 ha) have been identified, based on the following additional criteria:
- An area of at least 200 ha;
- A distance of < 40 km from existing breeding sites;
- If >40 km from existing breeding sites: a distance of <40 km from other potential breeding sites;
- Suitability of current hydrological regime and vegetation;
- Synergies with ongoing projects;
- Agreement with other land use interests.

Restoration of Aquatic Warbler habitats in Belarus is illustrated by two examples: Zvaniec and Scara-Dabramysl.

Zvaniec (52°3.00'N, 24°51.00'E; see chapter 4.1 for a detailed description of the site) is located on the divide between two river systems (Dniepr and Buh). Large areas are still in a near-natural hydrological state. However, there are negative impacts on the water levels, especially because of the Dniaproŭska-Buhski Canal and a fish farm adjacent to the mire. In 2004–2006, water-regulating facilities were built to stabilize water levels (funded by the UK Darwin Initiative). Since these measures did not prevent flooding caused by increased precipitation, in 2009–2010 new facilities were built, which have allowed water flow in and out of the peatland to be regulated effectively (funded by United Nations Development Programme (UNDP), Global Environment Facility (GEF), and Royal Society for the Protection of Birds (RSPB)). The area is partly (c. 20% of the area yearly) used for haymaking. Aquatic Warbler monitoring began in 1995–2005 with basic counts on only two sample plots. The population size estimate for this period based on extrapolation from these plots was 3,000–6,000 males. A detailed random stratified survey began in 2006. Using this method, it will be possible to examine the impact of habitat restoration on the Aquatic Warbler. Relatively comparable data are available for 2006, 2009, and 2010. These indicated that the number of males was 4,223–5,159 in 2006, 2,896–5,798 in 2009, and 2,254–4,428 in 2010. It is hoped that current management activities will keep the Aquatic Warbler numbers stable and reduce encroachment by scrub and reeds. However, active vegetation management will be needed to maintain suitable habitat conditions (A. Kozulin pers. comm.).

Scara-Dabramysl (52°47'20.18"N, 25°46'4.54"E) is the only formely drained and agriculturally used peatland with Aquatic Warblers. It is located in the Scara river floodplain, near the village of Dabramysl, Ivacevicy district, and Brest region. Formerly, the area was a shallow percolation mire with sedges and *Phragmites australis*. It was drained in 1985 for agriculture. A complete renewal of the drainage system was scheduled for 2000. However, the adjacent 'Vyhanascanskaje' zakaznik lobbied against cleaning of the canals. As a result, the drainage system is currently ineffective and hydrological conditions are mainly determined by river levels. The hydrological regime is now near-natural, with water levels of –20 to +20 cm. The total area of the peatland is about 10 km$^2$. Out of them a 123 ha plot is currently suitable for Aquatic Warbler. The shallow peat layer bears a mesotrophic sedge reed mainly of *Carex riparia* and *C. acutiformis*. Aquatic Warbler population was estimated at 29–44 singing males in 2010. The plot is still partly used as a hayfield by the agricultural enterprise 'Brest-travy', which usually mows once per year in July. Another c. 200 ha of the site is overgrown with bushes due to abandonment of land use. The rest of the area is not occupied by Aquatic Warblers due to unsuitable vegetation structure (mainly with *Phalaris arundinacea* dominating).

Of the fens rewetted by the GEF and BMU-ICI projects, two have the potential for Aquatic Warbler colonization due to their size and location: Hrycyna-Starobinskaje (3,860 ha) and Horeuskaje (198 ha, overall area of the peatland complex 20,000 ha). The GEF project has shown that depending on the level of peat soil degradation and the water level after rewetting, the restoration process of degraded fens to suitable Aquatic Warbler habitat can take a period of c. 30 years from rewetting to suitability for Aquatic Warblers (Kozulin et al. 2010a).

### 4.2.2 Degradation stages and restorability of biodiversity

Fauna and flora of peatlands are effected at minimal degradation stages (Table 20). More degraded peatlands are more difficult to restore and require explicit attention to components that might not have been directly impacted, but that have degraded as the longer-term but inevitable result of degradation of other components. Recommendations for rewetting and additional restoration measures and for management after rewetting must take the degradation stage of a site into account.

The rate of vegetation change (initiated by restoration measures and equated to restoration success) varies with target habitat type, pre-exploitation habitat type, and previous land uses. Much quicker success has been reported from cut-away bogs after extraction than from milled bogs, which generally require additional measures, such as plant re-introduction. Fen vegetation re-develops towards target vegetation more easily after drainage reversal, with the greatest success in controlled rewetting regimes (as opposed to rapid, permanent inundation of sites). Transitional stages (e.g. shallow water bodies) can provide valuable habitats for waterfowl, but these are transient, and are not typical habitats for the target bird communities of most conservation concern. Amphibians, reptiles and invertebrates are only poorly studied in relation to peatland habitat change, but do seem to respond rapidly to changes in environmental conditions (see box 12).

### 4.2.3 Plants

Vegetation development on peatland (see chapter 3.5 for an extensive description) is strongly determined by peat properties (Okruszko 1956)

## 4.2 Relationship between peatland condition and biodiversity values

Table 20: Effects of various degradation stages on peatland components and 'restorability' of peatlands (modified after Schumann & Joosten 2008). White = not affected, grey = slightly affected, black = severely affected. Degradation Increases from top to bottom and 'restorability' decreases. Peat accumulation is possible only after minimal and minor (and possibly moderate) degradation.

| Degradation stage | Fauna/ flora | Vegetation | Water regime | Soil hydraulics | Form and relief | Peat deposits | Site characteristis |
|---|---|---|---|---|---|---|---|
| Minimal | | | | | | | undrained, natural spontaneous vegetation, only hunting and gathering |
| Minor | ▒ | | | | | | not/slightly drained, low-intensity grazing/mowing or forestry |
| Modest | ▒ | ▒ | ▒ | | | | freshly deeply drained and/or regular mowing/grazing |
| Moderate | ▒ | ▒ | ▒ | ▒ | | | long-term very shallow drainage, long-term use |
| Major | ■ | ■ | ■ | ■ | ▒ | ▒ | long-term deeply drained or inundated |
| Maximal | ■ | ■ | ■ | ■ | ■ | ■ | intensively drained |

and response to rewetting depends above all on the method of extraction. Milling leads mostly to monodominant vegetation of non-peatforming vascular plant species even 20 years after rewetting, whereas peat cutting allows re-development of the original peat-forming vegetation (Poschlod 1992). Generally, the restoration of bogs as self-regulating landscapes after severe anthropogenic damage is impossible within human time perspective (Joosten 1995).

After milling of bogs, the surface of peatlands is composed of strongly decomposed peat (black peat) that results in a low water storage capacity and unfavourable conditions for plant growth. Such conditions inhibit the development to target bog communities. After abandonment of peat extraction and depending on the water level, usually fen communities with *Molinia caerulea* and *Eriophorum* species, dry and wet heath, and forest may develop (Schouwenaars 1992, Sliva 1997, Lavoie et al. 2005b). The restoration of milled sites often needs intervention by re-introduction of plants and artificial establishment.

Restoration of bog vegetation after peat cutting is more likely (especially in wet climates; Poulin et al. 2005) and after initial establishment of *Eriophorum*, *Molinia caerulea* and *Sphagnum* species may spread rapidly (Nick & Weber 2001). The depth of peat mining is the most important factor for the direction of the succession (Gremer & Poschlod 1991) and peat pits may be rapidly filled with *Sphagnum* mosses (Lütt 1992, Buttler et al. 1996). A mean annual water level near the surface favours *Sphagnum* revegetation.

Fens drained for agriculture are usually strongly eutrophic or polytrophic after rewetting, with eutrophic and mesotrophic conditions only on higher parts and in the centre. Here, the reduction of nutrient availability is the most important part of post-rewetting vegetation management. Initially, flooded fen polders develop open water areas with low cover of *Typha angustifolia* or *T. latifolia* and develop later into reed beds of *Phragmites australis*, *Glyceria maxima*, and *G. fluitans*. Under continued succession, *Carex* sedge beds develop (e.g. Timmermann et al. 2006, Margoczi et al. 2007). Studies from northeastern Germany report vegetation development towards less nutrient-rich conditions within 4–8 years of 'controlled' rewetting (i.e. a slow raise of water levels; K. Vegelin pers. comm.). Top soil removal in combination with rewetting can lead to the restoration of soft-water pools and small sedge marshes within 5 years. The complete restoration of fen meadows takes longer (10–15 years) and is only possible if hydrological measures that counter prolonged inundation and reinforce the discharge of base and iron-rich ground water are carried out (Jansen & Roelofs 1996, Jansen et al. 2004). According to Klimkowska et al. (2010), the combination of topsoil removal, hay transfer and exclusion of large animals resulted in a community with highest similarity to the target vegetation.

Table 21: Breeding (B) and non-breeding (G) Birds Directive Annex I species recorded before and after controlled rewetting starting in 2000 at Randow-Rustow (modified after Vegelin et al. 2009). Asterisk indicates only partial mowing in the previous year.

| English name | Scientific name | 1992–1994 | 2000 | 2002 | 2004 | 2008* |
|---|---|---|---|---|---|---|
| Barred Warbler | *Sylvia nisoria* | – | – | B | – | – |
| Bittern | *Botaurus stellaris* | – | B | B | B | B |
| Black Kite | *Milvus migrans* | – | B | G | B | B |
| Black Tern | *Chlidonias niger* | – | B | B | G | B |
| Bluethroat | *Luscinia svecica* | – | B | B | B | B |
| Crane | *Grus grus* | – | B | B | B | B |
| Eurasien Golden Plover | *Pluvialis apricaria* | – | – | – | G | – |
| Kingfisher | *Alcedo atthis* | – | G | B | B | G |
| Marsh Harrier | *Circus aeruginosus* | B | B | B | B | G |
| Osprey | *Pandion haliaetus* | – | G | G | G | G |
| Red Kite | *Milvus milvus* | B | G | G | B | G |
| Red-backed Shrike | *Lanius collurio* | – | B | B | B | G |
| Ruff | *Philomachus pugnax* | – | – | G | G | G |
| Spotted Crake | *Porzana porzana* | – | – | – | – | B |
| Whiskered Tern | *Chlidonias hybridus* | – | – | B | – | G |
| White-tailed Sea Eagle | *Haliaeetus albicilla* | – | G | G | G | G |
| Wood Sandpiper | *Tringa glareola* | G | G | – | G | G |
| Number of breeding birds | | 57 | 63 | 63 | 65 | 55 |
| Total number of birds | | 57 | 98 | 98 | 110 | 87 |

## Box 11

### Experience from rewetted sites in Belarus – Chernobyl zone

Dmitry Zhuravlev & Franziska Tanneberger

In the Belarusian part of Chernobyl zone (close to the Ukrainian nuclear power plant at Chernobyl), large peatland areas have been deliberately rewetted following the 1986 nuclear plant accident, to reduce radionuclide release from the soil. This section summarises the results of two studies on bird populations in these areas.

Nikiforov et al. (1995) studied a range of habitats in the Belarusian Chernobyl zone in 1988–1992, including formerly drained agricultural land abandoned and rewetted in the late 1980s (the year and intensity of rewetting was not stated; but is likely to have started before 1988 and continued throughout the study period, with raised ground water levels but no total inundation). They showed that with slight rewetting and abandonment (i.e. overgrowth by scrub), the density of bird species characteristic of moist or wet scrub increased (Table 22). Some species, such as Corncrakes *Crex crex* and Lapwings *Vanellus vanellus*, benefited from occassional burning, whereas others (e.g. Sedge Warblers *Acrocephalus schoenobaenus* and Reed Buntings *Emberiza schoeniclus*) declined on recently burnt sites.

Zhuravlev (1999 and unpubl. data) studied bird densities at two former polders in the Chernobyl region, one was rewetted delibrately in 1994/1995, and one rewetted by beaver activity. After rewetting, the density of many bird species characteristic of wet peatlands, e.g. Spotted Crakes *Porzana porzana* and Reed Buntings, increased substantially (Table 23).

Generally, density data from Chernobyl region must be treated with care since they may be influenced by an overall reduced arthropod abundance. Pape Moller & Mousseau (2009) showed that the abundance of invertebrates in the Chernobyl region decreased with increasing radiation, after controlling for soil type, habitat and vegetation height.

## 4.2 Relationship between peatland condition and biodiversity values

Table 22: Selected breeding bird densities (individuals/10 ha) on drained, abandoned and slightly rewetted sites in the Belarusian Chernobyl region (from Nikiforov et al. 1995).

| English name | Scientific name | 1988 | 1990 | 1992 – unburnt | 1992 – burnt |
|---|---|---|---|---|---|
| Corncrake | Crex crex | 0.3 | 1.5 | 0.7 | 1.3 |
| Crane | Grus grus | 0 | 0 | + | no data |
| Lapwing | Vanellus vanellus | 0 | 0 | 0 | 0.9 |
| Marsh Harrier | Circus aerugineus | + | 0 | 0.2 | 0 |
| Meadow Pipit | Anthus pratensis | 2.3 | 9.6 | 7.4 | 3.4 |
| Montagu's Harrier | Circus pygargus | + | + | 0.2 | 0 |
| Reed Bunting | Emberiza schoeniclus | 0.3 | 0.7 | 4.4 | 0.9 |
| Sedge Warbler | Acrocephalus schoenobaenus | 0.8 | 2.2 | 7.0 | 1.3 |
| Yellow Wagtail | Motacilla flava | 0.7 | + | 2.2 | 0.9 |

Table 23: Selected breeding bird densities (individuals/10 ha) on drained, abandoned and rewetted sites in the Belarusian Chernobyl region (from Zhuravlev 1999).

| English name | Scientific name | Drained, abandoned grassland | Rewetted abandoned grassland | | | |
|---|---|---|---|---|---|---|
| | | | 1995 | 1996 | 1998 | 1999 |
| Bittern | Botaurus stellaris | 0 | 0 | 0 | 1.4 | 0.8 |
| Common Snipe | Gallinago gallinago | 3.8 | + | 0.4 | 0 | 0.5 |
| Corncrake | Crex crex | 3.2 | 0.2 | 1.1 | 0 | 1.0 |
| Lapwing | Vanellus vanellus | 3.2 | 0 | 0.7 | 1.4 | 0.6 |
| Little Crake | Porzana parva | 0 | 0.3 | 0.7 | 0 | 0.3 |
| Marsh Harrier | Circus aerugineus | 0 | 0.3 | 0.7 | 0.9 | 0.6 |
| Meadow Pipit | Anthus pratensis | 7.0 | 0 | 0 | 0 | 0 |
| Montagu's Harrier | Circus pygargus | 1.3 | 0 | 0 | 0 | 0 |
| Reed Bunting | Emberiza schoeniclus | 4.5 | 0.7 | 2.1 | 2.8 | 4.8 |
| Sedge Warbler | Acrocephalus schoenobaenus | 26.1 | 10.5 | 11.9 | 9.3 | 14.9 |
| Spotted Crake | Porzana porzana | 0.6 | 0 | 0 | 0.9 | + |
| Yellow Wagtail | Motacilla flava | 3.2 | 0.3 | 1.1 | 0.9 | 1.3 |

### 4.2.4 Birds, amphibians, and reptiles

Rewetting of bogs seems to result in an increase in species richness (Müller 1988, Mazerolle et al. 2006), which is expected to decrease with decreasing areas of open water. For fens, transient stages with large numbers of roosting ducks (Gierk & Kalbe 2001) and the occurrence of the three 'marsh tern' species (Chlidonias spp.) as regular breeding birds (e.g. Sellin & Schirmeister 2004) are described. In Northeast Germany, 'controlled rewetting' favours the occurrence of most Bird Directive Annex I species as well as the total number of breeding and non-breeding bird species (K. Vegelin pers. comm.; Table 21). Rewetted polders have a huge potential for breedings rails:

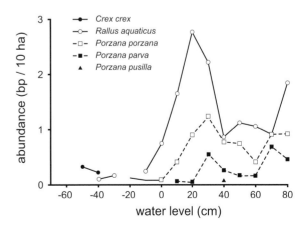

Fig. 47: Numbers of breeding rails in relation to water level (from Herold 2011). bp = breeding pairs.

### Box 12

**Biodiversity of GEF project rewetting sites**

Alexander Kozulin, Irina Viarshitskaya & Dmitry Zhuravlev

During the GEF project (see chapter 2.4; 2006–2010) flora and fauna monitoring was carried out before and after rewetting. Generally, wetland plant communities on project sites increased in area by 58–96% and the proportion of wetland bird species of the sites increased by 19–48%. Key monitoring outcomes are illustrated for two sites; the cutover fen at Barcianicha and the drained (for forestry) bog at Haloje.

At Barcianicha (54°06'N, 26°10'E), two monitoring plots show the vegetation development under slightly different conditions: On plot 1, bare or sparsely vegetated open peat prevailed before rewetting with almost 50% of the area inundated after rewetting. On plot 2 the reeds remained dominant after rewetting and the area of trees and meadow grasses decreased (Pugachevsky et al. 2009; Fig. 48). The number of 'forest' bird species decreased, whilst 'scrub' species increased. Among peatland species, Sedge Warbler and Reed Bunting densities increased (Pugachevsky et al. 2009; Table 24). Compared with bird communities, the effects on amphibians and reptiles were more pronounced. After rewetting, previously unrecorded species typical of peatlands were found (Moor Frogs *Rana arvalis* and Common Lizards *Zootoca vivipara*) and typical forest species declined from 54% to 31% (Pugachevsky et al. 2009; Table 25).

At Haloje (53°40'N, 28°25'E), the only forestry site of the GEF project where no peat extraction occurred, the change in vegetation composition was less pronounced than on the peat extraction sites. The two transect lines were dominated by shrubs such as *Ledum palustre* and *Vaccinium uliginosum* and mosses, both before and after rewetting (Pugachevsky et al. 2009; Table 26). However, the condition of the trees and the composition of mosses changed considerably. On plot 1, the condition of *Pinus sylvestris* and *Betula pendula* improved. In contrast, the condition of *B. pubescens* and *Populus tremula* declined (Pugachevsky et al. 2009). On plot 2 the cover of *Sphagnum* spec. mosses increased substantially, whereas the cover of *Polytrichum* spec. mosses decreased by over 50% (Table 26). On a study transect outside the plots, the cover of *Calluna vulgaris* decreased substantially after rewetting (i.e. by 30% in 2007–2009; Verzhitskaya et al. 2010).

The effects of rewetting on bird as well as on amphibian and reptile communities were rather pronounced even one year after rewetting. The number of 'forest' bird species declined, whilst the number of 'peatland' species increased. Even more pronounced, the

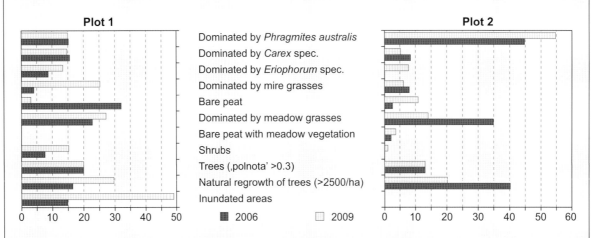

Fig. 48: Cover of vegetation types in the herb layer, trees/shrubs and inundated areas (%) at monitoring plots 1 and 2 at Barcianicha in 2006 (before) and 2009 (after) rewetting (modified after Pugachevsky et al. 2009). 'Polnota' = proportion of the plot covered by trees.

## 4.2 Relationship between peatland condition and biodiversity values

proportion of 'forest' amphibians and reptiles decreased and 'peatland' species increased (Pugachevsky et al. 2009; Table 27).

Generally, monitoring data of the GEF project indicate that the effects of rewetting on vegetation depend on:
- The hydrological regime;
- The depth and type of the residual peat layer;
- The chemical composition of ground water and peat;
- The method of peat extraction;
- The presence and location of man-made areas (peat mass storage areas, stumps, mounds for temporary roads, dams, etc.);
- The extent and intensity of fires;
- The activities performed after completion of peat extraction (agriculture, forestry, creation of water resources; after Kozulin et al. 2010a).

According to Kozulin et al. (2010a), the GEF project sites can be divided into three groups: peat extraction sites with fen peat, peat extraction sites with bog peat, and bogs drained for forestry. For each type, a scheme of vegetation communities typical of different water levels has been developed (see chapter 3.5). The highest floristic diversity was found on extracted sites with fen peat (Verzhitskaya et al. 2010). The rate of development and longevity of vegetation communities needs further research.

Table 24: Bird community composition based on habitat types at Barcianicha in 2006 and 2009 with density of species (individuals/10 ha) on the monitoring transect (modified after Pugachevsky et al. 2009). + = species present outside transect.

| English name | Scientific name | 2006 | 2009 |
|---|---|---|---|
| Peatlands (%) | | 11 | 14 |
| Great Reed Warbler | Acrocephalus arundinaceus | 1.25 | 2.5 |
| Lapwing | Vanellus vanellus | + | 0 |
| Marsh Warbler | Acrocephalus palustris | 0 | 2.5 |
| Reed Bunting | Emberiza schoeniclus | 6.25 | 10.0 |
| Sedge Warbler | Acrocephalus schoenobaenus | 0 | 2.5 |
| Snipe | Gallinago gallinago | 2.5 | + |
| Shore and water (%) | | 8 | 5 |
| Sand Martin | Riparia riparia | + | |
| Settlements (%) | | 3 | 0 |
| White Wagtail | Motacilla alba | 1.25 | 0 |
| Dry open areas (%) | | 3 | 5 |
| Citrine Wagtail | Motacilla citreola | 0 | 2.5 |

| English name | Scientific name | 2006 | 2009 |
|---|---|---|---|
| Forests (%) | | 47 | 32 |
| Black Stork | Ciconia nigra | + | |
| Bluethroat | Luscinia svecica | 5 | + |
| Chaffinch | Fringilla coelebs | 1.25 | 0 |
| Common Buzzard | Buteo buteo | + | + |
| Garden Warbler | Sylvia borin | 1.25 | 0 |
| Green Sandpiper | Tringa ochropus | 1.25 | 1.25 |
| Honey Buzzard | Pernis apivorus | + | |
| Tree Pipit | Anthus trivialis | 3.75 | 0 |
| Whinchat | Saxicola rubetra | 1.25 | 0 |
| Whitethroat | Sylvia communis | 0 | 7.5 |
| Scrub (%) | | 28 | 44 |
| Scarlet Rosefinch | Carpodacus erythrinus | 1.25 | 0 |
| Willow Warbler | Phylloscopus trochilus | + | + |
| Woodpigeon | Columba palumbus | + | 0 |
| Yellowhammer | Emberiza citrinella | 0 | 7.5 |

Table 25: Abundance of amphibians and reptiles based on habitat types at Barcianicha in 2006 and 2009 (after Pugachevsky et al. 2009).

| English name | Scientific name | Abundance (ind./ha) 2006 | Abundance (ind./ha) 2009 |
|---|---|---|---|
| Open water (%) | | 40 | 41 |
| Fire-bellied Toad | Bombina bombina | 40–50 males, 6.4 (3 sites) | 40–50 males, 6.4 (4 sites) |
| Pool Frog | Pelophylax lessonae | 15.0 (4 sites) | 20.5 (6–8 sites) |
| Forests (%) | | 54 | 31 |
| Common Frog | Rana temporaria | 16.6 | 12.5 |
| Common Toad | Bufo bufo | – | + |
| Sand Lizard | Lacerta agilis | 10 | 8.5 |
| Slow Worm | Anguis fragilis | 2.5 | – |
| Smooth Newt | Lissotriton vulgaris | + (1 site) | + (3 sites) |
| Forests and peatlands (%) | | 0 | 10 |
| Moor Frog | Rana arvalis | – | 6.6 |
| Peatlands (%) | | 0 | 10 |
| Common Lizard | Zootoca vivipara | – | 6.6 |
| Peatlands and shore (%) | | 8 | 8 |
| Grass Snake | Natrix natrix | 3.3 | 5.5 |

Table 26: Vegetation types and selected plant species before (2006) and after (2009) rewetting at Haloje peatland (from Pugachevsky et al. 2009; data on water levels from Verzhitskaya 2010). 'Polnota' = proportion of the plot covered by trees.

| | Plot 1 2006 | Plot 1 2009 | Plot 2 2006 | Plot 2 2009 |
|---|---|---|---|---|
| Water level in June | -0.23 cm | -0.19 cm | -0.30 cm | -0.23 cm |
| **Cover of main vegetation types (%)** | | | | |
| Dominated by Ledum palustre and Vaccinium uliginosum | 100 | 100 | 100 | 100 |
| … with Sphagnum spec. | … 77 | … 90 | … 52 | … 79 |
| … with Polytrichum spec. | … 4 | … 3 | … 46 | … 21 |
| Natural regrowth of trees (>2500/ha) | 68 | 100 | 100 | 100 |
| Trees ('polnota' >0.3) | 38 | 38 | 0 | 0 |
| **Cover of selected plant species (%)** | | | | |
| Andromeda polifolia | 12 | 16 | 16 | 11 |
| Betula pendula | 11 | 4 | 20 | 23 |
| Calluna vulgaris | 0.1 | 0.1 | – | – |
| Eriophorum vaginatum | 11 | 9 | 2 | 2 |
| Ledum palustre | 20 | 19 | 38 | 49 |

| | Plot 1 2006 | Plot 1 2009 | Plot 2 2006 | Plot 2 2009 |
|---|---|---|---|---|
| Water level in June | -0.23 cm | -0.19 cm | -0.30 cm | -0.23 cm |
| Pinus sylvestris | 11 | 11 | 6 | 7 |
| Vaccinium uliginosum | 57 | 29 | 23 | 20 |
| Sphagnum spec. | 64 | 72 | 46 | 59 |
| Polytrichum strictum | 28 | 19 | 36 | 30 |

Table 27: Proportion (%) of bird and amphibian/reptile species based on habitat types at Haloje in 2006 and 2009 (modified after Pugachevsky et al. 2009).

| Group | Birds 2006 | Birds 2009 | Amphibia and reptilia 2006 | Amphibia and reptilia 2009 |
|---|---|---|---|---|
| Peatlands | 9 | 17 | 26 | 44 |
| Forests and peatlands | – | – | 21 | 42 |
| Forests | 55 | 33 | 53 | 14 |
| Trees and shrubs | 27 | 25 | – | – |
| Dry open areas | 9 | 17 | – | – |
| Shore and water | 0 | 8 | – | – |

Herold (2011) recorded high densities of Spotted Crakes *Porzana porzana*, Little Crakes *P. parva*, and the first breeding record of Baillon's Crake *P. pusilla* in eastern Germany after 90 years (Fig. 47). An increase in population size and occurrence of amphibians after rewetting has been reported for northern Germany (Holsten et al. 2001, K. Vegelin pers. comm.) and for Belarus (see box 12).

### 4.2.5 Invertebrates

In a rewetted coastal transgression mire in northeastern Germany, beetles (Coleoptera) reacted to the change in salinity in the first year after rewetting, spiders (Arachnida) in the 3rd and 4th year and vegetation only in the 5th year (Müller-Motzfeld 1997). The colonization of a rewetted area by spiders and beetles was slower than that of bugs (Heteroptera) and leafhoppers (Auchenorrhyncha), which may be due to differences in mobility and different vegetation associations (Lieckweg & Niedringhaus 2010). A positive influence of rewetting on occurrence and reproduction of dragonflies (Odonata) was observed in bogs (Benken 1988) and fens (Mauersberger et al. 2010). Mobile aquatic invertebrates readily colonize man-made pools after rewetting (van Duinen et al. 2003), though this is less so for more sedentary species. Large-scale rewetting causing a functional homogenization may lead to a decline in total species numbers of aquatic invertebrates (Verberk et al. 2010).

## 4.3 Target and indicator species

Rob Field, Franziska Tanneberger & Richard Bradbury

### 4.3.1 Introduction

Regardless of the tasks of rewetting or restoration of peatlands, it is desirable to evaluate the direct effects of this intervention activity on biodiversity. A robustly conceived and executed monitoring program can greatly enhance the value of a restoration programme in achieving its own goals and informing future activities. The aims of any monitoring programme are fourfold:
- To determine the desired future state of biodiversity – 'the target';
- To identify how progress towards that target will be measured – 'the indicators';
- To measure the progress of sites towards the target state, against data collected on indicators before restoration begins – 'the baseline', and
- If logistics allow, to compare progress at the intervention site to that at a 'control' site, similar in previous use to the intervention site, but where no intervention has occurred.

These will allow a rigorous and repeatable assessment of the extent to which a project has achieved its biodiversity goals.

This chapter provides guidance and suggestions on the identification of target habitats and species and on the selection of indicators for peatland restoration projects. Biodiversity monitoring should accompany monitoring of the geophysical and chemical consequences of restoration (e.g. Traxler 1997; see also chapter 3.4 and 9.2).

### 4.3.2 Baseline surveys

The methodology chosen for monitoring the development of biodiversity due to site restoration should encompass the groups and species likely to be influenced by the interventions. The choice of targets and indicators will be partly influenced by the resources available and the task of restoration, but also the results of the baseline surveys conducted prior to intervention. The detailed site-specific knowledge gained during the baseline survey will determine the likely colonists and characteristic species of that site, and therefore which species will be suitable indicators.

### 4.3.3 Indicators

Indicators are species, groups of species or communities that are predicted to respond to the environmental change induced by the intervention, over a period of time suitable for a monitoring regime (Caro 2010). A species' or community's suitability as an indicator is dependent upon a number of factors:
- The scale at which a species operate, e.g. its territory size compared to the scale of processes ecologically relevant to the questions in hand;
- The time-period over which it is capable of change; factors such as generation time, mobility and size are all important in colonization and spread of organisms;
- The target habitat, species or community of the restoration;
- The desired changes in the physical and biological environment.

When choosing indicators, there are no pre-conceived reasons why any one biological, physical or chemical indicator is more appropriate than any another. The selection of appropriate indicators depends on the aims of the habitat change being undertaken, in particular whether it is:
- Species or habitat conservation driven, or
- Driven by other ecosystem services, such as carbon conservation, water quality or control (flood prevention), pollution control, etc.

Any monitoring scheme, including any pre-intervention monitoring, should begin well in advance of restoration works. The duration of most rewetting projects means that only primary succession stages and pioneer species will establish during the course of the project. These may be indicators for future target conditions, and this will necessarily influence the choice of within- and post- project indicators. Different species groups will respond differently to environmental conditions. The lower down the food web a species is, the more likely it is to react rapidly and directly to environmental changes. For example, Warner & Chmielewski (1992) suggest microorganisms such as testate amoebae are useful indicators for drained and re-wetted peatlands, by virtue of their responses to change over both short, and prolonged, time periods. Ideally, a wide range of indicators and measurements of environmental change (likely to vary at spatial and temporal scales appropriate to a site and project) should be chosen for monitoring during peatland regeneration (Chapman et al. 2003).

The types of species or communities selected as indicators will depend on the nature of the change that is desired on a site (Coppolillo et al. 2004). In the case of peatland rewetting, the rewetting is likely to be carried out to meet several objectives: the improvement of environmental quality and the ecosystem services this offers (e.g. reduction of greenhouse gas emissions or carbon sequestration), the restoration or creation of specific habitats and species assemblages, and the restoration of larger scale landscapes, within which a particular site may reside. The choice of indicators in these cases will depend upon the relative importance of these factors. If environmental (and particularly water) quality is a key goal, then species sensitive to this should be chosen. They need not be scarce or ecologically interesting, and indeed it may be beneficial that they are not. Groups such as water beetles, aquatic plants and diatoms are all commonly used. Their key property should be that their abundance varies directly with changes in water quality. Likewise, in projects where the prime motivation is habitat or species conservation, indicators of progress or success should be the abundance or colonization of a site by species or communities that are, or are part of, that which is sought. These so-called 'flagship' habitats or species may also be the target of the restoration (see below). They may be indicative of conditions likely to favour such species or they may be part of a larger community or assemblage of species, or they may be an apex species at the top of a food web, whose presence indicates the full functioning of that food web. (So-called 'umbrella' species; see box 13). Some organisms are also the engineers of observed changes in an ecosystem's structure. For example, *Sphagnum* species are important in peatland regeneration. They are not only indicators of the development of the target of restoration, but also an integral part of it. As architects of a functioning mire, they are essential for the habitat and its processes, but are not solely so, and require a complex community to develop with them, especially if a suite of ecosystem services is the aim of restoration (Francez et al. 2000, Lindsay 2010).

Furthermore, given that project sites usually vary in size, location, topography, and restoration technique, it is appropriate that site-specific targets are set, covering the range of conditions currently present, and those that are expected to arise. Site-specific data can then be converted into consistent metrics of progress towards site-specific targets that can be compared across sites.

Some criteria for the selection of indicator species/groups:
- Use very mobile species that are known to colonize newly created habitats quickly (such as dragonflies, ground beetles, butterflies);
- Use groups that include species with a broad range of environmental requirements (and thus indicate a variety of peatland conditions) and that can be monitored with one method (such as dragonflies with collection of larval cases);
- Use existing indicator species schemes, if possible (e.g. within the national implementation of the Convention on Biological Diversity or national monitoring schemes);
- Use standards that a project aims to comply with, e.g. CCBA (Climate, Community and Biodiversity Alliance) standard;
- Use groups of species that are relatively easy to identify, that consist of not too many species, but not so few that they do not cover the range of conditions likely;

# 4.3 Target and indicator species

> **Box 13**
>
> **The Aquatic Warbler as an umbrella species**
>
> Franziska Tanneberger
>
> Aquatic Warbler conservation requires large contiguous areas of suitable habitat, because of its particular breeding system of multiple paternity and uniparental (female) brood care (Heise 1974). It can therefore be regarded as an umbrella species (sensu Simberloff 1998 and Roberge & Angelstam 2004) encompassing the needs of the flora and fauna of mesotrophic and slightly eutrophic peatlands. Conserving the umbrella species Aquatic Warbler therefore means much more than the protection of a single species; protecting its unique fen mire breeding habitat brings equal benefits to a larger community of less prominent species.
>
> Hesse (1910) describes beautifully the large range of birds that share the habitat and surroundings of the Aquatic Warbler on fens near Berlin in Germany: 'The vast fens impressed me deeply. What a richness in plant and animal life! And most admirable, the birds, for example in a warm May evening – let us just think of the most characteristic species. The cawing of terns, the screaming of Lapwings, the piping of Redshanks, the yodelling of Godwits, the bleating of snipes, the trumpeting of Cranes, the booming of the Bitterns, the gobbling of Black Cocks, the whispering of Pipits, the whirring of *Locustella* warblers, the rattling and whistling of Aquatic Warblers, the plain song of the Whinchat – moreover several skeins of flying ducks, a few heavy-weight Bustards, Short-Eared Owls crossing quartering their territories, and here and there harriers sculling gull-like through the air, and the vast endless realms lightened by the evening sun – what a wonderful piece of nature!' Among EU Birds Directive Annex I species, 16 species often occur together with Aquatic Warblers, e.g. Bittern *Botauris stellaris*, Marsh Harrier *Circus aeruginosus*, Greater Spotted Eagle *Aquila clanga*, Great Snipe *Gallinago media*, and Wood Sandpiper *Tringa glareola*. They share habitats with a number of Habitats Directive Annex II arthropod species such as Larger Copper *Lycaena dispar* and *Carabus menetriesi*, and amphibians such as the Great Crested Newt *Triturus cristatus*, and the European Fire-bellied Toad *Bombina bombina*.
>
> Aquatic Warbler breeding sites are also closely linked with rare plants and habitats. They include eight habitat types from the EU Habitats Directive, among them three priority types. Examples of especially valuable sites include:
> - Rozwarowo Marshes (Poland) holds the largest population of *Myrica gale* in Northwest-Poland (Jurzyk 2004) and *Carex pulicaris* which is threatened with extinction in Poland. Part of this habitat type is listed under Habitats Directive Annex I Code 6410;
> - Chelm Marshes (Poland) are the largest area of calcareous fens with *Cladium mariscus* (Habitats Directive Annex I Code 7210) in Poland;
> - Biebrza Marshes (Poland) comprise large areas of transition mires and quaking bogs (Habitats Directive Annex I Code 7140) and in total 921 species of vascular plants and a large number of moss species, many of them rare and protected, e.g. Habitats Directive Annex II orchid *Liparis loeselii* and moss *Drepanocladus vernicosus*;
> - Key Belarusian breeding sites such as Zvaniec, Sporava, and Dzikoje comprise a large number of rare and threatened species (see chapter 4.1).

- Use groups for which the habitat requirements are well-known;
- Use species or species assemblages that are indicative of a landscape or habitat type ('landscape' or 'flagship' species);
- Use apex species that are indicative of a functioning food web ('umbrella' species).

Indicator plant species were suggested e.g. by Biewer et al. (1994) for fens in South-Germany, by Succow (1988) and Succow & Joosten (2001) for near-natural and moderately drained peatlands (as 'vegetation forms' which indicate ecological conditions, see box 6 in chapter 3.4), and may be also derived from Julve (2011) who lists vascular plant and bryophyte species of near-natural Holarctic peatlands with information on distribution and ecological requirements in a free online database. Similar resources exist for invertebrates or such information may be obtained by national/regional entomological specialists. Birds and amphibians are often used as indicator vertebrate species. Birds are easily recognisable, and wetlands and peatlands often have characteristic avifaunas (Tucker & Evans 1997). BirdLife International provides extensive country-specific information on distributions and conservation status of peatland specialist species throughout Europe (Heath & Evans 2000). Anderson et al. (1997) suggest important bird species for peatlands in the Peak District, UK. 'Landscape' species are likely to be site-specific, but are often apex predators of high mobility, so may also be useful as indicators of site connectivity within highly altered landscapes (Sanderson et al. 2002; Stroh & Hughes 2010). Amphibians are well suited as indicator species for rewetted peatlands since many of them react rapidly and directly to water level and water quality (see also box 12 in chapter 4.2). Many are protected at high national and/or international levels in Belarus (Drobenkov et al. 2005).

### 4.3.4 Targets

Target species or habitats are the desired future state of biodiversity on a site that is to be restored. In choosing target species or communities, it is important to recognise that in some cases it is not possible to regain exactly what has been lost due to site degradation. However, it is also important to recognise that restoration processes can be very slow. Target conditions can be derived from unaltered reference habitats, preferably within the same bio-geographic region and of the same type (e.g. Dyrcz 2010 and see chapter 4.1). Alternatively, targets can be derived from generic local conditions as the result of multi-site surveys, combining characteristics of restored and pristine habitats (e.g. Kratz & Pfadenhauer 2001, Kratz 1994, Coppolillo et al. 2004). Targets may be either (or both) individual species or community-types, and the choice of which will be governed by the drivers of restoration, and the likely result of that restoration. The target flora of a site may be mire species, both for the biodiversity conservation value of such an ecosystem and its component parts, but also for the carbon store that such a system will become once peat formation has restarted. At higher trophic levels, other species of conservation concern (particularly habitat specialists, e.g. Crane *Grus grus*, Aquatic Warbler *Acrocephalus paludicola*, Greater Spotted Eagle *Aquila clanga*, Great Snipe *Gallinago media*, Eurasian Curlew *Numenius arquata*) may be chosen, either because of their own problematic conservation status (e.g. Aquatic Warbler) or because they, despite their still favourable status, indicate the presence of an ecosystem type with a less favourable status (e.g. Crane). In this way a species may be both target and indicator of success.

The following references provide suggestions for choice of target animals and plants in peatland projects:
- Schopp-Guth (1999) provides 'Leitbilder' (= model or concept species) for the restoration of peatlands in Germany;
- Kratz & Pfadenhauer (2001) and Kratz (1994) suggest target conditions and a system of target species with hierarchical levels for fens in North-Germany;
- Rosenthal (2003) elaborates target plant species to evaluate the success of fen rewetting in North-Germany based on ecological requirements and gives successional series for levels of restoration;
- Biewer et al. (1994) establish target plant species for fens in South-Germany.
- Wagner & Wagner (2005) describe 'Leitbilder' and target conditions for fens in South-Germany, and distinguish between levels of anthropogenic change;
- Veen (1992) provides a system of peatland habitats with associated plant and bird species for various stages of rewetting at Dutch peatlands;
- Julve (2011) lists vascular plants and bryophytes of near-natural Holarctic peatlands and their ecological requirements.

### 4.3.5 Monitoring strategies

The spatial scale at which the change of indicator populations is measured should be appropriate to both the species and the site. For example, when monitoring vegetation types and change across a site, a variety of scales of measurement can be used (Table 28), depending upon the vegetation types chosen as indicators and targets. In reality, several scales of monitoring are likely at any one site, since the indicators of habitat change are likely to operate at a variety of scales, and should have been chosen partly for this reason. This will apply across groups, as well as within them. For example, within the BMU-ICI project, monitoring of vegetation types, ground beetles and birds has been done at several scales; vegetation (fine, broad and land-

Table 28: Vegetation monitoring approaches at different scales (after Stroh & Hughes 2010).

| Method | Scale | Summary | Advantage/Disadvantages |
|---|---|---|---|
| Fixed quadrats | Fine | • Quadrats located along a fixed transect<br>• Percentage cover, vegetation height per species recorded | Pros:<br>• small changes detected<br>• repeatable<br>• link to small scale hydrology measures<br>Cons:<br>• small area covered<br>• intensive survey effort required<br>• difficult to relocate sites over a long time period |
| Random quadrats | Fine | • Randomly selected quadrat locations<br>• Percentage cover, vegetation height per species recorded | Pros:<br>• small changes detected<br>• repeatable<br>Cons:<br>• small area covered<br>• intensive survey effort required<br>• difficult to relocate sites over a long time period |
| Transect surveys | Broad | Coarse abundance scales used over a defined walking route | Pros:<br>• can cover larger areas<br>• can be combined with surveys for other groups<br>• can detect broad community scale changes<br>Cons:<br>• small changes in community composition will be missed<br>• changes will take longer to be detected<br>• small species may be missed |
| Remote sensing | Landscape | Aerial photography or satellite imaging to detect landscape scale changes in vegetation communities | Pros:<br>• potential to assess changes over huge areas,<br>• efficient<br>Cons:<br>• expensive<br>• specialist skills required<br>• cannot determine community composition at anything other than very broad scale<br>• subtle changes may not be identified<br>• requires calibration by smaller scale surveys |

scape), ground beetles (fine and broad scale – 10 m to 100 m) and birds (broad and landscape scale – 100 m to 1 km).

Furthermore, the timing of surveys is also important, particularly for groups whose activity is seasonal or varies with time of day. When organising the bird surveys for the Belarus rewetting program, for example, bird surveys were conducted at three different times of day, to ensure that monitoring for different groups of birds was conducted when each is most active, and therefore detectable. Transect surveys were conducted during the morning and evening to detect species most active at those periods (e.g. small song birds in the morning; Aquatic Warbler, Spotted Crake *Porzana porzana*, Woodcock *Scolopax rusticola*, Corncrake *Crex crex*, and owls in the evening) and a raptor survey conducted at the hottest, midday period, so detecting birds of prey soaring on thermals above each site.

## 4.4 Peatland rewetting and biodiversity management

Rob Field & Franziska Tanneberger

### 4.4.1 General principles

Chapter 4.2 details the available information from studies of the effects of peatland restoration on biodiversity. A generic decision pathway for developing site management plans for biodiversity and ecosystem services as part of peatland restoration is provided in Table 29 (adapted from McBride et al. 2010). In general, the scale of planning decreases as progress is made through the pathway. It is important to consider the landscape context of a site during the initial stages of planning its management (McBride et al. 2010). Some of this planning will have been done as part of the selection of candidate rewetting sites, prior to intervention, but more detailed work will be site-specific, and needs to be elaborated after

## 4 Peatlands and biodiversity

Table 29: Checklist of key stages in deciding on appropriate management for (rewetted) peatland sites (after McBride et al. 2010).

| Stage | Actions |
|---|---|
| Site landscape context | Information gathering/research into:<br>Cultural, management, landscape, catchment, land use context of peatland |
| Site survey to assess current state | Species and habitat monitoring<br>Soil and geological survey<br>Hydrological assessment |
| Nutrient assessment | Nutrient status assessment |
| Past and future changes | Identify past changes in, and current factors influencing, flora and fauna, hydrological regime and nutrient status and assess how these might change in the future |
| Identify management constraints | Identify any constraints imposed by, or associated with:<br>Legal protection (site, local, regional)<br>Archaeological significance<br>Land ownership |
| Decide on site goals | Establish objectives:<br>Identify target habitats and species<br>Identify target hydrological regime<br>Identify target ecosystem services |
| Compare existing and target regimes and identify issues and problems | Establish how target conditions differ from existing species/habitats, hydrology etc. |
| Identify necessary changes | Establish what is needed to achieve target regime/habitat or address problems |
| Identify suitable techniques to achieve changes | What vegetation, water and nutrient management techniques are appropriate?<br>Research/draw on experience elsewhere (e.g. see box 13) |
| Evaluate suitability of techniques | Which constraints limit choice of management options<br>Do you have control over these constraints<br>Assess the costs of proposed management<br>Assess whether the required management is realistically achievable |
| Develop and implement action strategy to achieve objectives | Identify future funding sources, particularly if funding for restoration is short-term – site development and management may continue indefinitely |
| Monitor outcomes | Set up appropriate monitoring strategy for target and indicator species and habitats, see chapter 4.3 |
| Adaptive management strategy | Assess whether management is achieving desired objectives. If so, maintain current management, if not, restart the process by revisiting the hydrological assessment |

the site has begun to respond to intervention. It is also important to understand that such a process should be iterative, and flexible enough to respond to site-specific changes as the ecology and hydrology of a site develops. Once the landscape, social and land use contexts have been understood, the primary step is to understand current hydrology, how it has changed in the past and how it is likely to develop under the proposed management regime (McBride et al. 2010, Joosten & Clarke 2002). Previous and current drainage structures are important in this, as are their use by neighbouring land managers. The next step is to consider the eco-hydrology, i.e. how the hydrology interacts with the current flora and fauna, and how these will interact in the future.

It is generally beneficial to increase rather than decrease heterogeneity (van Duinen et al. 2003, 2004), and also to take account of any likely effects of climate change when planning for peatland biodiversity (Heijmans et al. 2002, Aerts et al. 2009, Breeuwer et al. 2009). The scale of management is also important, and management may need to include land uses outside the restored site in order to succeed, dependent upon the target species' or communities' scale. Dallimer et al. (2010) showed that agriculture at the landscape scale in UK uplands is integral to maintaining species richness and composition of peatland bird assemblages and they stress that management should not be limited to project sites alone. Such factors should not only inform the proposed choice of on-site management, but also feature in decisions on site choice before intervention (see also chapter 4.3).

Re-colonization of degraded sites by target species or communities is only likely to succeed

## 4.4 Peatland rewetting and biodiversity management

**Box 14**

**Managing a rewetted fen for climate and biodiversity protection – a case study from Belarus**

Rob Field, Susanne Bärisch, Annett Thiele & Wendelin Wichtmann

The project site Poplau Moch (55°02'777"N, 30°52'429'"E) is a fen peatland with a former bog lens in Viciebsk region. The peatland has a total area of 415 ha of which 350 ha have been destroyed by peat extraction (milling). The former bog lens has been milled down to the transitional peat layer. Drainage and peat extraction started in the 1950s and lasted until the 1990s. Before rewetting c. 78% of the peatland area was forested. In July 2010, building works to raise the water level were completed. In May 2011, the former extraction fields were largely flooded and partly overgrown with reed (Fig. 49 Colour plates II).

The rewetting of this site has lead to a mosaic of open channels, *Phragmites australis* dominated reed beds, a mix of dry and wet scrub, and drier birch and Alder *Alnus glutinosa* forest. This site will be managed in the future to maintain the GHG benefits of the vegetation changes due to raised water table, whilst also maintaining and enhancing the biodiversity value of the site. There are potential socio-economic benefits to be gained by combining habitat management with biomass extraction and sale.

The potential biomass yield from the reed beds lies in the lower range of the productivity potential of *Phragmites australis* which ranges between 3.6 and 43.5 t dry matter per ha and year (Timmermann 2003). However, even small yields per ha could generate benefits depending on the marketing potential and available infrastructure for harvesting and pre-processing of the biomass, as well as on the total available area for harvesting. Without harvesting, parts of reed beds will be colonized by willows so that the open aspect of some parts of the project area will get lost. Since large parts of Poplau Moch are already forested, it needs to be assessed whether the available sites with reed are sufficient for a feasible utilization. Furthermore, active cultivation of *Phragmites* could be an opportunity to increase yields.

Parallel to the main ditches, 50 m x 1.5 km long strips, which are not inundated, are dominated by alder regrowth. In this area the older stands of birch should be removed to support the further development of alder. The hydrological conditions are optimal for alder production and long-term management could yield high quality alder wood for timber and veneer production. On drier, forested areas, tree clearing could decrease water losses by transpiration and therefore GHG emissions. However, this would also decrease the tree carbon sequestration potential of the peatland. A calculation of this carbon trade-off is sensible before performing any clearing activities.

The above measure for maintaining GHG benefits and biomass extraction does not impinge on the biodiversity aims for the site. Prior to rewetting the site was dominated by a mixture of bare peat, heather, cotton grass, dry acidic grassland, and birch scrub. The topography of the site (comprising large areas of extraction blocks, milled to different depths, plus machinery tracks and drainage channels) means that even prior to rewetting, there was a degree of heterogeneity of vegetation. With the raising of the water table, the succession on this site is likely to tend towards wetter vegetation communities, although some dry scrub and forest areas will remain (see above). The management strategies proposed above will affect substantial areas of the site in different ways, but are unlikely to involve the entire site. The avifauna of the drier areas will remain the same, comprising typical bird species of birch and alder forest, common in this part of Belarus. However, with the maintenance of large areas of *Phragmites* reed bed, particularly by cutting for biomass, a typical avifauna of this habitat will establish. The complexes of reed bed, open water, and wet scrub that will occur and be maintained by biomass management following rewetting should support a wide range of passerines and other wildlife typical of reed beds and wet scrub in Belarus. The restored areas of peatland should also form a component of the territories of Lesser Spotted Eagles *Aquila pomarina*. In order for these co-benefits to accrue and be maintained, cutting regimes for both scrub and reeds for biomass should be carefully considered to fall outside the breeding season and possibly to be on a multi-year rotation in blocks to ensure suitable habitat is available in every year.

In other areas of the peatland, around the deeper extraction sites and drainage channels, a vegetation community more reminiscent of the original raised bog is likely to develop, though very slowly. This community is unlikely to be influenced by site management unless hydrology is changed, though care should be taken to avoid damage from extraction techniques and machinery. The range of bird species found on raised bogs that are of highest conservation value in Belarus are breeding waders. These are restricted to largely open areas containing pool complexes in a small number of fairly pristine raised bogs, which most commonly occur towards the wetter, outer edges of these mires. Since the rewetting is likely to result in areas of mixed wet heath, with some more *Sphagnum*-dominated areas, and patches of birch woodland, it is likely to provide suitable conditions for some of the more widespread breeding bird species of wet heath and areas of scattered woodland (Black Grouse *Tetrao tetrix* and Common Crane *Grus grus*). Removal of trees may help re-create more open areas, and make them more attractive to breeding waders. However, if water levels are successfully raised to sufficient levels to prevent establishment of coniferous forest and create area more suitable for breeding waders, then this is likely to kill existing trees anyway.

Any removal should be done manually (because of inaccessibility and damage caused by using mechanical removal, and also because trees are so thinly scattered).

Other flora and fauna are likely to benefit equally from the increase in wet fen habitats on the site, with relatively rapid colonization of the site (to suitable areas dependent on topography) of amphibians (typically, Pool Frog *Pelophylax lessonae*, Moor Frog *Rana arvalis* (Fig. 50 Colour plates II), and Fire-bellied Toad *Bombina bombina*) and reptiles, from the already wet abandoned milling areas and drainage channels. Wetland-typical invertebrates, such as White-faced darter *Leucorrhinia dubia* (already present on the site and of conservation concern in western and northern Europe) will expand their range and population on the site rapidly with increased water table.

The maintenance of high water tables and blocking of drainage channels of the exploited areas of Poplau Moch are likely to protect other areas of this site, not subject to extraction, from further damage. Small areas of high biodiversity value still remain in the area, and are negatively influenced by surrounding drainage. Restoration of high water tables in hydrologically connected areas will prevent degradation of intact areas and help preserve these areas and help them act as reservoirs of biodiversity to re-colonize rewetted areas.

if negative (often anthropogenic) impacts are removed. This does not preclude the establishment of (often more traditional, and more extensive) human management. Indeed, these may be an integral part of a management regime, maintaining favourable conditions. For instance, such activities may alter nutrient loads, suppress non-focal species or improve population continuity of target species. It can be desirable to re-establish traditional grazing or burning management, but where this is not possible, modern techniques with similar effects (e.g. mowing or mulching) can be applied. The same activity (e.g. grazing) can be either harmful or beneficial, depending on context, and site-specific evaluation of any method should be undertaken before widespread use (Schumann & Joosten 2008, McBride et al.

2010). Brooks & Stoneman (1997), Benstead et al. (1997), Joosten & Clarke (2002), McBride et al. (2010) provide useful guidance on the management of various peatland habitat types.

### 4.4.2 Key measures to restore biodiversity in peatlands

An overview on possible measures for peatland biodiversity restoration is presented in Table 30 (for more detail, see Schumann & Joosten 2008 and McBride et al. 2010). Key measures are:
- Installation, control, and maintenance of rewetting facilities:
  – This is usually more important in sites with sophisticated drainage;

## 4.4 Peatland rewetting and biodiversity management

Table 30: Summary of measures to restore biodiversity in peatlands (after Schumann and Joosten 2008).

| What | | Why | How |
|---|---|---|---|
| Keep in | Species | Support remaining populations or individuals of priority species by | Improving habitat conditions |
| | | | Reducing impact of herbivores/predators |
| | | | Reducing human impact |
| | Water | Prevent water losses from focal site to provide adequate water levels for focal species by | Maintaining catchment water table |
| | | | Raising catchment water table |
| | | | Decreasing evapo-transpiration |
| | | | Increasing water level on site |
| | Peat | Maintaining remaining peat as habitat and substrate of focal species by | Keeping up permafrost |
| | | | Reducing aerobic decomposition |
| | | | Reducing erosion |
| | Management | Support remaining populations or individuals of priority species by | Continuing traditional management |
| Keep out | Damage | Reduce damaging impacts of humans and animals to focal species by | Reducing human impact |
| | | | Reducing impact of animals |
| | Water | Prevent excessive water levels (harmful to focal species) by | Reducing water surplus |
| | Noxious substances | Reduce input of unwanted substances (incl. sediments) that may harm focal species by | Providing water of desired quality |
| | | | Reducing erosion |
| | Fire | Prevent damage to focal species and peat losses due to fire by | Preventing (expansion of) fire |
| Get in | Species | Introduce individuals of focal species (incl. nursing species) by | Introducing focal species |
| | | Encourage reproduction of focal species by | Improving habitat conditions |
| | | | Suppressing unwanted species |
| | Water | Bring in enough water to provide adequate water levels to support focal species by | Increasing water level on site |
| | | | Raising catchment water table |
| | | Supply water of desired quality to re-establish and support focal species by | Providing water of desired quality |
| | | | Raising catchment water table |
| | Peat | Re-establish peat accumulating species by | Improving habitat conditions |
| | | | Introducing keystone species |
| Get out | Biomass/ litter | Improve habitat conditions of focal species (creating germination niches) by | Improving habitat conditions |
| | | | Reducing unwanted substances |
| | Species | Reduce non-focal species that cause harm to focal species by | Suppressing non-focal species |
| | | | Improving habitat conditions |
| | Water | Draw off water surplus to establish adequate water levels for focal species by | Reducing water surplus |
| | Noxious substances | Dispose noxious substances (poison, nutrients) by | Improving habitat conditions |
| | | | Reducing noxious substances |

- And less so in easily rewetted sites (e.g. formerly poldered fens)
- Removal of scrub and trees:
  - Scrub tends to succeed to woodland and exacerbate drying through increased evapotranspiration (Brooks & Stoneman 1997);
- Reducing nutrient levels:
  - Nutrient levels have a significant effect on peatland vegetation, biodiversity, and nature conservation value;
  - Nitrogen, phosphorus, and potassium are most significant since they are the major nutrients that limit plant growth (McBride et al. 2010);
  - In fens and wet grasslands, the conservation of specialist plant species requires the restoration and conservation of P-limited ecosystems (Wassen et al. 2005);
  - A major threat to fens and fen biodiversity is eutrophication from sources such as:
    ◦ Nutrient loaded ground, surface and atmospheric water inputs;
    ◦ Intensive agriculture;
    ◦ Drainage of organic soils and resultant peat decomposition;
    ◦ Pollution of river waters by industry and sewage outputs.

### 4.4.3 Measures that can improve habitat conditions for target species

Further, additional measures that benefit target conditions, habitats or species are listed below (see also Schumann & Joosten 2008). Use and extent of these measures are likely to be site and context specific.
- Reduce nutrient availability by:
  - Mowing and removing or burning of biomass;
  - Grazing (Bakker and Olff 1995);
  - Preventing influx of eutrophic water.
- Regulate pH by;
  - Liming;
  - Controlling pH of influx water
- Increase site heterogeneity by:
  - Reducing competition from non-target species by:
    ◦ Grazing, selective removal (mechanical or chemical);
    ◦ Trapping.
  - Creating microhabitats, e.g. small pools;
  - Stimulating colonization by target species e.g. by creating artificial breeding sites, hay strewing;
  - Establishing nursing species to provide shelter for target species.
- Increase site connectivity to allow natural colonization by target species by:
  - Establishing habitat connections within and around the site (e.g. water courses, grazing animals (seed dispersal), controlled flooding);
  - Introduction of target species by e.g. planting, translocation, seeding.
- Reduce negative impacts, like:
  - Anthropogenic e.g. intensive agriculture, exploitation, drainage;
  - Over-grazing.

### 4.4.4 Guidance on land use effects

Vegetation succession schemes are presented in chapter 3.5 for both temperate bogs and fens reflecting effects of abandonment and rewetting. They have been developed to assess both greenhouse gas (GHG) emissions and biodiversity values of the successional stages during the Global Environment Facility (GEF) and BMU-ICI projects in Belarus. More detailed assessments on land use effects are needed for fens: Even if the target species of potential restoration are present in the seed bank, successful establishment is not certain, since degraded fens can be nutrient-enriched and therefore not suitable for such species growth. Furthermore, target species are unable to cope with the smothering and shading from the vigorous growth of nutrient tolerant species (Keddy et al. 1997). For these reasons, grazing, burning and mowing can help with the re-establishment of nutrient poor species (Keddy 1989, Weihe & Neely 1997).

Mowing, by hand scything, and the use of litter as bedding material for cattle was once commonplace on fen mires (Kotowski 2002, Joosten & Clarke 2002), but this practice has been completely abandoned in most of Europe (albeit more recently in central and eastern Europe), or replaced by more intensive mowing or grazing practices. This, along with nutrient enrichment, has led to vegetation and nutrient changes that have caused loss of habitat for many plant and animal communities (Kotowski 2002, Hodgson et al. 2005, Wassen et al. 2005). Middleton et al. (2006) review the effects of grazing, fire, and mowing on biodiversity in fens. Over-grazing can result in a permanent reduction in biodiversity, therefore the use of grazing as a management tool should be approached with caution. Fire is the most com-

## 4.4 Peatland rewetting and biodiversity management

mon and successful management tool in North America, although it is not effective in removing larger shrubs. In Europe, mowing has become an important management tool, and has been successful in maintaining species richness, particularly on sites where traditional mowing has been practiced for centuries. On suitable fen mires, biomass may be harvested and used in energy production ('paludiculture', Wichtmann et al. 2010a; see chapter 6.4), increasing economic sustainability and aiding site management. Useful information on mowing, grazing and burning of rewetted fens can also be found in Wagner & Wagner (2005), Kratz & Pfadenhauer (2001), Benstead et al. (1997), and McBride et al. (2010).

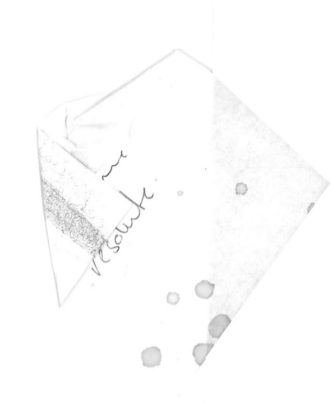

# 5 Driving forces and funding options

A wide range of drivers and funding options exists for rewetting degraded peatlands in Belarus and elsewhere. Legally, peat extraction companies in Belarus are currently obliged to restore peatlands after peat extraction has finished (see chapter 5.1). For various reasons, however, these restoration measures are usually insufficient.

Recently, the acknowledgement of peatlands in international conventions has grown substantially. The long road to sensitizing global conventions for climate change mitigation by peatlands is presented in chapter 5.2. Already in periods when peatland rewetting was still largely fameless, rewetting measures were implemented with funding from governmental and non-governmental grants. Examples from Belarus – e.g. the large rewetting project of the United Nations Development Programme (UNDP) and the Belarusian Ministry of Forestry mainly funded by the Global Environment Facility (GEF) – are presented in chapter 2.4. Along with the increasing recognition of peatlands in international conventions, opportunities for project-based funding of rewetting projects have improved. It also opened the way to 'carbon funding': Selling the emission reduction from rewetting peatlands (see chapter 3.4 for an emission monitoring tool) either on the voluntary carbon market (chapter 5.3) or on the compliance market (chapter 5.4) may yield revenues that can be used for managing rewetted peatlands (to enhance climate and biodiversity benefits) and/or for further peatland rewetting.

The legal framework for such carbon projects may differ substantially between countries. Belarus is listed under Annex I to the United Nations Framework Convention on Climate Change (UNFCCC) but was not listed under Annex B to the Kyoto Protocol when it was originally negotiated in 1997. Belarus was added to Annex B in 2006, but for this amendment to take effect three quarters of the Parties to the Kyoto Protocol have to ratify it. This has not happened yet. Selling emission reductions from peatland rewetting in Belarus is therefore – for the time being – only possible on the voluntary carbon market. The legal framework for voluntary carbon projects in Belarus is explained in chapter 5.5.

## 5.1 Legal obligations for the restoration of degraded peatlands in Belarus

Nina Tanovitskaya

### 5.1.1 Legal sources

In Belarus, landowners are legally bound to recultivate depleted and other degraded peatlands, i.e. to transform them into a condition suitable for after use in line with conditions agreed upon when the land was granted. The legal basis is given by:
- The Act 'On environmental protection' of 26th November 1992;
- The 'Regulations on the recultivation of land, degraded as a result of the extraction of minerals and peat, of geological, construction and other works' (1997);
- Two state standards (GOST 17.5.1.02 and 17.5.3.04); and
- Presidential Decree No 667 'On distraint and granting of land' of 27 December 2007;
- A technical code 'Rules and order of determination and change of the designation of land use of depleted and other types of degraded peatlands' (TKP 17.12-01-2008).

### 5.1.2 Types of recultivation

Until 1997 three types of land use after peat extraction were permitted in Belarus: agriculture, forestry, and as water body. The 'Regulations on the recultivation of land, degraded as a result of the extraction of minerals and peat, of geological, construction and other works' (1997) increased this number to seven approved types of after-use for all landowners, land users (including tenants) and other entities engaged in the surficial and subsurface extraction of minerals and peat, any construction works (e.g. industry, railway, other transport, energy, water management), and any

geological activities resulting in a degraded soil surface. The following main types of after-use are discerned:
- Agriculture (all activities that make a site suitable for agriculture, e.g. as arable land or grassland);
- Forestry (all activities that make a site suitable for establishing forest, e.g. anti-erosion forest, forest to protect adjacent water bodies, parks, and industrial plantations);
- Nature conservation (all activities that make a site suitable for rewetting and revegetation);
- Fishery (the creation of water bodies for fishery);
- Water management (the creation of water bodies for various purposes, e.g. for fire protection, irrigation, and cattle watering);
- Recreation (the creation of recreational zones), and
- Construction (all activities that make a site suitable for industrial and civil construction works).

Currently, depleted peat deposits and other damaged peatlands are designated for five types of after-use: nature conservation, water management, forestry, agriculture, and recreation.

The type of after-use is determined during the planning stage of peat extraction on the basis of key characteristics of the degraded peatland such as post-extraction relief, hydrology, type of bedrock, geographical and economic conditions, technical and social factors, etc. Recultivation of depleted peatlands must be implemented not later than one month after completion of peat extraction works.

Landowners and land users who extract peat are required to remove, store and use the fertile layer of the peat soil. The uppermost layer must be removed and used for composting or as soil improver for low productive soils. The thickness of the layer to be removed is determined on the basis of large-scale soil surveys and typically between 20 and 40 cm.

Recultivation activities are planned as part of the general engineering design plan for peat extraction. The recultivation plan must be agreed with the main land owner, the main land user, as well as with the land administrational authority and with the authorities for environmental protection and natural resources. Costs must be borne by the land owners and land users who extract peat. During the past 15 years, the main type of after-use of depleted peat deposits has been nature protection, e.g. rewetting usually with further transfer of lands to the forest fund. However, in practice recultivation activities of depleted peatlands are often not implemented by the land users because funds are lacking.

## 5.2 Sensitising global conventions for climate change mitigation by peatlands

Hans Joosten

### 5.2.1 Introduction

Whereas the general role of peatlands in the global carbon balance was familiar to peatland specialists (cf. overview in Joosten & Clarke 2002), the extent of greenhouse gas emissions from drained peatlands (see chapter 3.2) became known only recently and this knowledge only spread slowly in climate policy. This chapter describes how awareness on the climatic role of peatlands developed in global environmental conventions, especially in the UN Framework Convention on Climate Change (UNFCCC) and its Kyoto Protocol (see also chapter 5.4 and 7.5).

### 5.2.2 Peatlands in the Ramsar Convention

The first convention to pay attention to peatlands was, naturally, the Ramsar Convention on Wetlands of International Importance. In the 1990s the Convention became aware that peatlands were underrepresented in the Ramsar protection system. This resulted in special peatland oriented recommendations and resolutions of its Conference of Parties (COP). COP 8 (2002) Resolution-VIII.3 on 'Climate Change and Wetlands: Impacts Adaptation and Mitigation' called for managing wetlands adaptively in response to global climate change and recognised the role of peatlands in mitigating impacts of climate change. The resolution called on relevant countries to minimise degradation and promote restoration of peatlands as significant carbon stores. Resolution VIII.17 of the same COP adopted the 'Guidelines for Global Action on Peatlands' and called for establishing a Coordinating Committee for Global Action on Peatlands (CCGAP).

The activities of the CCGAP partners during COP 9 in Kampala, Uganda (November 2005) raised the attention to peatlands considerably. CCGAP stressed countries to incorporate peatland 'wise use' issues into climate change and desertification policies, as well as in the work of the UNFCCC and the Convention on Combating Desertification. Climate aspects of peatlands were explicitly recognized in COP 10 (2008). Resolution X.24 'Climate change and wetlands' mentioned the carbon storage function of wetlands and took note of the global 'Assessment on Peatlands, Biodiversity and Climate Change' and decisions of the 9th conference of parties of the Convention on Biological Diversity (CBD COP 9; see below). It advocated urgent action to reduce degradation, promote restoration, and improve management of peatlands and other wetland types that are significant greenhouse gas (GHG) sinks. The resolution furthermore called on Ramsar Administrative Authorities to provide guidance and support to their UNFCCC focal points in order to reduce anthropogenic greenhouse gas emissions from wetlands such as peatlands and to encourage study of the role of wetlands in carbon storage and sequestration, in adaptation to climate change (including flood mitigation and water supply), and in mitigating the impacts of sea level rise.

### 5.2.3 Peatlands in the Convention on Biological Diversity

Peatlands were firstly brought on the agenda of the CBD in 2004 by the adoption of CBD resolution VII/15, mentioning peatlands as valuable ecosystems, as habitats, and for carbon storage and sequestration. In the following years, a team of peatland experts coordinated by Wetlands International (WI) and the Global Environment Centre (GEC) produced the global 'Assessment on Peatlands Biodiversity and Climate Change' (Parish et al. 2007, 2008). This Assessment was endorsed by the CBD's Subsidiary Body on Scientific, Technical and Technological Advice (SBSTTA) in July 2007. In May 2008, CBD COP 9 in its decision on Biodiversity and Climate Change recognized the importance of the conservation and sustainable use of the biodiversity of wetlands and, in particular, peatlands in addressing climate change. The decision noted with appreciation the findings of the Assessment and encouraged Parties and other Governments to strengthen collaboration with the Ramsar Convention and to implement the Guidelines for Global Action on Peatlands and the actions listed in the global Assessment. Also CBD COP 10 in October 2010 had the retention of carbon stores as a strong message running through various elements of its adopted Strategic Plan and Thematic Decisions, lending considerable weight to peatland conservation and restoration.

### 5.2.4 Raising awareness in the UNFCCC

As wetlands and as carriers of important biodiversity, peatlands are clear objects of attention of the Ramsar and Biodiversity Conventions. For the UNFCCC, in contrast, peatlands were a marginal phenomenon, somewhere hidden in the soil compartment of the contested Land Use sector (see chapter 5.4). This was the reason that peatland interest groups started with raising the profile of peatlands in the Ramsar and Biodiversity Conventions, in the hope that via these conventions with an easier audience the message would trickle down to the Climate Convention more effectively. Instrumental in creating access to the UNFCCC were:
- The increasing cooperation between major global peatland stakeholder groups (International Peat Society (IPS), International Mire Conservation Group (IMCG), WI, GEC), initially in the Ramsar CCGAP;
- The globally important emissions from peatland conversion in Southeast Asia since the 1990s, especially from the catastrophic 1997/1998 El Niño fires (Page et al. 2002);
- The growing knowledge on the extent of degraded peatland and associated emissions (Joosten & Clarke 2002);
- The attention of the Intergovernmental Panel for Climate Change (IPCC) in its Guidance and Guidelines of 2003 and 2006 (IPCC 2003, 2006);
- The start of major peatland rewetting and restoration projects that provided the capacity for concerted action on the global level.

A first publication (1998) addressing the problems in Southeast Asia was Safford & Maltby's 'Guidelines for integrated planning and management of tropical lowland peatlands'. The UNFCCC was first explicitly addressed by the 1999 joint position paper 'The role of peatlands in man-induced climate change' (the 'Freising Statement', www.imcg.net/docum/freising.htm) of the IPS and the IMCG (see box 15).

Indonesian peatlands were the subject of the 2001–2005 'Peat Swamp Forest Carbon Sequestration and Management Project' funded by the Canadian International Development Agency (CIDA) and implemented by the University of Waterloo, Wildlife Habitat Canada and Canadian government advisors (Environment Canada and Canadian Forest Service) in cooperation with Indonesian agencies and Wetlands International. The project focused on demonstrating community-based management of tropical peat/swamp forests, fire suppression, peatland restoration, and carbon sequestration.

2001 was also the starting year of the Global Peatland Initiative (GPI), a platform programme promoting the wise use and conservation of peatlands, funding relevant projects, and channeling their results to international policies and conventions. GPI (founded by WI, Alterra, IMCG, IPS, and International Union for Conservation of Nature (IUCN) Netherlands, and funded by the Dutch Ministry of Foreign Affairs) managed to generate and support over 50 small but effective projects.

In May 2003 a global programme on 'Integrated Management of Peatlands for Biodiversity and Climate Change' started, a joint project of WI, GEC, IMCG and Wildlife Habitat Canada in conjunction with the governments of Russia, China, Indonesia, and other countries in the ASEAN region, and funded by the Global Environment Facility (GEF) through the United Nations Environment Programme (UNEP-GEF) and the governments of Canada and the Netherlands. Next to implementing various peatland rewetting projects in the partner countries, the project created the influential 'Assessment on Peatlands, Biodiversity and Climate Change'.

In autumn 2004 the Global Environment Facility agreed to finance a project presented by the Belarusian Ministry of Forestry through the United Nations Development Programme (UNDP-GEF) for restoring 17 degraded peatlands (42,000 ha; by 2010 28,000 ha on 15 sites were rewetted, see chapter 3.4) across Belarus. Next to biodiversity, the project was explicitly aimed at reducing the huge $CO_2$ emissions from drained peatlands and peat fires. A project on 'Vulnerabilities of the Carbon-Climate System: Carbon Pools in Wetlands/Peatlands as Positive Feedbacks to Global Warming' (led by GEC and the Global Carbon project, 2005–2006) brought experts from the Asia-Pacific Region and beyond together to share knowledge on the importance and vulnerability of tropical peatland carbon stores and their potential impact on future climate change (Parish & Canadell 2006).

Building on the success of the Global Peatland Initiative, the 'Central Kalimantan Peatland Project' (CKPP) started in December 2005. CKPP was designed to demonstrate the economic and ecological feasibility of peat swamp conservation and restoration through innovative financing schemes related to climate and carbon, livelihood development, community empowerment and improved policies and coordination between government levels. The large, integrative project was implemented by a consortium of non-governmental organisations (Borneo Oranguntan Survival Foundation (BOSF), CARE International Indonesia, Wetlands International – Indonesia Programme and World Wildlife Fund for Nature (WWF Indonesia), the University of Palangka Raya and the Government of Central Kalimantan). All these projects, that were followed by many others in later years, were instrumental in raising the profile of peatlands among global policy makers.

In 2005 peatland emissions also became subject of discussion in IPCC. Already in its 1996 guidelines (IPCC 1997) IPCC had noticed that 'histosols or peat soils …can lose massive amounts of carbon over a sustained period upon drainage and cultivation' but many countries neglected this fact in their national emissions reporting to the UNFCCC. The new 2006 Guidelines (IPCC 2006) aimed to improve overall consistency and completeness in reporting of the entire Agriculture, Forestry and Other Land Use (AFOLU) sector by integrating the chapters Agriculture (Chapter 4) and Land Use, Land-use change and

---

**Box 15**

**Excerpt from the IPS/IMCG 1999 Freising statement**

The UNF Convention on Climate Change should include actions designed to:
- Promote the maintenance of existing carbon stores in peatlands;
- Prevent the uncontrolled release of carbon from peatlands;
- Maintain the carbon sequestering role of pristine peatlands;
- Promote the restoration of disturbed peatlands for carbon sequestration;
- Reduce, by wise use, the emissions of greenhouse gases from peatlands currently being used;
- Promote further studies on carbon balance in peatlands and its role in global climate change.

## 5.2 Sensitising global conventions for climate change mitigation by peatlands

Forestry (Chapter 5) of the Revised 1996 IPCC Guidelines (IPCC 2007) and the Good Practice Guidance for Land Use, Land-Use Change and Forestry (GPG-LULUCF, IPCC 2003). In the chapter 'Wetlands' of the 2006 IPCC Guidelines, for the first time guidance was given for peatlands under extraction, whereas in the other land use chapters more attention was paid to $CO_2$ and $N_2O$ emissions from organic soils. During the preparation of the Wetlands chapter other peatland issues, such as peatland fires and rewetting of drained wetlands/wetland restoration, were discussed, but this did not result in concrete guidance, because adequate methodologies were not yet available. The 2006 Guidelines advised countries where such activities were planned to consider research to assess their contribution to greenhouse gas emissions or removals.

### 5.2.5 UNFCCC 2006 Nairobi

UNFCCC COP 12 (November 2006, Nairobi, Kenya) was the first UNFCCC meeting with attention to peatlands. Two booths of WI and GEC informed on the role of peatlands in climate change. Two well-attended side events focussed on the immense emissions from peatland drainage and fires in South East Asia and promoted the interim key findings of the global Assessment on Peatlands, Biodiversity and Climate Change. During the Subsidiary Body for Scientific and Technological Advice (SBSTA) meeting on deforestation, GEC made an intervention to stress that peatlands are found in more than 100 developing countries, covering 60 million ha and storing over 80 Gt Carbon. Emissions from degradation of forested peatlands in developing countries are the largest single source of emissions from deforestation globally. When the decision was adopted by SBSTA, Singapore emphasised the need to look at peatlands within the framework of work on deforestation.

One of the major peatland events in Nairobi was WI and Delft Hydraulics presenting the results of the PEAT-$CO_2$ research project (Hooijer et al. 2006). The report showed that drained peatlands in Indonesia annually emit 600 million t of $CO_2$ by peat decomposition and 1,400 million t through fires. Local demands for economic development and poverty reduction as well as the global demand for hardwood, paper pulp and palm oil are the driving forces behind peatland destruction. The report concluded that deforested and drained peatlands in Southeast Asia are a major obstacle to meeting the aim of stabilising greenhouse gas emissions and recommended that international action be taken to help Southeast Asian countries, especially Indonesia, to conserve their peat resources and to restore high water tables.

### 5.2.6 Developments after Nairobi

In June 2007 the 3rd International Conference of the Michael Otto Foundation on Wetland Protection and Climate Change in Belarus adopted a resolution on peatlands, supported by the Ministry for Natural Resources and Environmental Protection of Belarus, the National Academy of Sciences of Belarus, APB-BirdLife Belarus, UNDP, the Michael Otto Foundation, The Royal Society for the Protection of Birds (RSPB), and IMCG. With this resolution the Conference Participants, among others, agreed:
- To aim at the restoration of 260,000 ha degraded peatland in the first phase, and a larger area in the longer-term;
- To try and raise funds for the restoration and long-term sustainable management of re-wetted peatlands in Belarus through selling high quality carbon credits on the voluntary market;
- To develop standards and methodologies for peatland carbon projects in Belarus;
- To develop a monitoring system for verifying greenhouse gas sinks and avoided emissions in restored peatlands;
- To advocate for an inclusion of greenhouse gas emissions from peatland under the Kyoto Protocol post 2012.

The December 2007 UNCCC COP took place in Indonesia, the country with massive ongoing deforestation and peatland drainage. The PEAT-$CO_2$ report (Hooijer et al. 2006) and a workshop on 'Carbon-Climate-Human Interactions on Tropical Peatlands" (Yogyakarta, Indonesia, August 2007) had just shown that because of peatland emissions, Indonesia was the third largest greenhouse gas emitter of the world. The resolution 'Reducing Emissions from Deforestation in Developing countries: Approaches to Stimulate Action' adopted at this COP explicitly refers to forest carbon stocks, which include carbon stored in the associated forest soils. A large group of stakeholders was attracted by the especially organised Wetland Day, which emphasised peatland. The Executive Summary of the Assessment on Peatlands, Biodiversity and Climate Change was officially launched by Achim Steiner, UN Under-Secretary General and Executive Director of UNEP, Ahmed Djoghlaf, Executive Secretary of the CBD, Marcel Silvius, Programme Manager of WI and Faizal Parish, Director of GEC. They called for the international community to take urgent action to protect and restore peatlands, or as Achim Steiner put it: 'Just like a global phase out of old, energy guzzling light bulbs or a switch to hybrid cars, protecting and restoring peatlands is perhaps another key "low hanging fruit" and among the most cost-effective options for climate change mitigation.'

### 5.2.7 Addressing the Kyoto Protocol: Wetland restoration and management

2008 brought the first direct intervention on peatland in the UNFCCC by a background paper submitted by Iceland to the Kyoto Protocol Ad Hoc Working Group (AWG-KP 6) at its meeting in Accra, August 2008. The paper noted that the high emissions from degraded peatlands are disproportionate to their size and that, whereas the problem is large, it is also concentrated and may therefore be easier addressed than many other emissions sources.

Referring to the sizable emissions from drained peatlands in its own country (almost half of the combined emissions from fossil fuel use and industrial processes) Iceland proposed to add 'Wetland restoration and degradation' as a new Land Use, Land-use change and Forestry (LULUCF) activity under the Kyoto Protocol.

The proposal attracted much attention as shown by the supportive submissions to the next Bonn meeting (April 2009) of, among others, Australia, Belarus, Japan, New Zealand, Switzerland, the EU, and a whole series of non-Annex 1 countries, such as Indonesia and Singapore. Some countries (e.g. Japan and China) pointed at the hitherto unsolved accounting problems.

What followed was one and a half year of discussions on content and language (cf. chapter 7.5; Joosten 2009a, b, c, 2010a, b) that resulted in a definition at the 2010 COP in Cancun that was unanimously accepted by all LULUCF expert negotiators: '"Rewetting and drainage" is a system of practices for rewetting and draining on land with organic soil that covers a minimum area of 1 ha. The activity applies to all lands that have been drained and/or rewetted since 1990 and that are not accounted for under any other activity as defined in this appendix, where drainage is the direct human-induced lowering of the soil water table and rewetting is the direct human-induced partial or total reversal of drainage.' (see chapter 5.4).

### 5.2.8 Concluding remarks

This chapter described some steps on the 'long and winding road' (Joosten 2009b) that a multitude of stakeholders had to go to raise awareness on peatland emissions in the UNFCCC and to bring peatland rewetting as an accountable activity under the Kyoto Protocol. Whereas in Cancun (December 2010) unanimity was reached among LULUCF negotiators on the definition and content of such activity the aims have not fully been reached yet. Discussions still have to follow on the voluntary or mandatory character of the activity (see chapter 5.4) and on the concrete role of peat soils and peatlands in other UNFCCC mechanisms, such as the Clean Development Mechanism, REDD+ and other mechanisms under development. The pursued parallelisation among UNFCCC mechanisms may lead to a mutual control of rules and modalities and a modification of their content. And the unpredictable element of politics beyond rationality will always remain and may still frustrate the outcome. What will remain beyond dispute is recognition of the important role that peatlands play in global emissions and that peatland rewetting and restoration can play in climate change mitigation. Peatlands have arrived in the UNFCCC deliberations and they are there to stay.

## 5.3 Selling peatland rewetting on the voluntary carbon market

Robert O'Sullivan & Igino Emmer

### 5.3.1 Introduction

Emission reductions that cannot be accounted for under the Kyoto Protocol and thus cannot be credited and transferred under Joint Implementation or International Emission Trading (see chapter 5.4) can be sold on the voluntary carbon market. The voluntary market is divided into two segments, the 'over-the-counter' (otc) market and the exchange market. The otc market consists of bilateral (and typically confidential) sales of carbon credits under some type of emission reductions purchase agreement (ERPA). In contrast the exchange market functions via facilities that are especially constructed for the purpose of trading (i.e., 'exchanges'). Whereas the exchange market has been significant in past years, it collapsed with the Chicago Climate Exchange (CCX) in 2010 (Linacre et al. 2011).

Corporate social responsibility and public relations, i.e. the wish to compensate inevitable emissions on a voluntary basis, are leading motivations to buy carbon credits on the voluntary market (44%). Other motives are resale investment (22%) and anticipation of future regulations (20%). Private firms are the predominant buyers next to governments, NGOs and individuals. The majority of buyers were located in Europe in 2010 (41%) followed by the US (37%), Latin America (5%) and Asia (4%) (Peters-Stanley et al. 2011).

### 5.3.2 Volume and prices

The voluntary market grew rapidly up to the time of the global financial crisis in 2009. It has started to recover in 2010, but is still (mid-2011) significantly less than its 2008 high (Table 31). While market value is not insignificant, it is dwarfed by the compliance market that was valued at US$ 142 billion in 2010 (Linacre et al. 2011).

In 2010 an estimated 128 million carbon credits were transacted on the otc market, up from 55 million in 2009 (Peters-Stanley et al. 2011) and 54 million in 2008. Credits from terrestrial carbon (forestry and agriculture) have grown in significance in recent years. In 2010 credits from forest protection projects (Reducing Emissions from Deforestation and forest Degradation (REDD)) represented 29%, from other forestry projects 13%, and from agricultural soils 5% (Peters-Stanley et al. 2011). This is a big increase compared to 2008 when terrestrial carbon credits made up only 5.6 million credits or 11% (Hamilton et al. 2009). Wetland projects were until now completely absent from the market, which was primarily due to voluntary market standards only being recently recognizing wetland projects as being eligible to create carbon credits.

Market wide the (volume-weighted) average price paid for carbon credits on the voluntary market has declined over the last few years. In 2008 it was US$ 7.34 per credit (Hamilton et al. 2009). This dropped to US$ 6.5 in 2009 and US$ 6.0 in 2010 (Peters-Stanley et al. 2011). Concrete prices are, however, strongly variable depending inter alia on the buyer, the type of project, the co-benefits it may have, the standards chosen for the project, whether the credits are sold wholesale or retail, the volume of credits sold, and the contractual terms under which the credits are sold (e.g. whether it is a payment on delivery contract, pre-payment etc.). No transactions of carbon credits from peatland or wetland projects have been recorded to date, but other terrestrial carbon projects can provide an idea of the price ranges. For example, the highest recorded price for credits in the entire voluntary market in 2010 came from improved forest management (US$ 136 per credit), whereas the average price from this project type was around US$D 6 per credit. REDD also showed significant variation with some smaller volumes transacting for close to US$ 24 per credit, but high volume sales at a low price brought the average for REDD down to around US$ 5 per credit, and also depressed overall market averages.

### 5.3.3 Carbon credit sales

Options for structuring the sale of carbon credits include forward sales, spot sales, option sales, or a combination of these.

Forward sales involve signing a contract to sell carbon credits before they have been generated. Payments are normally made upon delivery of the credits, though advance payments for not yet delivered credits are also sometimes possible. Smaller payments to cover some project development costs such as validation may be made in advance without providing security, but larger payments associated with project implementation typically need to be secured. Forward sales typi-

Table 31: Value of voluntary carbon market (US$ million). 2007 and 2008 data from Hamilton et al. 2009, 2009, and 2010 data for World Bank estimates from Linacre et al. (2011), for Ecosystem Marketplace estimates from Peters-Stanley et al. (2011), respectively.

| 2007 | 2008 | 2009 | | 2010 | |
|---|---|---|---|---|---|
| | | World Bank | Ecosystem Marketplace | World Bank | Ecosystem Marketplace |
| 335.3 | 704.8 | 357.8 | 415 | 393.5 | 424 |

cally command reduced prices compared to spot transactions because they carry an enhanced project and regulatory risk. Since forward sales deal with carbon credits that do not yet exist, an inherent risk remains that the credits will not be generated. This is reflected in a price discount. At the same time, forward prices reflect an expectation of further market developments, which may lead to additional corrections of the price. Defining time and place of delivery is essential for forward sales. For Verified Carbon Standard (VCS) projects that involve the sale and delivery of Verified Carbon Units (VCUs) delivery can only occur after the project has been registered as a VCS project and verification and issuance of VCUs have taken place. Many forward sale agreements for afforestation or reforestation projects run for anywhere from five to 20 or more years. Forward sales in other project categories tend to be shorter, but still may run for several years from the date of execution.

Spot sales involve the sale of credits after they have been generated and issued. This sale may either occur through a bilateral otc contract or on an exchange. Prices and volumes are fixed in advance. Payment and delivery occur within a few days after execution for otc transactions or instantaneously on exchanges. Payment may also occur before delivery, or be mediated by a broker that releases payment and credits simultaneously.

Option sales involve a contract between two parties for a future sale of carbon credits at a reference price. The buyer of the option gains the right, but not the obligation, to engage in that transaction, while the seller has the corresponding obligation to fulfill the transaction.

In 2010 otc sales were dominated by forward sales (55%) and spot transactions (22%). Other transaction types such as pre-payment sales were less common (18%) (Peters-Stanley et al. 2011). All sales contain a number of key terms with varying levels of risk and reward for the seller, including payment, price structures, volume, delivery requirements, and defaults and remedies. Choosing the most appropriate option is the key component of any marketing strategy. Recently some REDD projects have been exploring financing projects (in return for transferring title to credits) through innovative 'green bonds' as an alternative model to traditional sale and purchase contracts.

### 5.3.4 Marketing strategy

Key components in a marketing strategy for the voluntary market include:
- Choosing a selling strategy. Assess whether a project should sell its credits under one or more forward contracts, via spot transactions, or a combination of both. This assessment should involve an evaluation of how much and what sort of risks the project is willing to bear;
- Telling the story. Once a selling strategy has been identified project developers need to 'tell the story' of their project. Most voluntary market buyers are driven by public relations or corporate social responsibility goals, so a project that tells a good story will help attracting buyers and maximizing the price. The story should explain all positive aspects of the project, including co-benefits such as biodiversity conservation, scientific or research value, as well as social benefits of the project. It should also highlight the standard(s) that the project intends to apply;
- Developing steps for otc transactions. A term sheet setting out the key terms (price, volume, delivery terms, defaults and remedies, governing law) under which the VCUs will be sold needs to be developed. It should be accompanied by the project's story as well as a technical description of how the emission reductions are calculated, including baseline assumptions and emission reduction estimates. The term sheet is then circulated amongst potential buyers in the target market who are given a certain time period to respond with their offer for the credits. The most appealing offer is selected, an exclusivity agreement between buyer and seller is often entered while the buyer conducts additional due diligence, the parties negotiate a sale and a purchase agreement is negotiated and executed. Due diligence does not occur in all transactions, unless there is an advance payment involved in which case it is almost always conducted by the buyer to assess the project's risks.

The costs to implement the marketing strategy contain two components:
- Project costs associated with addressing legal issues associated with the project (e.g. agreements with national governments) that need to be resolved prior to the sale of credits; and
- Transaction costs associated with drafting term sheets, identifying buyers, responding to due diligence requests (though the actual costs of

## 5.3 Selling peatland rewetting on the voluntary carbon market

due diligence are normally borne by the buyer), negotiating contracts (for otc forward and spot transactions), and/or identifying brokers, paying broker fees or registering to sell on exchanges (for spot exchange transactions).

The project costs need to be estimated by the project developer. The transaction costs will vary depending on the parties involved and how the provision of these services is structured, but can range from a percentage of the transaction value to fixed prices for discrete services or a combination of these. The costs will depend on the size and complexity of the transaction. An additional brokerage fee may be charged directly to the buyer or shared between buyer and seller.

### 5.3.5 Standards

The voluntary market is populated by a plethora of competing standards to demonstrate the integrity of projects. In 2008 96% of voluntary credits were third party verified, making third party verification a de facto starting point for any project or standard to be seen as credible. VCS was the dominant standard on the otc market, accounting for 34% of recorded transactions in 2010 (down from 48% in 2008). The Climate, Community, and Biodiversity Standard (CCB Standard) increased from 3% of otc transactions in 2008 to 19% in 2010, even though it is limited to land based carbon projects (Peters-Stanley et al. 2011 for 2010 data and Hamilton et al. 2009 for 2008 data, respectively). It must be noted that the CCB Standard assesses co-benefits of a project and does not quantify emission reductions so it usually is combined with an additional standard such as the VCS.

The next most common standards in the otc market in 2010 were the Climate Action Reserve (16%) and the Gold Standard (8%). A total of 17 other standards made up the rest of the market (23%), though eight of these 17 accounted for less that 0.1% combined (Peters-Stanley et al. 2011). Most standards do not recognize peatland restoration projects as an eligible project category. Peatland projects are currently only eligible under the VCS, International Standards Organisation (ISO), and CCB standards, and potentially under the Social Carbon standard. The VCS only recently (March 2011) adopted a new 'peatland rewetting and conservation' category. The ISO standard was designed as a neutral standard that could be applied to any greenhouse gas accounting standard. The Social Carbon standard, similar to CCB, focuses on social and other environmental aspects (and on its website also recommends to be combined with VCS), but unlike CCB also develops and applies carbon quantification standards if needed. To date there have not been any projects in Belarus registered under the VCS, CCB, or Gold Standard.

A number of standards obtain a premium on the voluntary market because they certify co-benefits or are associated with a particularly attractive jurisdiction. The VCS, as the de-facto benchmark standard for the voluntary market, does not attract a premium. The CCB, as the leading quality assurance standard for land based carbon projects, attracts a premium that, dependent on the buyer's priorities and willingness to pay, ranges between US$ 1–3, and may be as high as US$ 6 or more per credit (Neeff et al. 2010). The cheapest carbon was associated with the CCX and the American Carbon Registry with average prices of less than US$ 1 per credit (Hamilton et al. 2009). CCX credits were particularly devalued over ongoing concerns of their environmental integrity.

### 5.3.6 A new standard: VCS Peatland Rewetting and Conservation (PRC)

As peatland projects are acceptable under the VCS since March 2011 (due to significant efforts of the BMU-ICI project) and taking into account the broad recognition of that standard, VCS is the best option to create carbon credits from peatlands. This should, if possible, be combined with CCB in order to realize the premium end of the voluntary market.

The purpose of the VCS is to assist project proponents, project developers, methodology developers, and validation/verification bodies in developing and auditing projects and methodologies (VCS 2011). The VCS-PRC standard provides guidance on, amongst others, eligible project categories, greenhouse gas (GHG) sources and carbon pools, baseline determination, leakage calculation, and GHG emission reductions and removals calculation. Eligible agriculture, forestry and other land use (AFOLU) project categories include Afforestation, Reforestation and Revegetation (ARR), Agricultural Land Management (ALM), Improved Forest Management (IFM), Reduced Emissions from Deforestation and forest Degradation (REDD), and Peatland Rewetting and Conservation (PRC) (see Table 32).

Project development and implementation are addressed at three different levels: 1) The AFOLU

Table 32: Types of PRC activities that may be combined with other AFOLU project categories.

| Baseline scenario | | Project activity | Applicable guidance |
|---|---|---|---|
| Condition | Land use | | |
| Drained peatland | Non-forest | Rewetting | RDP |
| | | Rewetting and conversion to forest/revegetation | RDP+ARR |
| | | Rewetting and paludiculture/erosion avoidance | RDP+ALM |
| | Forest | Rewetting | RDP |
| | Forest with deforestation/degradation | Rewetting and avoided deforestation | RDP+REDD |
| | Forest managed for wood products | Rewetting and improved forest management | RDP+IFM |
| Undrained peatland | Non-forest | Avoided drainage | CUPP |
| | Forest | Avoided drainage | CUPP |
| | Forest with deforestation/degradation | Avoided drainage and deforestation | CUPP+REDD |
| | Forest managed for wood products | Avoided drainage and improved forest management | CUPP+IFM |

requirements define how projects and methodologies can comply with the VCS standard. 2) Methodologies explain step-by-step how emission reductions or removals are to be estimated in line with the requirements following accepted scientific good practice. 3) Project description or design documents provide information on how a specific project complies with the AFOLU requirements and how it applies the methodologies. The PRC requirements must be applied across all AFOLU categories when they occur on peatland, e.g. ARR on peat, REDD on peat. In addition, PRC project activities can exist stand alone, e.g. rewetting of drained peatland or avoided drainage of non-forested peatland.

### 5.3.7 Specific PRC requirements

In peatlands, GHG emissions and carbon stock changes largely depend on hydrological conditions. Therefore, most PRC requirements relate to hydrology or to soil moisture-dependent processes (VCS 2011).

Projects aiming at the conservation of undrained or partially drained peatland must demonstrate that there is either no hydrological connectivity to adjacent areas, or, where there is, that a buffer zone is established to ensure that adjacent areas will not significantly affect the project area, such as causing the water table in the project area to drop, resulting in higher GHG emissions.

PRC projects must furthermore demonstrate that their peat carbon stock is 'permanent'. The maximum quantity of GHG emission reductions that may be claimed by the project is limited to the difference in peat carbon stock between the project and the baseline scenario after 100 years. This limit is established because in peatlands that are not fully rewetted, the peat will continue to oxidize leading to GHG emissions and possibly to an eventual complete depletion of the peat.

Biofuel crop production activities on drained peatland or on peatland cleared of, or converted from native ecosystems are not eligible. Biofuel crop production on rewetted peatland must follow the PRC requirements. Some forms of biomass production on peatland (i.e. paludicultures with mosses, alder, papyrus, reeds, sedges, and willow) are compatible with rewetting and may even lead to peat accumulation in the long run (see chapter 6).

Rewetting of Drained Peatland (RDP; Table 32) concerns practices establishing a higher water level on drained peatland. A clear relationship between GHG emissions and water level has been established in scientific literature with most changes occurring at water levels close to the surface (see chapter 3.1 and Couwenberg et al.

2011). Afforestation, Reforestation and Revegetation (ARR) project activities that involve nitrogen fertilization or active lowering of the water level, such as draining in order to harvest, are not eligible, as they are likely to enhance net GHG emissions.

Conservation of Undrained or Partially Drained Peatland (CUPP) concerns activities that avoid drainage in undrained (or further/deeper drainage in partially drained) peatlands that are threatened by drainage. These activities aim at reducing $CO_2$ emissions by avoided peat oxidation and/or by avoiding increased fire incidence. Projects that continue or maintain active drainage are not eligible.

Due to the extensive local, regional, and global demand for peat, projects that avoid peat mining are likely to suffer significant (potentially 100%) leakage (i.e. an increase in emissions outside the project boundary that occurs as a consequence of the project activity's implementation) and are therefore not eligible. Project activities that serve the demand side and avoid peat mining by providing alternatives for peat as fuel or substrate, are outside the scope of AFOLU but may qualify under another VCS sectoral scope, e.g. energy.

GHG emissions for both the baseline and project scenarios can be assessed using water level or another justifiable parameter as a proxy. Emissions of $CH_4$ from drained peatland are negligible and may conservatively be neglected in the baseline. Transient peaks of $CH_4$ after rewetting, however, necessitate the inclusion of $CH_4$ in the project emissions calculation. $N_2O$ emissions also must be included. A methodology establishes the criteria and procedures by which the $CH_4$ and $N_2O$ sources may be deemed insignificant (for which VCS has set specific rules) or may be conservatively excluded (based on a quantitative assessment or by using peer-reviewed literature). GHG accounting has to take the peat depletion time into account. The peat depletion time is the moment in the baseline scenario that all peat would have been disappeared due to oxidation or other losses and after which thus no GHG emissions from the peat would take place anymore. The peat depletion time has to be assessed on the basis of peat depths, water levels, and associated subsidence rates. No emission reductions can be claimed beyond the peat depletion time.

Methodologies for Rewetting of Drained Peatland (RDP) projects explicitly addressing anthropogenic peatland fires must establish procedures for assessing the baseline frequency and intensity of fires in the project area.

PRC project activities must account for leakage due to activity shifting (e.g. continued deforestation and associated drainage outside the project area, a shifting of agricultural practises; see chapter 7.2). In addition, 'ecological' leakage may occur in PRC projects by changes in GHG emissions in ecosystems that are hydrologically connected to the project area (e.g. forests that die off outside the project area as a result of rewetting of the project area).

## 5.4 Selling peatland rewetting on the compliance carbon market

Hans Joosten

### 5.4.1 Introduction

Better management of ecosystem carbon stocks (reservoirs) and fluxes (emissions and removals) can substantially contribute to reducing atmospheric greenhouse gas concentrations (Trumper et al. 2009, Blaustein 2010). Climate policy has become increasingly aware of this potential, although the important role of peatlands in climate change mitigation has until recently not sufficiently been acknowledged. This chapter explains how the UN Framework Convention on Climate Change (UNFCCC) and its Kyoto Protocol (KP) deal with land use, presents an overview of the latest peatland-oriented developments, and discusses the possibilities of generating and trading carbon credits under these international agreements.

### 5.4.2 The Climate Convention and its Kyoto Protocol

The UNFCCC aims at achieving 'stabilization of greenhouse gas concentrations in the atmosphere at a level that would prevent dangerous anthropogenic interference with the climate system' (art. 2 UNFCCC). Progress with respect to this goal is monitored by means of greenhouse gas inventories that all countries have to submit. Reporting for the land use sector is done on the basis of 'land use categories' (forest land, cropland, grassland, wetlands, settlements, other lands), using guidelines prepared by the Intergovernmental Panel on Climate Change (IPCC).

The Kyoto Protocol is the only legally binding mechanism within the Climate Convention. Under the Protocol, 37 industrialized countries and countries in transition to a market economy (the 'Annex I countries' of the UNFCCC) have legally binding emission limitation and reduction commitments. They have obliged themselves to reduce their emissions in the first commitment period (2008–2012) collectively by 5.2% compared to the reference year 1990. Main object of the Kyoto Protocol is to decrease the emissions of greenhouse gases (GHG) from the sectors 'energy', 'industrial processes', 'solvents and other products', 'agriculture', and 'waste' (GHG sources). Simultaneously the possibility was opened for compensating these emissions by GHG sinks in the so-called LULUCF ('land use, land-use change and forestry') sector. Accounting for the LULUCF sector is done on the basis of pre-defined 'activities'.

The activity 'afforestation, reforestation and de-forestation' (ARD) was considered most important as a GHG sink and most easily to monitor and was made mandatory under the Kyoto Protocol (art. 3.3 KP). Mandatory means that countries *must* account for carbon stock changes and GHG fluxes from ARD. In contrast, accounting for all other land use activities ('forest management', 'cropland management', 'grazing land management', and 'revegetation') was made voluntary (art. 3.4 KP). This means that countries may *choose* whether to account for an activity or not.

### 5.4.3 Peatlands in the Kyoto Protocol: the current situation

As GHG emissions from drained peatland may be substantial (see chapter 3.2), peatland rewetting would be a beneficial practise under each land use activity. Some 40% of the drained peatlands in the Kyoto Protocol countries have been drained for agriculture, some 30% for forestry. The KP activities 'forest management', 'cropland management' and 'grazing land management' would thus cover 70% of drained peatlands, whereas the activity 'revegetation' additionally covers the largely bare drained peatlands under/after peat extraction. So the current KP already allows accounting for rewetting of the vast majority of drained peat soils, provided that the relevant activities are chosen. The freedom of choice is exactly where the bottleneck is: hardly any country has chosen 'cropland management', grazing land management' and 'revegetation' for the first commitment period (Table 33). The reasons are clear: Choosing an activity (e.g. grazing land management) implies that the country has to account *all* effects (e.g. $N_2O$ emission from manure, carbon emissions from the soil, etc.) of *all* practises (e.g. grazing, fertilizing, occasional tillage, etc.) on *all* lands subject to that activity (= all land used for grazing cattle). This requires information about the emissions associated with these practices (e.g. how much $N_2O$ emission is associated with nitrogen fertilization) and knowledge about the area subject to these practices (e.g. how much land is fertilized how much). If, for example, Germany would want to choose 'grazing land management' in order to claim emission reductions from rewetting grassland on peat soil, it has to monitor and account all emissions from grasslands. This does not only apply to the 600 km$^2$ of grassland it has rewetted, but also to the remaining 6,000 km$^2$ of grassland on drained peatland and to the 60,000 km$^2$ of grassland on mineral soil. Countries that want to account for rewetting thus also have to monitor and account for much larger areas of land with much less emission reduction options or even with increasing emissions. This principle generally makes 'the calf larger than the cow'. Other reasons for not choosing 'cropland management' and 'grazing land management' are that emissions from these activities are difficult to monitor and that the emission factors (especially for $N_2O$) have a large uncertainty range. The most important reason is, however, that croplands and grasslands may also be net GHG sources (because of fertilization induced $N_2O$ emissions and because of soil carbon losses) instead of GHG sinks and that it is not clear how the balance has changed in comparison with the reference year 1990. So instead of including a risky line in their GHG budget, countries prefer not to deal with these activities at all.

The only LULUCF activity that many countries have chosen is 'forest management'. The monitoring of this activity is considered simple because the GHG budget is largely approached by monitoring changes in wood stock, for which century long experience in commercial forestry exists. Monitoring of forest management is also *made* simple, because most countries simply forget or, with wrong and dubious arguments, neglect the emissions of $CO_2$ and $N_2O$ from organic soils (Barthelmes et al. 2009). 'Forest management' is, however, first and foremost popular because this activity (and similarly ARD) is accounted in a completely different way than the

## 5.4 Selling peatland rewetting on the compliance carbon market

Table 33: Choice (for the first commitment period) of art. 3.4 activities by Annex I countries with a significant extent of peatlands (FM= forest management, CM= cropland management, GM= grazing land management, RV= revegetation).

| Country | FM | CM | GM | RV |
|---|---|---|---|---|
| Belarus | Not elected | Not elected | Not elected | Not elected |
| Canada | Not elected | Elected | Not elected | Not elected |
| Denmark | Elected | Elected | Elected | Not elected |
| Estonia | Not elected | Not elected | Not elected | Not elected |
| Finland | Elected | Not elected | Not elected | Not elected |
| France | Elected | Not elected | Not elected | Not elected |
| Germany | Elected | Not elected | Not elected | Not elected |
| Iceland | Not elected | Not elected | Not elected | Elected |
| Ireland | Not elected | Not elected | Not elected | Not elected |
| Latvia | Not elected | Not elected | Not elected | Not elected |
| Lithuania | Elected | Not elected | Not elected | Not elected |
| Netherlands | Not elected | Not elected | Not elected | Not elected |
| Norway | Elected | Not elected | Not elected | Not elected |
| Poland | Elected | Not elected | Not elected | Not elected |
| Romania | Elected | Not elected | Not elected | Elected |
| Russian Federation | Elected | Not elected | Not elected | Not elected |
| Sweden | Elected | Not elected | Not elected | Not elected |
| UK | Elected | Not elected | Not elected | Not elected |
| Ukraine | Elected | Not elected | Not elected | Not elected |

other LULUCF activities and the other sectors in the Kyoto Protocol. Whereas all others use 'net-net accounting', i.e. compare the emissions in the commitment period 2008–2012 with those in the reference year 1990, the Kyoto forest world uses 'gross-net accounting', i.e. only looks at emission changes within the commitment period itself. As a result of this double standard, the forestry sector seems to perform much better than it does in reality ('as the atmosphere sees it'). It generates carbon credits that are counted as equal to those from other activities and sectors, but in reality are inferior. Forest accounting thus unrightfully makes countries richer than they are.

### 5.4.4 Peatlands under a future Kyoto Protocol

If peatland rewetting wants to become attractive for climate change mitigation, the rules of the Kyoto game have to be changed. Peatland rewetting on cropland and grassland has little chance as long as the activities 'cropland management' and 'grazing land management' are only voluntary. Peatland rewetting on forest land does not give credit as long as the emissions from drained forest soils are neglected. Rewetting 'wetlands' (i.e. the land category under which also peat extraction sites are classified) is even completely discouraged as no activities (except 'revegetation') are eligible for 'wetlands'.

There are three major options to facilitate peatland rewetting under a future Kyoto Protocol:
- Adopting a land-based approach: Land-based accounting is the full accounting of all GHG

fluxes on all (managed) land in a country. It provides a complete ('wall-to-wall') picture of what is happening across the entire land use sector, rather than, as in the current LULUCF practice, countries choosing which land use activities they want to account. Land-based accounting precludes perverse selection and closes loopholes from (unaccounted) displacement of emissions between sectors and land categories. A notorious loophole is the cultivation of 'biofuels' on drained peat soils, where the reduced emissions from using 'biofuels' are accounted under the energy sector, but the (much larger) increased emissions from the soil are neglected in the land use sector. Land-based accounting would simply allow to account for peatland rewetting, in whatever land category it would occur. The political support for direct implementation of such comprehensive approach is, however, limited. Many countries sympathise with the option, but argue not yet to be able to manage the necessary inventory and monitoring. (This is a somewhat odd argument, because since 2005 the Annex I countries already report all these emissions annually to the UNFCCC). Some countries propose to go for full land-based accounting in the third commitment period (after 2018/2020), but this has not yet materialized in concrete proposals. The last 'proposal of the chair' of the Kyoto Protocol Working Group (AWG-KP) of 10 December 2010 (FCCC/KP/AWG/2010/CRP.4/Rev.4) reads: 'The Conference of the Parties serving as the meeting of the Parties to the Kyoto Protocol, agrees that it is desirable to move towards complete coverage of managed lands when accounting for the land use, land-use change and forestry sector';

- Increasing the number of mandatory activities: As long as a land-based approach is not within reach, countries could account for peatland rewetting under relevant land use activities. There is indeed political support in UNFCCC for increasing the number of mandatory activities and to move activities from art. 3.4. (the voluntary part) to art. 3.3 (the mandatory part) of the Kyoto Protocol. The largest chance exists for forest management. Also cropland management (with its relevance for soil carbon) is under consideration but there the reservations of developed countries are much larger. The chance that all current voluntary activities will become mandatory is virtually nil;

- Creating a new activity focussed on peatland rewetting: This last option under the Kyoto Protocol has been most widely discussed in the past two years. In Cancun (December 2010), unanimity was reached among LULUCF negotiators in the AWG-KP on the definition and content of such activity (see also chapter 7.5).

### 5.4.5 The proposed new activity 'rewetting and drainage'

The definition of the proposed new activity "Rewetting and drainage" is a system of practices for rewetting and draining on land with organic soil that covers a minimum area of 1 ha. The activity applies to all lands that have been drained and/or rewetted since 1990 and that are not accounted for under any other activity as defined in this appendix, where drainage is the direct human-induced lowering of the soil water table and rewetting is the direct human-induced partial or total reversal of drainage.' (FCCC/KP/AWG/2010/CRP.4/Rev.4).

The phrasing 'rewetting *and drainage*' was chosen to reach a balance in the definition: if climatically positive activities are accounted for, also their negative counterparts must be accounted (cf. deforestation in ARD).

The definition refers to 'organic soil' because a recent IPCC workshop (Eggleston et al. 2011) has concluded that, in contrast to a few years ago, now sufficient new science is available for IPCC to develop methodological guidance for reporting GHG fluxes in peatlands. For mineral soil wetlands the IPCC workshop was not so sure. The focus on organic soils is appropriate as organic soils are the global hotspots of emissions in the land use sector. In the European Union (EU), for example, cropland on organic soil (= 2% of the total cropland) is responsible for 43% of all cropland emissions, whereas cropland on mineral soil (88% of the total cropland) is responsible for only 5% (J. Grassi pers. comm.). The focus on organic soils makes it unnecessary to monitor lands with much less relevant emissions.

With 'since 1990' the definition reflects that 'rewetting and drainage' follows the normal Kyoto net-net accounting and that the activity only applies to lands where a lowering or a raising of the water level has taken place since 1990. Areas where the water level was changed before 1990 and where this lower or higher water level has been maintained continuously from then onward are not subject of the activity.

## 5.4 Selling peatland rewetting on the compliance carbon market

The epithet 'direct human-induced' stresses the anthropogenic character. Spontaneous rewetting (e.g. by beaver dams) or drainage (e.g. by river incision) do not fall under the activity and cannot be accounted for. Special attention was required to define 'rewetting' in a way to exclude 'flooding'. This restriction was politically necessary because some countries were afraid of having to report and account the (sometimes huge) methane emissions from water reservoirs (especially for hydro-electricity). Rewetting is defined as the partial or total reversal of drainage, and drainage as the lowering of the soil water table. This phrasing implies that the activity only concerns land that originally was 'wet', subsequently has been drained, and now is made wet again. Areas that are flooded but never have been drained do not fall under the activity '*re*wetting'. Also drained areas that have been flooded to the extent that the mean water level (fluctuations) by far exceed that of the area before drainage, do not comply with the activity.

'Rewetting and drainage' may become an effective instrument to facilitate peatland rewetting, as long as land-based accounting has not been achieved or not all art. 3.4 activities have become mandatory. 'Rewetting and drainage' allows for accounting the rewetting of lands that currently fall outside mandatory and voluntarily elected activities. 'Rewetting and drainage' furthermore allows for closing the emerging loophole between the unaccounted LULUCF and the accounted energy sector in the case of biofuels. If forest management and cropland management (the activities with the largest biofuel loophole risk) do not become mandatory, only mandatory 'rewetting and drainage' will close this loophole.

### 5.4.6 KP emission trading and peatland rewetting

The emission reductions, to which the contracting Parties to the Kyoto Protocol are committed ('compliance'), do not have to be totally implemented in these countries themselves. The Kyoto Protocol provides 'flexible mechanisms' that enable Parties to offset their emissions by acquiring emission reductions from other countries by – what is popularly called – 'carbon trading'. These mechanisms were introduced to lower the costs of achieving the targets, because the cost of reducing emissions varies considerably among regions and sectors, whereas the benefit for the atmosphere is the same, wherever on Earth a reduction is realized.

The Kyoto Protocol has currently three mechanisms for emissions trading:
- International Emissions Trading (IET, art. 17 KP) in which a country sells its surplus in 'carbon credits' to a country with a deficit. (In climate policy various concepts related to emission rights and emissions reductions are distinguished, including Assigned Amount Units (AAUs), Emission Reduction Units (ERUs), Removal Units (RMUs) and Certified Emission Reductions (CERs) in the Kyoto Protocol, and various types on the voluntary market. For simplicity reasons we call them all 'carbon credits'.) In 'economies in transition' (former East block states) the sales of 'hot air' (emission reductions caused by the economic collapse since 1990) can be 'greened' via so-called Green Investment Schemes (GISs). In a GIS the sale proceeds of IET in 'hot air' are reinvested in emission reduction projects or other projects beneficial to the environment. Such greening allows the buyer to use 'hot air' for his compliance purposes without running the reputational risk of 'non-additionality', i.e. of using credits that are not generated through genuine emission reduction policies. GISs are not part of the Kyoto Protocol, but are self-imposed commitments from a seller towards the buyer country (Simonetti & De Witt Wijnen 2009). In Ukraine, for example, peatland rewetting projects are envisaged as possible GIS-cover of national hot air sales;
- Joint Implementation (JI, art. 6 KP) relating to projects in which a developed country finances a GHG emission reduction project in another developed country and in return receives the carbon credits achieved by that project;
- Clean Development Mechanism (CDM, art. 12 KP) relating to projects in which a developed country finances a GHG emission reduction project in a developing country and in return receives the carbon credits achieved by that project.

In order to comply with its collective KP obligations, the European Union started in 2005 with the European Union Emissions Trading Scheme (EU ETS), a company-based cap-and trade system that covers the major energy and industrial companies. The EU ETS has since 2005 developed to become the largest emissions trading scheme in the world (Fig. 51). Similar, but much smaller regional regulated markets exist in, for example, the USA (Regional Greenhouse Gas

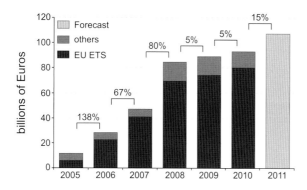

Fig. 51: Development of the global carbon market (in billions of Euros). Source: http://carboncapitalist.com/global-carbon-market-value-set-to-increase-by-15-in-2011/. EU ETS = EU Allowance Units in the European Union Emissions Trading Scheme; others = e.g. Certified Emission Reductions (CERs) purchased directly from CDM projects or traded on exchanges, Regional Greenhouse Gas Initiative in ten Northeastern and Mid-Atlantic states of the United States (RGGI), and Assigned Amount Units (AAUs) according to art. 3 and 17 of the Kyoto Protocol. The value for 2011 is forecasted.

Initiative RGGI), Australia (New South Wales Greenhouse Gas Abatement Scheme GGAS), and Canada (Alberta's Specified Gas Emittors Regulation SGER). Since the start of the Kyoto Protocol (2005) more than 345 billion (=345,000 million) Euro of carbon sales have been realised, of which almost 100 billion Euro in 2010 (Fig. 51, Hamilton et al. 2010). The compliance market is fully dominated by EU ETS (82.3%), followed on large distance by CDM (14.1%), RIGGI (1.9%), IET (1.4%), and JI (0.2%) (Hamilton et al. 2010, Kossoy & Ambrosi 2010, 2009 figures). The voluntary market with less than 1% of the global carbon trade (Hamilton et al. 2010, see chapter 5.3), dwarfs in comparison.

These huge figures have, however, hardly any relevance for peatland rewetting. EU ETS and RIGGI currently do not deal with land use at all, whereas CDM only allows afforestation and reforestation in its land use portfolio. IET could be relevant with respect to peatland rewetting if countries would select art. 3.4. activities. JI allows LULUCF and therewith peatland rewetting projects, but JI is not very popular, because the European Union has excluded JI from its EU ETS market and in combination with uncertain post-Kyoto rules this leads to a negligible demand for JI credits. From a carbon credit generation point of view, International Emission Trading, i.e. work-

ing for the national budget, is more interesting, as the least requirements are adhered to this mechanism. Joint Implementation and Clean Development Mechanism (similarly to the Verified Carbon Standard) have stricter criteria (e.g. with respect to reference level, additionality and permanence) so that with the same interventions less marketable carbon credits are generated. In comparison to the voluntary markets, the compliance market, furthermore, generates (much) higher prices for the same emission reduction.

### 5.4.7 Peatland rewetting on the future compliance market

The future of peatland rewetting on the compliance market will strongly depend on the future of the Kyoto Protocol system, on which considerable differences in perspectives exist. Countries like Japan, Russia and Australia have stated that they will not continue with a new legally binding Protocol after 2012, if major emitting countries that are currently outside the Protocol (USA, China, Brazil, South Africa) do not accept similar reduction obligations. Also the USA, currently not a KP Party, expects binding emissions reduction targets from the major developing countries. Anyway: the discussions in the past few years have brought considerable progress with respect to land use. Compared to 1997 when the Kyoto Protocol was negotiated, the scientific basis has become much more robust and the political will has increased to address the huge GHG emissions from the land use sector. The EU ETS and CDM currently consider to include LULUCF in their systems (depending on the outcome of the discussions on a post-Kyoto framework) to increase flexibility.

The awareness of the importance of peatlands has strongly increased (see also chapter 5.2 and 7.5) and peatland rewetting may get its fair place among the activities under a successor of the Kyoto Protocol. A strong link with peatland conservation and rewetting also exists in the REDD+ mechanism (Reducing Emissions from Deforestation and Forest Degradation). REDD+ offers great opportunities for reducing the massive $CO_2$ emissions from drained and degraded peat swamps in countries like Indonesia, Malaysia and Papua New Guinea. An important question will be to what extent the rules for developed countries under the Kyoto Protocol will also apply to developing nations under REDD+ and vice-versa.

Discussions in and outside the Kyoto Protocol seem to strengthen the case for a stronger commitment to peatland rewetting within the UNFCCC. Such a commitment will exert a strong influence on the future compliance market with respect to peatlands.

## 5.5 Voluntary emission reduction projects – how to start in Belarus

Irina Voitekhovitch & Alexander Grebenkov

### 5.5.1 Introduction

Any emission reduction project for the voluntary market must be recognized and approved at the national level, i.e. go through a national voluntary emission reduction (VER) project cycle, in order to generate, determine, and verify voluntary greenhouse gas emission reductions.

The Belarusian national voluntary emission reduction project framework is based on two legal acts:
- The Decree of the Council of Ministers of the Republic of Belarus No. 466 dated 14 April 2009 (Decree 466) 'On Approving the Provision on the Procedure of Submission, Review and Monitoring of Voluntary Emission Reduction Projects';
- Presidential Decree No. 625 dated 8 December 2010 (Decree 625) 'Some issues on reduction of greenhouse gas emissions'.

### 5.5.2 The domestic project cycle

The domestic project cycle (Fig. 52) comprises eleven steps. Based on the above-mentioned documents, the national procedure of submission, review, control, and monitoring of voluntary greenhouse gas emission reduction projects starts with identification of a project initiator (step 1), an organisation that initiates and/or implements a VER project. The project initiator elaborates a project idea note (PIN) or VER Project Proposal (step 2) and submits it (step 3) to the:
- Republican government bodies and other state-run agencies subordinate to the Government of the Republic of Belarus (if the project initiator is a governmental organisation or non-governmental organisation (NGO) with shares in the authorized capital owned by the Republic of Belarus and managed by these governmental bodies);
- The oblast executive committees and the Minsk municipal executive committee (if the project initiator is a municipal unitary enterprise subordinate to an executive committee or a non-governmental legal entity with shares/stakes in the authorized capital owned by the municipal authorities or NGOs, registered on their territory).

The relevant governmental body or executive committee reviews the PIN and transfers it to the Ministry of Natural Resources and Environmental Protection (step 4) in the format approved by the Ministry. The Ministry of Natural Resources and

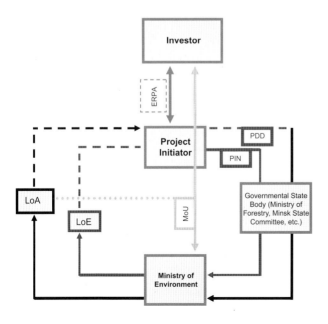

Fig. 52: The domestic project cycle in Belarus. ERPA = emission reductions purchase agreement; PDD = project design document; PIN = project idea note; LoA = letter of approval; LoE = letter of endorsement; MoU = memorandum of understanding.

Environmental Protection (henceforward referred to as 'Ministry') shall review a VER project proposal within 30 days after its submission.

After reviewing a proposal, the Ministry issues a Letter of Endorsement (step 5) of a VER project proposal in accordance with the format approved by the Ministry or a justified denial of issuing such a letter.

Receipt of the letter of endorsement of a VER project proposal is the ground for elaboration of a full package of documents including the Project Design Document (PDD; step 6) of a VER project and other documents required by the legislation (the format and content of these documents are determined by the Ministry). The elaborated VER PDD must be forwarded to the Ministry for approval.

The Ministry must review a VER PDD within 45 days after its submission and issue a Letter of Approval (step 7) of a voluntary emission reduction project in the approved format to the relevant governmental body or project initiator or a written notification about rejection of projects with respective justification.

The Letter of Approval is the starting point for the international project cycle, although the latter may also run parallel with the domestic project cycle (some international carbon standards, however, ask for the Letter of Approval on the implementation of the project signed by the government to register the VER project). The Letter of Approval is the foundation for concluding a Memorandum of Understanding (step 8) between the Ministry and a foreign investor concerning the terms of execution of VER projects and the transfer of voluntary emission reduction units. After the Memorandum of Understanding is signed, the PDD together with the Letter of Approval must be submitted to the independent entity (e.g. under Verified Carbon Standard: Validation/Verification Body (VVB)) for international assessment (determination) and confirmation of all information contained in the documents (step 9).

The project initiator must monitor the voluntary emission reduction project and simultaneously provide the respective state governmental body and Ministry of Environmental Protection with a project monitoring report according to the project monitoring plan. The report is elaborated within the format approved by the Ministry of Environmental Protection and is based on the monitoring data to confirm the correspondence to baseline and project emissions estimation within the VER project. Within a month after the submission of the monitoring report the Ministry shall perform its review, evaluate its effectiveness, and issue an Approval of Monitoring Report (step 10) for the relevant state body or project initiator.

Once the monitoring report has been approved, the project initiator submits it to the independent entity. Within 10 days after the receipt of independent entity's approval the governmental body or project initiator must inform the Ministry about this. Based on the approval issued by the independent entity and taking into account the terms of the Memorandum of Understanding and financing of the VER project, a confirmed (verified) amount of emission reduction resulting from the implementation of these projects is then transferred (step 11) to the foreign investor.

# 6 Land use options for rewetted peatlands

Wendelin Wichtmann &
Franziska Tanneberger

## 6.1 Overview on land use options after rewetting

### 6.1.1 Introduction

Peatland restoration measures are stimulated by environmental and economical as well as conservational needs. Rewetting aims to stop peat mineralization and thus the release of $CO_2$ into the atmosphere (chapter 3.2), as well as ground- and surface water pollution. Future land use of rewetted peatlands should contribute to the reduction of greenhouse gas (GHG) emissions, being in line with sustainability demands (Wichtmann & Wichmann 2011a). It should also recognise the goals of international instruments for biodiversity conservation. This is achievable by very different management schemes available for rewetted peatlands. All have in common that the mean water level is near the soil surface (i.e. paludicultures, see below).

There is a great potential for biomass use from wet and rewetted peatlands which covers the same range of opportunities like biomass production on mineral soils – food, fodder, fuel, and fibre (Table 34). Besides traditional agricultural uses for food, fodder, bedding and fertilisation (chapter 6.2), biomass can be used as a raw material for industry (chapter 6.3). It can substitute fossil fuels for energy generation (chapter 6.4) and be a feedstock for the synthesis of pharmaceuticals and cosmetics (see chapter 6.2). Last but not least it can be used as growing media in horticulture (see box 17).

Although still largely unrecognised, rewetted peatlands are generally well suited for biomass use in the energy industry and manufacturing sector since they do not (or rarely) harbour rare

Table 34: Examples of biomass utilization from wet peatlands in temperate Europe (changed after Wichtmann et al. 2000). Q = demand for quality: ++ = high. + = medium, 0 = low).

| Utilization | | Vegetation | Harvest | Q* |
|---|---|---|---|---|
| Agricultural | Ex situ fodder (hay, silage) | Wet meadows, reeds | Early summer | ++ |
| | In situ fodder (grazing) | Wet meadows, reeds | Year-round | ++ |
| | Litter | *Carex* meadows, reeds | Summer/autumn | 0 |
| | Compost | Wet meadows, reeds | Late summer | 0 |
| Industrial | Roofing material | Reeds | Winter | ++ |
| | Form-bodies | Wet meadows, reeds | Autumn/winter | + |
| | Construction/insulation | *Phragmites* reeds | Winter | ++/0 |
| | Paper (cellulose) | *Phalaris-Phragmites* reeds | Winter | + |
| | Basket-ware | Willow shrubs | Autumn | ++ |
| | Timber/furniture/veneer | Alder swamps | Frost | ++ |
| Energetic | Direct combustion and gasification | Alder/willow swamps, reeds | Autumn/winter | 0 |
| | Fermentation | Wet meadows, reeds | Early summer | + |
| | Liquid 'sun fuels' | Wet meadows, reeds | Year-round | 0 |
| Other | Officinal | Natural mires/plantations | Early summer | ++ |
| | Food | Natural mires/plantations | Summer/autumn | ++ |
| | Growing media | Peat moss stands | Year-round | ++ |

plant and animal species and are often overloaded with nutrients. By taking biomass off the site, depending on the harvesting regime, significant amounts of nutrients can be removed. As a result, the rewetted sites can become more valuable habitats for characteristic mire species that can thrive only under rather nutrient-poor conditions (see chapter 4.2). On highly degraded nutrient-rich sites, planting of reeds or trees prior to rewetting can speed up the establishment of desired stands.

Land use on rewetted peatlands has a range of advantages. It helps to combat climate change (e.g. by substitution of fossil fuels, chapter 6.4), supports biodiversity (e.g. by maintenance of threatened flora and fauna habitats and by preventing the sites from overgrowth with shrubs, box 8 in chapter 4.1) and provides social benefits (see chapter 6.5 for more detailed information).

### 6.1.2 What is paludiculture?

Paludiculture (lat. 'palus' = swamp), the cultivation of biomass on wet and rewetted peatlands, is an innovative alternative to conventional drainage-based peatland agri- and silviculture (Wichtmann & Joosten 2007, Wichtmann et al. 2010a). The precondition for paludiculture is that the peatlands are sufficiently wet, that peat is conserved and peat accumulation is maintained or re-installed.

---

**Box 16**

**The touristic potential of peatlands**

Nina Tanovitskaya

In Belarus, at several depleted peatlands, hunting areas and eco-tourism facilities were established. The establishment of protected areas on depleted peatlands is based in the Act of the Republic of Belarus 'On specially protected natural territories' of 20 October 1994 and in the Decree of the Ministry of Natural Resources and Environmental Protection of the Republic of Belarus 'On approval of rules for announcement, preparation, conversion and cancellation of protected areas' of 10 May 2001. These sites are often surrounded by slightly disturbed parts of the peatland. Since Belarus is famous for its vast peatlands in near-natural conditions, there is a considerable demand for individual, or small-group 'peatland' tourism. This demand gives Belarus a chance to present its peatlands as a destination for this kind of tourism and to value the educational and recreational potential of peatlands. Important conditions for the further development of this kind of tourism to meet the specific needs of individual and small-group tourism consumers are:
- Undisputed preservation of natural peatland complexes in a natural or near-natural state: The attractiveness of peatlands in Belarus for domestic and foreign eco-tourists depends primarily on a high degree of naturalness and characteristics derived from it (difficult access, isolation from economically developed territories, their specific biodiversity);
- Availability of basic services and equipment like ecological trails and marked routes, observation towers and hides, equipped parking facilities, information support and service infrastructure (accommodation, catering, special transport, hire of equipment);
- Designation of these destinations as protected areas of highest national and international status, e.g. as Ramsar Sites, United Nations Educational, Scientific and Cultural Organisation (UNESCO) Biosphere Reserves and World Heritage Sites, Important Bird and Plant Areas.

Some steps in this direction are already taken in Belarus. Fourteen touristic routes have been developed in the reserves 'Srednya Prypiac' and 'Sporauski zakaznik'. For example, the one-day trip to the reserve 'Srednya Prypiac' from Pinsk to the waterworks 'Kachanovichi' can be done on the waterway, by bike, or on foot. The touristic route in the Sporauski zakaznik is 10 km long and runs along the river Jasielda from the village Staromlyny to Lake Sporauski. However, a touristic development that is based on the natural and cultural values of peatlands in Belarus, requires more systematic and strategic certainty and planning. The development of individual and small-group tourism is hampered by a number of alternative peatland uses that involve large-scale development and transformation of natural peatlands in Belarus.

## 6.1 Overview on land use options after rewetting

Paludiculture uses that part of net primary production (NPP) that is not necessary for peat formation (which may amount to 80–90% of NPP). In temperate, subtropical and tropical zones, where plant productivity is high, peat is generally formed by roots and rhizomes. Here mires by nature hold vegetation of which aboveground parts can be harvested without harming peat formation. The quintessence of paludiculture is to cultivate plants that:
- Thrive under wet conditions;
- Produce biomass in sufficient quantity and quality; and
- Contribute to peat formation.

Harvesting biomass from rewetted peatlands allows sustaining their typical ecological functions and combines the following positive effects:

- Establishment of alternative, site-adapted land use;
- Adoption of new land use concepts with minimal damage to the environment;
- Mitigation of $CO_2$ emissions;
- Fostering of peat-forming plant species and restoration of the sink function of the peatland, e.g. for carbon and nutrients;
- Restoration of habitats for typical mire species;
- Production of raw materials for energy and industrial use;
- Revitalisation of traditional land uses combined with new ways of processing.

Among the various possible paludicultures, winter reed cutting for thatch has probably the longest tradition (Rodewald-Rodescu 1974, Häkkinen 2007). Also the mowing of wet meadows for litter

---

**Box 17**

**Sphagnum as a renewable resource**

Greta Gaudig & Sabine Wichmann

The most important raw material in professional horticulture is slightly decomposed peat called 'white' or 'blond' peat, which has over time developed from peat mosses (*Sphagnum*) in living bogs. About 30 million m$^3$ of 'white' peat are globally used for this purpose annually. The use of peat presents two main problems:
- Peat extraction destroys the important functions of bogs for e.g. biodiversity, climate regulation (carbon storage), and water balance as well as archives of the landscape;
- Peat is a finite resource. In most countries of western and central Europe the deposits of fossil 'white' peat are nearly exhausted.

Therefore, a non-polluting alternative ensuring a lasting and sustainable supply of high-quality raw material for professional horticulture has to be developed. This alternative is the cultivation of peat mosses (*'Sphagnum farming'*). Fresh peat moss biomass has similar physical and chemical properties as 'white' peat and enables plant cultivation without a loss of quality.

Since 2004 the University of Greifswald has been studying optimal conditions for *Sphagnum* growth in greenhouse and field experiments within four projects that were conducted in cooperation with several research as well as industrial partners. They were funded by the German Agency of Renewable Resources (FNR) and the German Federal Ministry of Economics and Technology (BMWi). A pilot field study on 1,200 m² demonstrated the feasibility of *Sphagnum* farming on a rewetted cut-over bog in northwestern Germany as long as continuously high water levels are maintained. For up-scaling of these promising results a new field experiment on 3 ha of former bog grassland was successfully installed in spring 2011. Additionally, in order to use inundated cut-over bogs for *Sphagnum* farming an artificial floating mat was developed. Plant cultivation experiments conducted by the Horticulture Research Station Hanover (Chamber of Agriculture in Lower Saxony) and Humboldt University of Berlin showed that Sphagnum biomass – even up to a proportion of 100% – is suitable as a raw material for growing media in horticulture. Investigations on *Sphagnum* farming for horticultural purpose are also going on in other countries such as Canada, Finland, and Chile. In addition to the production of a renewable alternative to fossil peat, *Sphagnum* farming enables a non-polluting and sustainable after-use of abandoned cut-over bogs (see www.uni-greifswald.de/~sphagnumfarming/).

is a traditional form of paludiculture performed over centuries before issues such as GHG emissions from peat soils became recognized. In the last years, paludicultures have been tested within several pilot projects. Common Reed *Phragmites australis* was cultivated as a renewable resource was tested in rewetted peatlands in northern Brandenburg (Wichtmann 1999a). Cattail *Typha latifolia* and *T. angustifolia* cultivation was subject of a research and development project in constructed wetlands on peat soils in the Danube floodplains in Bavaria (Wild et al. 2001). Alder *Alnus glutinosa* was planted for timber and furniture production on a pilot area in the rewetted Trebel valley in Mecklenburg-Western Pomerania (Schäfer & Joosten 2005). Recently, new paludiculture activities were implemented within several projects in Poland and Germany. They were motivated by nature protection but aim at sustainable use of the harvested biomass (Lachmann et al. 2010, Wichtmann & Wichmann 2011a). An overview on recent paludiculture projects is given in Wichtmann & Wichmann (2011b).

### 6.1.3 Land use options after rewetting in Belarus

In Belarus land users are obliged to recultivate depleted and other degraded peatlands, i.e. to transform them into a condition suitable for a follow-up use in line with the conditions agreed in the granting of the land (see chapter 5.1). This is obligatory for any surface and subsurface extraction of minerals and peat, any construction work (e.g. for industry, railway, other transport, energy, water management), as well as all geological activities resulting in a degraded soil surface. The permitted land use types of depleted peatlands include:
- Agriculture;
- Forestry;
- Nature conservation;
- Fishery;
- Water management;
- Recreation;
- Construction.

Among these land use options, only nature conservation use is suited for rewetted peatlands. These uses may be combined, e.g. use as a nature protection area is largely open to other soft uses (e.g. hunting, fishing, and eco-tourism, see chapter 6.2.2) or even requires them (e.g. mowing as habitat management). Recently some tourism has developed, providing opportunities for hunting or eco-tourism, facilitating all kinds of bird watching and wilderness experiences (see box 16; N. Tanovitskaya pers. comm.).

## 6.2 Biomass use for food and fodder

Wendelin Wichtmann

### 6.2.1 Introduction

Peatlands can be used for the production of comestible goods or animal fodder, which affects them in different ways. Some uses like hunting, gathering of mushrooms or berries have little impact on the peatland itself because they do not require any changes in hydrology. Principally, all other land uses on peatlands can be distinguished corresponding to the intensity of drainage (see also chapter 2.2). Drainage intensity again corresponds with the extent of peat mineralization that is initiated by aeration of the or-

---

**Box 18**

**The berry perspective**

Nina Tanovitskaya

A promising outlook for the use of depleted cutover peatlands after rewetting is berry cultivation: American Cranberry *Vaccinium macrocarpon*, Northern Cranberry *Vaccinium oxycoccos*, Bog Bilberry *Vaccinium uliginosum*, and Lingonberry *Vaccinium vitis-idaea*. For the cultivation of American Cranberry the central and southern areas of Belarus are most favourable. Recommended sites are recently exhausted cutover bog and transitional peatlands with a peat layer of at least 30–50 cm and a medium acidic reaction (pH 3.0–5.5). On depleted peatlands, after certain restoration activities it is also profitable to cultivate Bog Bilberries and Lingonberries. There is already positive experience in planting Northern Cranberries and Bog Bilberries in Belarus. An experimental station of the National Academy of Sciences is located in the Hantsavichy area for the purpose of cultivating Northern Cranberries and Bog Bilberries. There are also several successful production sites for cultivating Northern Cranberries in various regions of Belarus.

ganic soil connected with problems like leaching of nutrients into the groundwater and greenhouse gas (GHG) emissions into the atmosphere (Zak et al. 2008). Peat conserving methods for land use on organic soils require water levels that stop further mineralization. Even by low intensity management of grasslands this precondition is not realised. This implies that even low input land use schemes on peatlands give no guarantee for sustainability (Wichtmann & Wichmann 2011a) unless the water table remains sufficiently high. In Central Europe, approximately 14% of peatlands are currently used for agriculture (Joosten & Clarke 2002).

## 6.2.2 Soft uses

An old and widespread form of peatland utilization is the gathering of (parts of) plants. Especially in the boreal zone of Eurasia a wide variety of wild edible berries and mushrooms are collected and dried for provision of food and vitamins during the winter period (Joosten & Clarke 2002). Berries from *Vaccinium* species, *Empetrum*,

---

**Box 19**

**The value of cranberry collection from Marocna raised bog**

Sviataslau Valasiuk

A valuation study has been conducted with the research topic 'valuation of peatlands' ecosystem services' by APB-BirdLife Belarus aiming at a cost-benefit analysis of various management scenarios concerning Marocna raised bog (4,200 ha) located in Stolin district, Brest region. Assessed values of cranberry resources used by local communities have been derived from a preliminary experts' opinion survey. A questionnaire was handed out to 10% (in total 149) of the households in the villages Haradniany and Hlinka in Stolin district in December 2009. The questionnaire was designed to discover the socioeconomic characteristics of households, their annual cranberry collection and their willingness to pay (WTP) for the conservation of Marocna in its near-natural state. The annual value of the relevant ecosystem service was estimated as the difference between total income of the households that collect and sell cranberries and their monetary costs.

According to the survey, in 2009 the mean cranberry yield (Fig. 57 Colour plates III) amounted to 125.5 kg per household. Around 60% of the households reported more or less regular use of Marocna's cranberry resources. The yield is mainly sold except for 15% that is used for own consumption. The average annual income per 'collector' household in Haradniany and Hlinka made up around 392,000 BYR (c. 105 Euro, 2009 prices; see Table 35).

Annual income per ha raised bog made up 69,400 BYR, which is less than half of the preliminary assessed value in the experts' opinion survey.

The following conclusions can be derived from the study:
- Cranberry collection contributes substantially to the income of many local households, but less than previously estimated by expert opinion;
- Conservation and sustainable use of the raised bogs' ecosystem services is economically at least competitive with peat extraction;
- The majority of respondents declared their willingness to pay for conservation of Marocna.

Table 35: Income from collecting cranberries in two Belarusian villages.

| Village community | Total number of households | Number of 'collector' households | Annual income per household (thousand BYR) | Total annual income (thousand BYR) |
|---|---|---|---|---|
| Haradniany | 752 | 422 | 392 | 165 474 |
| Hlinka | 675 | 379 | 392 | 148 530 |
| TOTAL | 1427 | 801 | 392 | 314 004 |

> **Box 20**
>
> **Medicinal plants and medicine**
>
> Nina Tanovitskaya
>
> Exhausted peat deposits show potential for the cultivation of medicinal and melliferous herbs native to peatlands. In Belarus more than 50 species of peatland plants are used as raw materials for the pharmaceutical industry or are good melliferous herbs. These officinal plants are either gathered from natural stands or can be cultivated in rewetted peatlands. Some understanding of officinal plants has been gained with experiments on Bogbean *Menyanthes trifoliata*, Hemp Agrimony *Eupatorium cannabinum*, Meadowsweet *Filipendula ulmaria*, Sweet Calamus *Acorus calamus*, and European Bugleweed *Lycopus europaeus* in northeastern Germany. They can be successfully cultivated on land with a high water table (Kersten et al. 1999).
>
> In case depleted peatlands after industrial peat extraction still hold peat of adequate quality, this peat can be used for balneology, medicine, and cosmetic purposes. Large positive effects on human health can be reached by using very small volumes of peat. Peat can help to cure diseases of e.g. joints, the musculoskeletal system, viscera, eyes, and skin.
>
> Medical use is probably the most promising area of peat usage, which will continue to develop even if people stop using peat in many other ways, such as for fuel or fertilizer.

and *Ribes* (Joosten & Clarke 2002) currently still play an important role in some peatland regions in Belarus (see box 18 and box 19). In other regions of the world plants like Wild Rice *Zizania aquatica* (North America), *Menyanthes trifoliata*, *Acorus calamus,* and *Hierochloe odorata* (all Europe) and Sago Palm *Metroxylon sagu* (Malaysia) have been or are still used for human nutrition or medical use (Joosten & Clarke 2002, see also box 20). Also eco-tourism and hunting of mammals and birds are traditional uses of mires and peatlands (see box 16). All these soft forms of land use are connected with some unavoidable disturbance like noise (hunting), waste disposal and some infrastructural measures, like building of footbridges, pathways and places for resting. At least the impact of these soft land use forms on the site, especially on biodiversity, is dependent on the intensity of its implementation.

### 6.2.3 Arable farming on peatlands

The percentage of peatlands used for arable farming is difficult to assess. Traditionally in Central Europe, ploughing for cultivation of crops has been widely implemented to reclaim peatlands after drainage. At the beginning of the 20th century, in Germany, the first cultivated crop on newly drained fen peatland was Hemp *Cannabis sativa* (Scheel 1937), because of its ability to compete with weeds. Nowadays potatoes or even wheat can be found as arable crops on peat soils all over Europe, either on cutover bogs or on fens especially drained for agriculture (Fig. 53 Colour plates III). The latest development is the cultivation of Maize *Zea mays* as raw material for biogas production (Wichtmann et al. 2009). All these land use options require water tables of more than 35 cm below soil surface in spring when agricultural activities on site start. This often results in mean annual water levels of >80 cm below soil surface (Wichtmann et al. 2010a).

### 6.2.4 High intensity grassland

Intensive grassland utilization on peat soils requires similar interventions in hydrology as arable agriculture. This kind of high input land use on fen grasslands with three to four cuts per year has lost its importance in Germany because of severely decreased soil productivity (but is still standard in other regions). High productivity grass species sowed on these degraded peat soils, damaged by previous intensive land use, cannot form stable grassland communities but are out-competed by Twitch Grass *Elymus repens* after four to five years. The resulting stands show lower total performance, quality, and decreased digestibility for ruminants (Käding et al. 1990). This necessitates regular ploughing (or a complete killing of the sward by herbicide application) and reestablishment of the desired high productivity grasses (Succow & Joosten 2001).

Modern breeds of dairy cows with their ever increasing milk production require more and more high quality fodder that can only be produced on mineral soils. This has led to the agricultural abandonment of vast areas of peatland or to a switch to very low input management schemes, dependent on subsidies.

## 6.2.5 Low intensity grassland

In Central Europe low intensity grassland management was widespread. The use of wild plants as straw (litter) or fodder (hay) is traditionally common. Peatlands serve as wild pasture for domestic animals. Only slightly drained peatlands are used for the production of litter or low quality hay (Fig. 55 Colour plates III). However, this kind of low impact peatland use has been drasticly decreasing during recent decades resulting in a widespread abandonment of the sites (Fig. 54 Colour plates III).

In most peatland areas of Germany and Poland, low input use often followed a period of more intensive agriculture, that had been given up because of various problems concerning amelioration systems, decreasing carrying capacity of the soils, plant species composition changes and decreasing productivity of the sites (Succow & Joosten 2001). Correspondingly the peat soils of these sites are deeply affected by mineralization, shrinking, and soil compaction. Usually these sites are managed by cattle grazing (c. one livestock unit/ha) for meat production, cut once or twice a year for haymaking or some combination of both. The hay is used e.g. for suckler cows or young cattle with low performance. Fertilisation and phytosanitary measures are not common. In Germany these land use forms are only competitive because of subsidies. Nevertheless this low input management is dependent on deep drainage as cattle as well as the heavy machinery used are dependent on water table heights to a maximum of 40–60 cm below soil surface. This leads to the same problems that can be observed in intensive grassland management on peat soils (Wichtmann & Koppisch 1998). An exemption is grazing with Water Buffaloes *Bubalus bubalis* for meat and milk production (Fig. 56 Colour plates III). They accept water levels near surface and forage biomass with low energy contents. Wet grassland management with buffaloes is an upcoming approach performed in several nature conservation projects (Wiegleb & Krawczynski 2010).

## 6.2.6 Conclusions

Most land use forms on peatlands are not sustainable, because they require drainage (Wichtmann & Wichmann 2011a). New concepts for site-adapted land use of peatlands favour the utilization of the harvested biomass outside traditional agricultural application, e.g. as construction material (see chapter 6.3) or as a fuel (see chapter 6.4).

## 6.3 Biomass use for raw material

Franziska Tanneberger & Wendelin Wichtmann

### 6.3.1 Introduction

Aside from food and fuel, there are several other possibilities to use biomass from natural and rewetted peatlands that have commercial potential. Besides its use as fodder and fertiliser other traditional uses are in construction, insulation and as raw material for manufacturing. The use of biomass as a raw material for construction or handicraft should always be favoured its use as a fuel. In many cases the biomass can substitute fossil resources and the carbon can be sequestered during the life cycle of the product. The lifespan of such products could be anything from one year e.g. for materials used in floristics, up to fifty or more years e.g. for construction materials. The key parameter is the half-life of carbon in each end use product (e.g. use of wood in house construction c. 70–100 years, furniture 30 years, and paper one year). The half-life is the time after which half the carbon placed in use is no longer in use. The disposition of carbon after use includes recycling, disposal in landfill or dump, or emission to the atmosphere by burning, with or without energy produced (Skog & Nicholson 2000).

### 6.3.2 Use as raw material

The use of peatland biomass for thatch is well established in many European countries such as UK, Germany, Poland, and Romania (see also box 24 in chapter 6.5). In Germany the demand for high quality Common Reed *Phragmites australis* as roofing material (Fig. 58 Colour plates III) and for manufacturing of mats cannot be satisfied by inland reed harvest. Therefore, the material needs to be imported from southern and eastern Europe, Turkey and even China in order to cover the demand (Schäfer 1999). In the UK home-based production of reed meets about 25% of the demand, with the remainder also imported (McBride et al. 2010). Thatching is the most widespread use referring to mass of Common Reed in Europe (Lautkankare 2007).

Weaving of Common Reed is a traditional way of manufacturing it to make several kinds of products for construction and insulation. In Iraq it is even used for the construction of complete houses (box 21). Depending on the product en-

> **Box 21**
>
> **Traditional construction with Common Reed**
>
> Wendelin Wichtmann
>
> In the floodplains of Euphrat and Tigris in southeastern Iraq Common Reed *Phragmites australis* is traditionally used for the construction of private houses, mosques, and other buildings. This tradition, as well as that in the Mesopotamian marshes, was destroyed by the Iraqi Bath party regime before and during the Iraqi war in the 1990s. After the war some of the former inhabitants came back to the marshes and revived this tradition. By binding cylindric barrels and connecting them to each other complete houses are constructed. Such as a mudhif, a traditional Marsh Arab guesthouse, which is made entirely out of reed and which has been in construction since c. 5,000 years (Fig. 60 Colour plates III).

visaged the culms (stems) are put in single layers for weaving into plaster porter mats, insulation for garden plants (against frost) or privacy screens. For weaving thicker mats, thicker layers are put onto the loom. Such thick mats, enmeshed with stainless steel netting are used as insulation material. Woven mats can be fixed to walls and ceilings and by that serve as plaster porters. Such mats can also be used as screens fixed to fences or as insulation material protecting frost sensitive plants in the garden (e.g. roses). A use as construction boards, fire protection panels, plasters and insulation boards is currently in development e.g. in northeastern Germany in the research and development project 'Vorpommern Initiative Paludikultur' (VIP) at Greifswald University. This project also involves the development of clay plaster production with digested reed fibres. Ministry of Natural Resources and Environmental Protection of the Republic of Belarus Recently developed products like form bodies for packaging material or holders for floral arrangements seem to be economically interesting (Wichtmann 1999b).

There is long-term experience in the production of paper pulp from reed biomass in different parts of the world, like China, Russia, Sweden, and Romania. During socialist times large factories were constructed e.g. in the Volga and Danube deltas. They processed Common Reed harvested from vast reed beds of several thousand hectares for paper production (Rodewald-Rodescu 1974).

### 6.3.3 Harvesting of biomass from wet peatlands

The harvesting process should always be in line with the anticipated use of the biomass and correspondingly adjusted to it. Some pre-processing is possible during harvesting like chopping, combing, bundling, or baling. But processing is not necessarily needed for every use. High quality biomass (for thatching or for weaving mats) and unspecific biomass (for plasters and form-bodies) can be distinguished.

In general, biomass harvesting and processing in wet peatland management faces two main challenges: the machinery should be adapted to wet sites with low carrying capacity and should have a high acreage performance. The carrying capacity of peatlands is determined by the humidity of the soil (thus the ground water level) and vegetation cover (plant species, density of the sod). After Prochnow & Kraschinski (2001) the trafficability of peatlands is not a technical but rather an economic problem. Very light and specially equipped machinery adapted to harvesting wet sites is available, but usually associated with poor performance, increased time consumption and high costs for purchasing as well as for execution of harvesting. The carrying capacity of the plant sods will decrease by repeated passing over. Additionally, for the use of light and site-adapted machinery, it may be necessary to fortify tracks or access points that are frequently used to avoid structural damages in peat soils.

There are several mature technical solutions for harvesting and processing of unspecific biomass on large wet peatlands. The main basic machine types are (Wichtmann & Tanneberger 2009):

- Caterpillars: crawler-mounted, tracked vehicles, available from several supplier companies, have large bearing surfaces so their weight is distributed very well over the ground, but they can cause damage to the vegetation by tight manoeuvres;
- Seigas: light vehicles with two or three axels and low pressure balloon tyres. Originally built by the Danish 'Seiga' company, they are no longer produced by them, but now built by several small companies. They are able to float but

are hard to manoeuvre which in some cases can cause damage by wheel-spin;
- Machinery usually used in dryer grassland (wet meadow) management or on frozen soils.

These machines have in common cutting devices which are mounted on the front of the machine and any transport facilities constructed on the rear or on a trailer. The cutting device can be arranged in different ways, according to the anticipated utilization of the biomass: As a sickle bar mower or a horizontal or vertical rotating mower, possibly combined with a shredder (Fig. 59, Fig. 62 Colour plates III). Harvested biomass with culm (stem) lengths of 1–15 cm can be taken for most uses in the industry and handicraft sector (see below) as well as for energy purposes (see chapter 6.4).

### 6.3.4 Quality standards

Regarding reed, there are specific quality standards for thatching and weaving, which must be considered before harvesting. The reed culms must be harvested unfolded and in full length. This necessitates using sickle bar mowers combined with binding facilities which makes transport in bundles possible. After harvesting the material should retain its properties for decades. For this reason the biomass should not be too nutrient rich as it should not decompose for a long time. If rewetted sites are eu- or hypertrophic due to strong degradation of the peat soils before rewetting this will be reflected in productivity and nutrient contents of the standing biomass. This implies that not every reed stand is suitable for harvesting so selection would be necessary.

The quality standards regarding raw material for manufacturing with weaving looms are the same as for thatching: Unfolded long culms of reed are transported in bundles to the facility and stored in piled bundles until further processing. The bundles are opened and the culms are then threaded onto the loom for weaving.

For other uses of biomass from wet peatlands, as raw material, plant species composition and their ingredients play a subordinate role. In this context species composition may only be of interest as different species promise different yields. Because of its high yield potential Common Reed *Phragmites australis* is the most attractive species to be harvested. For most purposes the demands on quality are low (see Table 34 in this chapter), but the biomass should be dry and not decomposed in order to be stored before processing. The biomass can be compacted for transport to manufacturing sites e.g. as big bales. After unpacking these big bales the loose biomass is chipped or milled for most of the utilization schemes.

## 6.4 Biomass use for energy

Wendelin Wichtmann & Jenny Schulz

### 6.4.1 Introduction

Several sustainable land use alternatives have been tested recently, combining positive environmental effects of peatland restoration with attractive biomass yields: After rewetting, nutrient-rich fen grasslands can be converted into highly productive stands of Common Reed *Phragmites australis* (Timmermann 1999), Reed Canary Grass *Phalaris arundinacea* (Wichtmann & Succow 2001), and Cattail *Typha latifolia*, *T. angustifolia* (Wild et al. 2001) or planted with Black Alder *Alnus glutinosa* (Schäfer & Joosten 2005). Less common is the use of sedge *Carex* spp. reeds and Grey Willow *Salix cinerea* shrub land (Timmermann et al. 2006) (see chapter 3.5). The development depends on hydrology, trophic status, former land use, and diaspore potential.

These plant communities develop spontaneously after rewetting, usually with heterogeneous species composition, limiting their use for manufacturing. Their low nutritional content restricts their value as animal fodder. However, there is the option to use the biomass as fuel. This requires high yields in order to be economical since the unprocessed biomass has little added value. Average yields of about 10 t dry biomass per hectare per year can be achieved (Table 36).

### 6.4.2 Harvesting

Harvesting of biomass from wet peatlands for energy production has to follow the same rules as other uses of unspecific biomass (see chapter 6.2). The material can already be processed during harvesting and compressed for transport or storage. The machine purchased within the BMU-ICI project in Belarus has been developed using experience gained from vegetation management in the Biebrza peatlands in Poland (Fig. 62 Colour plates III, box 22). In this instance the biomass is chopped and blown to the transport bunker on the trailer. The harvester transports the biomass

Table 36: Productivity of different wetland species (from Timmermann 2003).

| Species | Productivity t ha⁻¹ yr⁻¹ |
|---|---|
| Common Reed (*Phragmites australis*) | 3.6–43.5 |
| Cattail (*Typha latifolia*) | 4.8–22.1 |
| Reed Canary Grass (*Phalaris arundinacea*) | 3.5–22.5 |
| Reed Sweetgrass (*Glyceria maxima*) | 4.0–14.9 |
| Great Pond Sedge (*Carex riparia*) | 3.3–12.0 |
| **for comparison:** Fallow wet grassland  High-intensity grassland | 6.4–7.4  8.8–10.4 |

to the edge of the peatland as soon as the bunker is full. To prevent sward damage, transport during harvest should cover the shortest distance possible. This can be achieved by using additional vehicles for transporting, driving alongside the mowing machine.

### 6.4.3 Processing of biomass for energy production

Processing the harvested biomass is usually a precondition for energy use. End products include loose litter, silage, big bales, briquettes, pellets, and charcoal. They can be used on their own or in mixtures with other types of biomass like wood or energy crops. The fresh biomass has to be removed from the site and can either be processed during harvesting, or afterwards. In case of winter harvest the humidity of the standing biomass is low enough to allow storage or combustion. Loose and dry biomass is usually stored in barns. Bales can be stored in open stocks, covered by tarpaulin or in a barn. Required moisture content of the biomass varies according to the end product. The 'Herlt technique', where big bales are gasified, requires less than 20–25 volume%, the production of briquettes and pellets less than 15%. Briquettes have to be stored at a moisture content of <18%. In Belarus briquetting and pelleting are important as the peatlands are often located far from retail markets, resulting in long transport distances and potentially high costs. Briquettes also have the advantage that they can be used like any other solid fuel (peat briquettes, wood) in heating systems that most people use for domestic space heating. Pellets are suitable for furnaces with an automatic feeder system.

In most cases the biomass has to pass a shredding or milling procedure before compression. Pelleting Common Reed and Reed Canary Grass in trials using a stationary pelleting device did not show any disadvantages compared to other types of biomass (e.g. wood). The finished products are shown in Fig. 64 Colour plates III. Also Eder et al. (2004) recommended pelleting as a suitable preprocessing technique for the thermal utilization of reed. Pelleting machines process 100–220 kg/h. The performance can be increased by running several devices in parallel. The use of mobile pelleting devices allows moving pellet production equipment from one biomass harvesting site to another, if these sites do not yield sufficient biomass for a full workload of stationary machines.

### 6.4.4 Sunfuels and combined techniques

Biomass can serve as a feedstock for the production of charcoal, which in turn can be converted into liquid fuels ('biomass to liquid', BTL), or used directly. This has the advantage of concentrating the material and reducing transport costs. The most important quality parameters for the biomass are a high content of organic matter and humidity lower than 15% (Choren AG pers. comm.). For the production of liquid fuels only pilot plants exist so far. To date there is no mature technology available. These techniques are promising for the future utilization of biomass from rewetted peatlands because demands on quality are low.

### 6.4.5 Biogas production

Biomass from rewetted peatlands can be used for the production of methane, fed directly into the gas grid. There is great potential in the use of biomass for space heating, especially via Combined Heat and Power (CHP) systems. Biogas is produced by anaerobic digestion of biodegradable materials such as biomass, manure or sewage, municipal waste, green waste, and energy crops. Biogas comprises primarily of methane ($CH_4$, 50–75%) and carbon dioxide ($CO_2$, 25–50%) and has an average caloric value of 6 kWh/m³ (4–7.5 kWh/m³ depending on methane content; Fachverband Biogas pers. comm.). Two different biogas production types can be distinguished:
- The mesophilous process that needs temperatures of c. 35–38°C;
- The thermophilous process that needs a temperatures >55°C.

Fig. 53: Wheat cultivation on peatlands in southwestern Belarus near Zdzitava (2009; photo: Wendelin Wichtmann).

Fig. 54: Abandoned near-natural grasslands in Jasielda floodplain, formerly used for litter production (2009; photo: Wendelin Wichtmann).

Fig. 55: Traditional haymaking in the Jasielda floodplain is limited to sites close to villages. The stand of Reed Canary Grass *Phalaris arundinacea* in the foreground is already abandoned (2009; photo: Wendelin Wichtmann).

Fig. 56: Water Buffaloes *Bubalus bubalis* in the winter corral, waiting for the grazing season in reed beds in the National Park 'Vorpommersche Boddenlandschaft' on Darss peninsula (March 2011; photo: Wendelin Wichtmann).

Fig. 57: Harvesting cranberries (photo: Sergei Zuenok).

Fig. 58: Reed used for thatching (photo: Franziska Tanneberger).

Fig. 59: Bar mower cutting device mounted on a Kässbohrer snowcat, constructed for use in the Peene Valley peatlands (photo: Achim Schäfer).

Fig. 62: Ratrak mowing machine with a trailer, each based on a snowcat, developed for mowing in the Biebrza peatlands, Poland (photo: Lars Lachmann).

Fig. 60: Inside a re-constructed guesthouse of the Marsh Arabs in Iraq (Iraqi mudhif interior) (photo: U.S. Army Corps of Engineers 2004, Public domain http://commons.wikimedia.org/).

Fig. 63: Big round bale from Common Reed *Phragmites australis* produced by common hay making techniques under frozen soil conditions (photo: Wendelin Wichtmann).

Fig. 64: Pellets from Reed Canary Grass *Phalaris arundinacea* (right) and Common Reed *Phragmites australis* (left; photo: Wendelin Wichtmann).

Fig. 67: Prof. Jürgen Augustin explaining equipment for greenhouse gas measurements in the rewetted former polder Zarnekow in Peene Valley, Northeast-Germany (photo: Franziska Tanneberger).

Fig. 68: The school team during installation of minimum-maximum water level measuring devices in Sporauski zakaznik together with APB/CIM project staff Nadzeya Liashchynskaya and Annett Thiele (photo: Ruslan Shaikin).

Fig. 70: Small-scale agriculture on former kolkhoz land in Northwest-Ukraine – a potato field on a peatland with abandoned kolkhoz buildings in the background (photo: Annett Thiele).

Fig. 71: Drained, abandoned peatland in Northwest-Ukraine. The typical peatland flora is absent while weeds and ruderal communities cover the degraded peat soil and bushes are encroaching the area. Since the demand for these abandoned sites is low, they have a good potential for rewetting (photo: Susanne Bärisch).

## Colour plates, Part III 123

Fig. 73: The central lake of Dalbeniski (2010; photo: Annett Thiele).

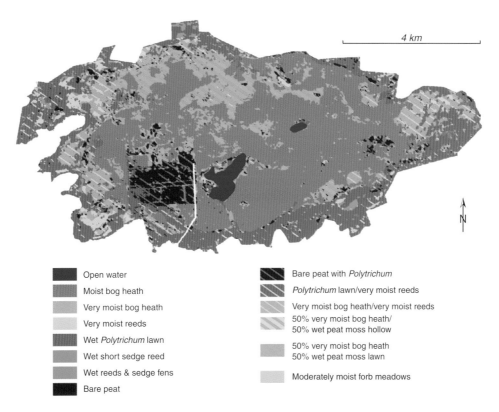

Fig. 74: GEST (greenhouse gas emission site type) map of project territory Dalbeniski, derived from maximum likelihood classifier (supervised classification, based on 129 training sites) (map: Mikhail Kudriakov & Annett Thiele).

Fig. 75: The burnt area with dwarf shrubs before rewetting of Dalbeniski (2010; photo: Annett Thiele).

Fig. 76: Very moist bog heath in the central area of Dalbeniski before rewetting (2010; photo: Annett Thiele).

## 6.4 Biomass use for energy

The core of a biogas plant is the fermenter. Residues from gas production can be used as fertilizer on farmland. Biogas is usually burnt in motors producing electricity in a generator and residual heat of 80–90°C. Profitability of a biogas plant often depends on to what extent surplus heat can be used (sold). More efficient techniques to produce electricity will be available in the future (FNR 2006). The most common and readily available method for biogas production is fermenting a fluid substrate like liquid manure with a co-ferment like silage from Maize or grass (wet fermentation). Recent developments include the use of silage only (dry fermentation) (B. Linke pers. comm. 2009).

20–40% of electricity produced is consumed by the plant itself and 30–50% of the thermal energy is used to heat the fermenter. The larger a biogas plant is the lower these percentages will be. Biogas production needs technically skilled and specialized operators and plants have to be designed for individual sites. There are Biogas facilities in Belarus e.g. in Brest and Zaslavl (Minsk region; Palkin 2008).

The biogas potential of a substrate depends on the content of total dry matter (DM) and on the proportion in the total dry matter of organic mass (C, H and O), nutrients (N, P, K) and organic contaminants (FNR 2006). Silage from highly productive grass provides biogas yields of 450–700 m$^3$/t DM, similar to those of maize. Grass from low intensity meadows has lower yields. For dry fermentation the required quality of the biomass used as co-ferment or sole feedstock equals that for high performance dairy cows. Therefore, biomass from rewetted peatlands is less suitable as a co-substrate for biogas plants. Only material from early summer cuts on peatlands for habitat restoration is of sufficient quality to be used in biogas plants.

However, research on how to use biomass of lower quality is in progress. Tests are conducted on straw and grasses with a crude fibre content of about 30% and a DM content of >70% using liquid manure as basic fluid. First results are promising and show satisfactory gas yields. It is estimated that in the near future the technology will be sufficiently mature to be available for the market. Further research is currently underway into fermentation of cellulose-rich material in combined wet/dry fermentation (Vogel & Ahlhaus 2009) and dry fermentation processes. Here the results are promising too (Loock Biogassysteme GmbH Hamburg 2009).

### 6.4.6 Combustion and gasification

For loose biomass combustion Danish REKA heating facilities can be used. These were originally developed for straw. Big bales can be burned directly in Herlt gasifiers (round bales) and cigar burners (rectangular bales) (Fig. 61). Most of the added value has biomass that is compressed into pellets or briquettes. Therefore, both the BMU-ICI project (box 22) and the EU project 'Wetland Energy' (box 23) aim at demonstrating the production of briquettes from biomass from wet peatlands. For combustion briquettes are better suited than smaller fractions (Kastberg & Burvall 1998). Homogenous coarse powder from Reed Canary Grass *Phalaris arundinacea* resulted in higher ash and silica contents (causing damage in the combustion chamber) than briquettes produced from the same species.

The major advantage of gasification is that the quality required is lower than that for combustion (humidity 20–25% for Herlt gasification). Lower water contents are suboptimal for the gasification process because the temperatures of about 500°C developing at the margins of the bales cannot be conducted into the core of the bale. Reed from winter harvest is very suitable for gasification because of its high cellulose content. The main quality requirements regarding biomass for combustion are (Wichmann & Wichtmann 2009):
- High heating value;
- Low water content;
- Low content of problematic substances (Cl, S, N, Si);
- Low ash content;

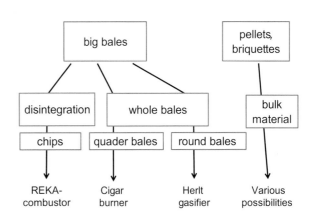

Fig. 61: Possibilities for heating with biomass from rewetted peatlands.

> **Box 22**
>
> **Fuel briquettes production in Sporauski zakaznik**
>
> Uladzimir Malashevich & Olga Chabrouskaya
>
> As for most Aquatic Warbler breeding habitats all over Europe, the problem of overgrowing is very urgent also for Sporauski peatlands (Belarus). This protected area held about 500–630 singing males in 2010 or 5% of the global population. The need of vegetation management was prioritised as essential in a management plan to conserve its unique biodiversity. Pilot works on active vegetation management started at Sporava in 2006. These pilot activities showed the technical possibility of mowing the mire even with usual agricultural equipment. However, such mowing with regular tractors equipped with normal wheels appeared to be very weather dependent and harmful for the delicate peat soils and was easiest to be performed during frost seasons. Monitoring of key bird species proved the effectiveness of such conservation measures.
>
> In order to find a solution for economical sustainability of mowing, a feasibility study on using biomass from wet peatlands in Belarus was carried out in 2009. Based on this study a business plan for a sustainable enterprise to achieve additional project measures in restored peatland was elaborated in 2010. According to this plan, the most cost-effective way of using biomass obtained from mowing of Aquatic Warbler habitats is the production of fuel briquettes and a biomass project could be realised at Sporava. To perform weather independent mowing on wet peatlands special machinery is required. Such caterpillar mowing devices were specially adapted to high water table conditions within the EU LIFE Project 'Conserving Aquatic Warbler in Poland and Germany', organised by the Polish Society for the Protection of Birds (OTOP). Such a ratrak machine was tested before in the Biebrza marshes and delivered to Belarus in autumn 2011.
>
> At Sporava, starting in 2012, mowing will be performed on approximately 500 hectares annually. During further processing, the biomass will be dried if necessary, milled and compacted into briquettes. Buildings for storing the manufactured biomass have been rented by APB-BirdLife Belarus. For the briquetting equipment buildings have been prepared and biomass processing equipment has been installed.
>
> After the pilot phase, the biomass briquettes will be sold as a substitute mainly for peat briquettes (widespread in use in rural areas of Belarus). The income from sales of briquettes will cover the costs of vegetation management making the whole cycle self-supporting. The costs of mowing and fuel briquettes production will be assessed during winter 2011/12. The fuel briquettes from plant biomass can be used as an alternative carbon-neutral fuel for heating. Biomass harvesting will simultaneously guarantee the maintenance of globally threatened fen mire biodiversity and will provide new opportunities for income generation in rural areas.

- High ash melting temperature;
- Appropriate conditioning;
- Low emissions.

### 6.4.7 Heating value

The heating value (caloric value, CV) of dry gramineous biomass is with 16.4 MJ $kg^{-1}$ (*Festuca arundinacea*) to 17.8 MJ $kg^{-1}$ (*Miscanthus* spec.) only a little lower than that of wood (18.2–18.8 MJ $kg^{-1}$) (Hartmann 2001). Reeds have heating values between 17.5 MJ $kg^{-1}$ (Eder et al. 2004) and 17.7 MJ $kg^{-1}$ (Barz et al. 2006). The heating value of Reed Canary Grass is about 16.9 MJ $kg^{-1}$ (Kastberg & Burvall 1998). A comparison of reed fuel from peatlands in Peene Valley, Germany with literature data for fossil fuels and other biomass fuels (Barz et al. 2006) showed that among the biomass fuels with lower heating value than fossil fuels, reed has one of the highest heating values. The content of problematic substances of reed was in the range of all other biomass fuels and only the ash content was significantly higher than that of wood.

## 6.4 Biomass use for energy

---

**Box 23**

**The Europe Aid project 'Wetland Energy'**

Wendelin Wichtmann

In January 2011 the project 'Implementation of new concepts for wet peatland management for the sustainable production of biomass-based energy (wetland-energy)' has started. This project aims at the sustainable use of re-wetted peatlands and the production of fuel briquettes from biomass from rewetted peatlands and from the maintenance of wet meadows in Belarus. Project partners in Belarus are the International Sacharov Environmental University and the Institute for Natural Resources (former Peat Institute) of the National Acadcmy of Sciences of Belarus.

With the experiences from the BMU-ICI project in Belarus in mind, the Michael Succow Foundation applied for a two-step tender from the Europe Aid thematic programme of the European Commission (EC) 'Environment and sustainable management of natural resources, including energy'. The first draft was successful and the Michael Succow Foundation was asked to submit a detailed project proposal.

By written confirmation of 30 July 2010 the project was approved proclaimed. After official registration of the project at the Belarusian authorities during the second half of 2011 first activities in Belarus could start with in-country activities.

Goal of the project is to build up a demonstration project at a pilot site in Belarus for testing new methods in wet peatland management and harvesting peatland biomass (paludiculture). In cooperation with regional stakeholders (zakaznik, peat factory, kolkhoz, local energy suppliers) the current ecologically and environmentally questionable management and utilization of peatlands and peat will be changed into a sustainable land use system. Fossil fuels will be substituted by biofuels. Biomass from rewetted peatlands, degraded by peat excavation, will be used to produce bio-briquettes where peat has been the raw material before. New options for income generation in rural areas will be opened up. Another aim of the project is to reinstall adapted land use on abandoned wet meadows for biofuel production to prepare them as habitats for e.g. Aquatic Warblers and hunting grounds for Greater Spotted Eagles.

---

### 6.4.8 Water content

The water content strongly determines the heating value (Rode et al. 2005) and should be generally ≤15%. With increasing water content the heating value is reduced (e.g. for reed by 24% at 20% humidity, by 60% at 50% and by 100% at 83%, Darroch-Thompson & Ash 2003) and the suitability for storage decreases as biological decomposition processes start at 16% humidity. This may result in substantial losses and formation of fungal spores (Rode et al. 2005). Several quality parameters (carbon monoxide, volatile organic compounds, dust content) for exhaust gases also increase with humidity (Weiss 2001). A low water content and thus higher heating value can be secured by harvesting in late winter. In the case where management for nature conservation demands harvesting in summer, the water content should be decreased by making hay or silage or by using biomass drying equipment. Under unfavourable conditions (wet soils, high rainfall frequencies) the drying process may require considerable efforts (Bellebaum 2004).

### 6.4.9 Problematic substances

The content of problematic substances such as Cl, S, N which cause fast deterioration (e.g. corrosion) of the mechanical components of the combustion plant is higher in gramineous biomass than in wood (Table 37). Biomass from Belarusian peatlands is assumed to have a slightly lower content of critical components such as chlorides than *Miscanthus* from fertilized mineral soils, which usually result in a higher mineral content (Table 37, Hartmann & Hering 2004), and also than the reed biomass from the Neusiedler See in Austria (Wichtmann & Tanneberger 2009).

### 6.4.10 Ash content

The ash content is considerably higher in gramineous biomass than in wood (Tab. 37). Comparing ash compositions of *Miscanthus* and Common Reed we find that values are more or less similar. Ash content of Common Reed from the Neusiedler See is about 5% of dry matter and exceeds that of wood pellets 20 times (Eder et al. 2004). However, Paist et al. (2007) analysed reeds mown in winter at Estonian lake shores and arrived at significantly lower values for total ash contents (Tab. 37). Burvall & Hedman (1998) showed for Reed Canary Grass that delayed harvesting (March) decreased contents of alkali, chloride, nitrogen, and sulphur considerably.

### 6.4.11 Ash melting

Ash melting temperatures of gramineous biomass are often lower than those of wood. This makes adapted techniques for combustion necessary in order to avoid slagging. Examinations with Common Reed from Hungary in co-operation with the University for Applied Science Stralsund, Germany, showed initial changes at the surface of the ash with temperatures higher than 1,350°C. Such temperatures are usually not reached during the combustion of biomass. Also Kask et al. (2007) found that ashes from winter mown reeds did not melt under 1,330°C. Burvall & Hedman (1998) showed for Reed Canary Grass that delayed harvesting (March) increased ash melting temperature from 1,070°C to 1,400°C. It is unlikely that the ash melting behaviour of Common Reed will lead to slagging if it is used for combustion in grid- or fluidized bed combustors (Hofbauer et al 2001, Eder et al. 2004).

### 6.4.12 Emissions

Emissions have been studied from combustion of Common Reed from sites at the Neusiedler See, Austria. Under optimal combustion conditions no emissions of CO occur and the release of unburned carbon hydrides is lower than 50 mg/m$^3$. No significant emissions of dioxines and furanes or nitrous oxides could be noticed. The recommendation was to abstain from special equipment for nitrous oxide reduction (Hofbauer et al. 2001). On the other hand the trials resulted in rather high $SO_2$ emissions of about 20–50 mg/Nm$^3$. By application of filter tissue and application of lime into the stream of smoke, $SO_2$ and HCl emissions could be neutralised (Hofbauer et al. 2001). Combustion trials with Reed Canary Grass powder by Kastberg and Burvall (1998) resulted in low NO emissions under stabilized combustion. Particle emissions in flue gas leaving the boiler was 0.7 g/m$^3$, which was unexpectedly low.

## 6.5 Benefits from land use on rewetted peatlands

Franziska Tanneberger & Wendelin Wichtmann

### 6.5.1 Introduction

Land use of rewetted peatlands has several effects on the site itself, on the environment, on biodiversity, and on the economy. These effects can occur separately or in combination. The negative effects of drainage-based agriculture on peatlands, requiring a mean water level significantly lower than 10 cm below soil surface, are described in numerous publications (Succow 1988, Succow & Joosten 2001, Kratz & Pfadenhauer 2001). However, also 'wet' agriculture and forestry (i.e. paludiculture) on rewetted peatlands with a mean water level higher than 10 cm below soil surface may have negative effects, e.g. inappropriate harvesting dates leading to losses in biodiversity or high dynamics in water levels which may result in high

Table 37: Content of substances problematic in combustion of *Miscanthus*, Common Reed, Reed Canary Grass, and wood (in % dry weight). Values from literature compared with mean values for sedge and reed biomass from wet peatlands in Belarus (BWP).

| Content | *Miscanthus* | *Miscanthus* | Common Reed | Common Reed | Reed Canary Grass | Wood from spruce | BWP |
|---|---|---|---|---|---|---|---|
| Source | Hohmann 1995 | Roll & Hedden 1994 | Eder et al. 2004 | Paist et al. 2007 | Kastberg & Burvall 1998 | Hartmann et al. 2003 | Wichtmann & Tanneberger 2009 |
| Chloride | 0.3 | 0.2 | 0.20 | 0.1 | 0.05 | 0.005 | 0.02 |
| Sulphur | 0.1 | 0.1 | 0.04–0.05 | 0.04 | 0.1 | 0.015 | 0.09 |
| Nitrogen | 0.4 | 0.8 | 0.24–0.30 | 0.3 | 0.6 | 0.13 | 0.51 |
| Ash | 5.4 | 6.9 | 5.12 | 3.2 | 8.0 | 0.6 | |

emissions of greenhouse gases (GHG). However, in general the positive effects prevail.

As in an example from Belarus, the implementation of regular vegetation management on wet and rewetted peatlands (see box 22 in chapter 6.4) may have the following benefits:
- Climate: Substitution of fossil fuels (especially peat briquettes) by a renewable fuel;
- Biodiversity: Habitat restoration and maintenance for threatened flora and fauna;
- Economy: Establishment of a self-repayment and self-financing model operation for the use of biomass from wet and rewetted peatlands and contribution to regional development.

In the case of cost-covering biomass production, the wet land use of peatlands causes lower mitigation costs than many other bio-energy options for $CO_2$ mitigation. For rewetted sites, the sale of 'carbon credits' from emission reductions by rewetting (as established by the BMU-ICI project, see chapter 7.1) can provide income in addition to the earnings from the biomass production for energy itself. Briquette production can help to maintain the site conditions created through the BMU-ICI project by rewetting: The management of newly developed vegetation can keep succession in a desired state, impeding the development of shrubs and thus preventing changes in water consumption by increased evapotranspiration and keeping appropriate conditions for target species (see chapter 4.3). In addition, regular land use may help to detect problems with the new rewetting facilities (e.g. dams, sluices) at an early stage.

### 6.5.2 Climate benefits

Drainage of peatlands for conventional agriculture, forestry and peat extraction and the use of peat for energy and growing media are currently, worldwide, responsible for 6% of the total anthropogenic $CO_2$ emissions (see chapter 3.2). Recent efforts to mitigate anthropogenic GHG emissions include substituting fossil fuels by biofuels, i.e. fuels produced from biomass with a short regeneration cycle. Drained peatlands are also increasingly used for the production of biomass fuels, claiming positive effects on climate protection. Examples for such cultivation are Oil Palm in Southeast Asia, Sugar Cane in Florida, Maize and *Miscanthus* in temperate Europe and part of the peatland forest wood in Scandinavia. This generally leads to much larger $CO_2$ emissions from the oxidising peat soil than can be saved by replacing fossil fuels (Couwenberg 2007, Wicke et al. 2008, Sarkkola 2008). Biogas from Maize cultivated on drained peatlands, for example, leads to emissions of some 880 t $CO_2$·per terajoule (TJ) produced energy, palm oil from peatlands to 600 t $CO_2 \cdot TJ^{-1}$ (Couwenberg 2007). This is much higher than the $CO_2$ emissions from combustion of fossil fuels like peat (106 t $CO_2\,TJ^{-1}$), coal (anthracite 98 t $CO_2\,TJ^{-1}$), oil (73 t $CO_2\,TJ^{-1}$), or natural gas (52 t $CO_2\,TJ^{-1}$; IPCC 2006).

Paludicultures on rewetted peatlands, in contrast, contribute to climate change mitigation in two ways:
- By reducing GHG emissions from drained peatland soils (Fig. 65);
- By replacing fossil resources by renewable biomass alternatives.

An example for the positive climatic effect of paludiculture is the cultivation of Common Reed *Phragmites australis* on rewetted peatlands. The rewetting as such results in a GHG emission reduction of some 15 t $CO_2$-eq. $ha^{-1}$ $year^{-1}$ (Fig. 65, compare chapter 3.4). With a conservative yield of 12 t DM $ha^{-1}$ $year^{-1}$ and a heating value of 17.5 MJ kg $DM^{-1}$, one hectare of reed can replace fossil fuels in a cogeneration plant that would otherwise emit 15 t $CO_2$. Assuming GHG emissions from handling (mowing, transport, storage, delivery and operation of the combustion plant) to amount to 2 t $CO_2$-eq. $ha^{-1}$, using reed biomass from paludiculture would thus avoid emissions of almost 30 t $CO_2$-eq. $ha^{-1}$ $year^{-1}$ (Wichmann & Wichtmann 2009).

### 6.5.3 Biodiversity benefits

Rewetting of drained peatlands is generally beneficial for nature conservation as vastly degraded peatlands are 'biodiversity deserts'. In paludiculture, beside target plants, non-harvestable species may establish. These are often species that have become rare and endangered because of the massive decline of their naturally wet and moist habitats (e.g. in *Sphagnum* cultivation plots 'weeds' such as *Drosera* spec. have been found). With regard to species and habitat conservation, rewetting of degraded peatlands and a subsequent use for biomass cultivation should generally be preferred over keeping the areas in a drained and degraded state (cf. Wichtmann & Joosten 2007).

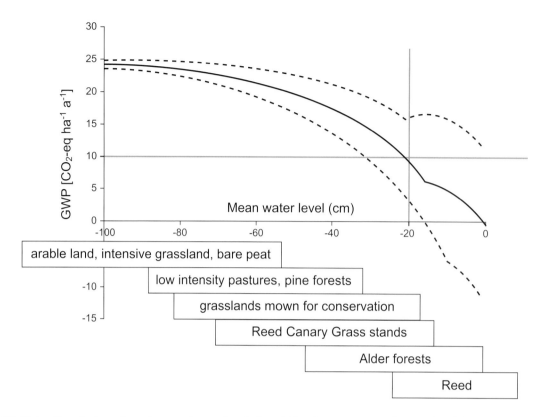

Fig. 65: Net soil emissions from temperate peatlands in relation to mean annual water level for different types of land use, expressed in Global Warming Potential (GWP) (from Wichtmann et al. 2010a). $N_2O$ emissions are conservatively neglected. The grey lines mark a reduction of 10–15 t $CO_2$-eq ha$^{-1}$ year$^{-1}$ as compared to conventional peatland agriculture and forestry.

Since agricultural land use and peat oxidation have enriched the soil with nutrients, rewetting often leads to high productive but species-poor vegetation. Regular harvesting of the biomass keeps the vegetation short and the litter layer thin. This reduces the trophic level and allows less competitive species to establish and remain. An example is the Aquatic Warbler *Acrocephalus paludicola* (see box 8 in chapter 4.1), a fen mire flagship species and the only globally threatened passerine species of continental Europe. Its breeding habitats have suffered from increasing drainage and eutrophication. Thus, the species became more and more land use dependent, because only regular mowing maintains the necessary open, sparse vegetation (Tanneberger et al. 2010a; see box 24).

In the case of areas designated as conservation sites, paludiculture can be considered as a cost-effective management option, instrumental but ancillary to conservation. On former heavily degraded sites, where any rewetting and management will increase biodiversity, climate benefits can prevail. Here, monitoring is recommended to track the appearance of protected species and habitats and to be able to modify management (see chapter 9.2). To prevent conflicts e.g. when early summer mowing for biogas production destroys breeding habitats or when winter harvesting leaves insufficient old-grown reed, it is necessary to establish and communicate clear priorities.

### 6.5.4 Economic benefits

In the temperate zone most drained peatlands were used as pastures or meadows. Nowadays large areas are abandoned as a result of progressive soil degradation, insufficient productivity and fodder quality for dairy cattle, as well as regional decline in livestock (Wichtmann et al. 2010a). In Germany, in general, grazing for meat production on marginal lands or protected areas, e.g. by suckler cows or lambs, generates deficits of several hundred Euro per hectare per year and fully depends on financial support by nature protection

## 6.5 Benefits from land use on rewetted peatlands

> **Box 24**
>
> **Birds and reed harvesting**
>
> Franziska Tanneberger
>
> An impressive example of the biodiversity benefits of paludiculture is the Aquatic Warbler *Acrocephalus paludicola* (see box 8 and Fig. 44 Colour plates II) breeding site in Rozwarowo Marshes, Poland. Here, the harvesting of reed from a slightly drained peatland for thatch is currently profitable without major subsidies and it benefits the Aquatic Warbler. The site is the stronghold of the particularly threatened Pomeranian population of this species, accounting for, in recent years, more than 50% of its population. A recent study (Tanneberger et al. 2009) observed three successful broods of Aquatic Warblers in reed cut in the previous winter and could not find differences in total biomass of potential prey between the regularly winter cut Rozwarowo breeding sites and other Pomeranian breeding sites.
>
> Biomass harvesting from rewetted peatlands has several effects directly relevant to biodiversity conservation:
> - Reduction of the thickness of the litter layer and of old standing vegetation;
> - Depletion of access nutrients from top soil mineralization and atmospheric deposition;
> - In the case of late summer or winter cuts of grass and sedge vegetation, maintenance of vegetation height and density; and
> - In the case of early summer mowing of grass and sedge vegetation, reduction of vegetation height and density as well as weakening of reed and favouring of sedge vegetation;
> - Reduction of the successional encroachment of bushes (e.g. willows) and trees.
>
> Among the management options of paludiculture, winter reed cutting for thatch has probably the longest tradition and its effects on biodiversity are best studied. With regard to plant conservation, winter cutting can maintain (Güsewell & Le Nédic 2004) or increase species diversity as it substantially reduces the ruderal and nitrophilous character of the vegetation (Gryseels 1989) and benefits shorter and more species-rich vegetation (Cowie et al. 1992). For typical reedbed birds, Kube & Probst (1999) showed low population densities e.g. of Bearded Tits *Panurus biarmicus* in brackish reedbeds in Northeast-Germany up to three years after cutting. In contrast, Hawke & José (1996) describe that the same species preferred recently cut brackish reedbeds (with rapid litter accumulation) in southwestern England above areas that had not been cut. Other characteristic reed species such as Sedge Warbler *Acrocephalus schoenobaenus* are apparently also able to rapidly recolonize cut areas (Hawke & José 1996). Often, the reduced abundance of potential prey for passerine birds is used as an argument against reed cutting. According to Ditlhogo et al. (1992) and Schmidt et al. (2005), appropriate management for invertebrate conservation in reedbeds and fen meadows would include rotational burning or cutting, with parts of the bed left unmanaged, forming a mosaic of ages and allowing recolonization by cutting-sensitive invertebrates. Bibby & Lunn (1982) generally see little conflict between harvesting activities and bird conservation, as long as patches of uncut reed remain.

and agricultural subsidies (Plachter & Hampicke 2010). The continued costs of drainage – with all external diseconomies – are, furthermore, largely borne by society, not by the individual user.

Paludiculture fosters income generation from primary production where, at least in the European Community, previously a subsidy-oriented, non-sustainable land use of drained peatland took place. Income e.g. from selling fuel briquettes may cover fixed and variable costs as well as costs for labour, relating to harvesting and processing of the biomass. Whether net margins are possible depends on many factors, e.g. the performance of workers and machines and the degree of mechanisation. Most activities have to be done in autumn or winter during periods when in agriculture only low workloads occur. In such a situation, employees may be kept in work and additionally jobs could be generated, dependent on the intensity of processing and creation of value.

Although special wetland-adapted harvesting machinery is required, thermal utilization of winter harvested *Phragmites* reeds in Northeast-Germany can fully compete with *Miscanthus* or straw from mineral soils without subsidies or payments for ecological services. For individual farms the perspectives of paludiculture are decisively determined by the agricultural subsidies that competitive (but unsustainable!) land use options receive (e.g. EU direct payments), not by objective economic costs and revenues (Wichmann & Wichtmann 2009).

# 7 The BMU-ICI project

## 7.1 Summary of the project

Irina Voitekhovitch, Viktar Fenchuk,
Zbig Karpowicz & Susanne Bärisch

### 7.1.1 Project overall context

The drained peatlands of temperate Europe (especially Germany, Poland, Belarus, Ukraine and Russia) constitute an important source of greenhouse gas (GHG) emissions and are – after Southeast Asia – the second most important global hotspot in this respect. Draining of peatlands is an important source of peat fires which can lead to vast amounts of additional emissions. In Belarus, peatlands cover 2.9 million ha, i.e. almost 15% of its territory. During Soviet times, more than half of the country's peatlands, or 1.5 million ha, were drained, most of them for agricultural usage and mining for fuel. After the breakdown of the Soviet Union, large areas of drained peatlands were abandoned and the remaining lands were often used ineffectively (see chapter 2.1 and 2.3).

Restoring degraded peatland can halt peat mineralization that in turn stops or greatly reduces the release of $CO_2$ and other GHGs. It can also decrease the number of peatland fires, and in the long run possibly produce sinks for carbon from the atmosphere. It has an immense, thus far untapped, emission reduction potential. Furthermore, peatlands serve as important habitats for flora and fauna. Fen mires and raised bogs are today some of the most threatened habitats in Central Europe. Restoring peatlands, therefore, can reduce GHG emissions as well as enhance biodiversity in Belarus.

### 7.1.2 Project partners

The project 'Restoring peatlands and applying concepts for sustainable management in Belarus – climate change mitigation with economic and biodiversity benefits' was coordinated by the Royal Society for the Protection of Birds (RSPB), UK, in collaboration with the Michael Succow Foundation (MSF), Germany, and APB-BirdLife Belarus. The project was financed in the framework of the International Climate Initiative (ICI) of the German Federal Ministry for the Environment, Nature Conservation and Nuclear Safety (BMU) through Kreditanstalt für Wiederaufbau Entwicklungsbank (KfW). It was carried out with support of the United Nations Development Programme (UNDP) in Belarus and the Ministry of Natural Resources and Environmental Protection of the Republic of Belarus. The project began in autumn 2008 and will be completed at the end of 2011.

**The Royal Society for the Protection of Birds (RSPB), UK**

RSPB has a proven capacity in the execution of wildlife conservation projects. The headquarters are situated in UK, supported by three national and nine regional offices. There are over 1,300 staff, and more than 13,000 volunteers support the organisation. The international division has about 45 staff. RSPB draws upon various European funding streams and programs for executing international projects (e.g. EC Life-Nature, EC 'Environment and Forest in developing countries', European Bank for Reconstruction and Development (EBRD), and European Commission Technical Assistance to the Commonwealth of Independent States (ECTACIS)), besides receiving funds from the UK government (e.g. Civil Society Challenge Fund (CSCF)) and the Global Conservation Fund (GCF). The RSPB is the largest wildlife conservation organisation in Europe, with a membership of over one million, together with over 195,000 youth members. RSPB manages 200 nature reserves covering 130,000 ha that are home to 80% of our rarest or most threatened bird species. Some further activities are to champion birds and the environment to decision makers, protect, restore and manage habitats for birds and other wildlife and to perform research into the problems facing birds and the environment.

RSPB was the interface between the project and KfW and was responsible for overall coordination and supervision. RSPB established a Project Implementation Unit (PIU) which was comprised of members from each partner organisation. The RSPB has worked successfully in Belarus since

# 7 The BMU-ICI project

> **Box 25**
>
> **The International Climate Initiative (ICI) in a nutshell**
>
> Federal Ministry for the Environment, Nature Conservation and Nuclear Safety
>
> The International Climate Initiative has been financing climate protection projects in developing and newly industrialising countries and in transition countries in Central and Eastern Europe since 2008. One major focal point of this funding is the G5 states (Brazil, China, India, Russia and South Africa). Through this, the German Federal Ministry for the Environment, Nature Conservation and Nuclear Safety is making an effective contribution to emission reductions and adaptation to climate change. This new form of environmental cooperation complements the Government's existing development cooperation.
>
> Funding of 120 million Euro per year is available for the ICI from revenues from the sale of emission allowances. A decision by the German Bundestag forms the basis of this worldwide investment in climate protection.
>
> When selecting projects, the Federal Ministry for the Environment, Nature Conservation and Nuclear Safety attaches great importance to the development of innovative and multipliable approaches that impact beyond the individual project itself and are transferable. Through targeted cooperation with partner countries the Climate Initiative provides important momentum for negotiations on an international climate agreement for the post-2012 period. The Climate Initiative also makes a valuable contribution to international discussions on structuring a climate financing architecture.
>
> The focus of the International Climate Initiative lies in the areas of:
> - Promoting a climate-friendly economy;
> - Promoting measures for adaptation to the impacts of climate change and;
> - Promoting measures for preservation and sustainable use of carbon reservoirs/ Reducing Emissions from Deforestation and Degradation (REDD).
>
> The funds from the ICI primarily flow into bilateral projects, but projects by multilateral organisations are also supported. More information on the key elements of the International Climate Initiative is available here: http://www.bmu-klimaschutzinitiative.de/en/theme

1998, providing a 'Country Programme' package of financial and technical assistance to APB-BirdLife Belarus. Its prime focus is conserving the two globally threatened species, the Aquatic Warbler *Acrocephalus paludicola* and the Greater Spotted Eagle *Aquila clanga*, and protecting Important Bird Areas as well as to help build capacity in biodiversity conservation in general.

**Michael Succow Foundation (MSF)**

The Michael Succow Foundation for the Protection of Nature (MSF) was founded in 1999 at the Institute of Botany and Landscape Ecology at Greifswald University, Germany, where it is still based. Currently there are about 20 employees who are supported by a broad network of expert volunteers. MSF has a partnership agreement with the Institute of Botany and Landscape Ecology, which is an internationally known centre for scientific research on mires and peatland restoration. MSF dedicates its efforts to the conservation of landscapes and ecosystems following the motto 'Preserve and Sustain'. The main focus of the Foundation's work lies in transformation countries, especially in Eurasia. Several MSF projects deal with providing support in the establishment of protected areas, such as biosphere reserves and national parks, in countries such as Russia, Kazakhstan, Uzbekistan, Azerbaijan, and Ethiopia.

The BMU-ICI project idea was developed by MSF. Within the project, MSF, together with Greifswald University, was primarily responsible for the scientific input to the project e.g. with regard to greenhouse gas emission assessments, monitoring, and development of a standard and methodology for the voluntary carbon market. Greifswald University also played a key role in United Nations Framework Convention on Climate Change (UNFCCC) advocacy (see chapter 7.5). Other responsibilities in the project like sustainable use of wet peatlands, capacity building (together with Leibniz Centre for Agricultural Landscape Research ZALF Müncheberg, Germany) and relations with BMU and other stakeholders in Germany were also with MSF. Greifswald University and later MSF have cooperated with institutions in Belarus since the 1990s.

## 7.1 Summary of the project

> **Box 26**
>
> **Projects managed by the Development bank of the Federal Republic of Germany (KfW Entwicklungsbank) within the International Climate Initiative (ICI).**
>
> Frank Mörschel
>
> KfW Entwicklungsbank is Germany's leading development bank and is an integral part of the KfW Bankengruppe. KfW helps to achieve the goals set by the German Development Cooperation: to reduce poverty, protect the climate, ensure peace, and impact globalisation in such a way that those living in the world's poorer regions will benefit, too. In financing, advising and guiding development projects throughout the world, KfW's efforts focus on the needs of their partner countries in order to improve livelihoods, within these countries.
>
> KfW Entwicklungsbank invested 4.5 billion Euro in 2010, of which around 64% was mobilised on the capital market. Approximately 250 million people worldwide benefit from these commitments. Activities aimed at combating and adapting to climate change and protecting the environment were funded with 2.6 billion Euro in 2010, which amounted to just under 58% of the total financing volume.
>
> KfW Entwicklungsbank works on behalf of the German Government, in particular the German Federal Ministry for Economic Cooperation and Development (BMZ), but also for other Ministries like the German Federal Ministry for the Environment, Nature Conservation and Nuclear Safety (BMU).
>
> Nature conservation is an important area within the activities of the KfW Entwicklungsbank. In 2010 alone, 1.5 million ha of endangered natural landscape areas, particularly tropical forests, were protected, and a contribution was made towards preserving the livelihoods of 400,000 people.
>
> In Belarus, Ukraine and Russia, KfW Entwicklungsbank finances pilot projects through the International Climate Initiative of BMU, aiming to restore degraded peatlands. Rewetting of degraded peatlands achieves several objectives at once, demonstrating a unique win-win situation:
> - Reducing the emission of greenhouse gases and thus contributing to climate protection;
> - Restoring the natural biodiversity of peatlands;
> - Improving the livelihoods of local people, such as through the alternative use of biomass and sustainable fuel production.
>
> Such pilot projects constitute best-practice examples that can, and should be, replicated throughout Eastern Europe.

**Achova Ptusak Backauscyny (APB) – BirdLife Belarus**

APB is the official partner of BirdLife International in Belarus which was registered by the Ministry of Justice in Belarus in June 1998. APB is a membership-based organisation with about 1,950 members. APB has 20 full-time staff and utilizes funding from the Frankfurt Zoological Society, Global Environment Facility (GEF), other BirdLife International organisations as well as UNDP, RSPB and the Michael Otto Foundation. The latter three organisations have funded a project on the implementation of urgent conservation actions in fen mires in Belarus, where APB has acquired intensive experience in rewetting peatlands. As a public organisation APB-BirdLife Belarus is also very active in awareness raising for the conservation of species and ecosystem diversity in Belarus.

Within the project, APB coordinated and managed all the in-country activities and in-country partners, performed technical planning of activities at the rewetting sites, the subsequent implementation of management measures after rewetting, and the implementation of a pilot enterprise focused on biomass harvesting and briquetting, to produce a sustainable alternative fuel on rewetted peatland sites, at Sporauski zakaznik.

**UNDP Belarus**

UNDP is the UN's global development network, an organisation advocating for change and connecting countries to knowledge, experience and resources to help people build a better life. UNDP operates on the ground in 25 countries and territories in Central and Eastern Europe, the Caucasus and Central Asia. It works closely with national partners – including governments, civil society and the private sector – supporting country needs and priorities that fall into UNDP's areas of expertise. UNDP started its activity in Belarus in 1992 having a mandate to support and supplement national efforts to struggle with the most important problems of economic development, to promote social progress and improvement of standards of people's life.

UNDP has outstanding experience in the execution of a five-year project on peatland restoration launched in 2006 (see chapter 2.4). Through this project, 15 degraded peatlands were rewetted, five of them with BMU-ICI funding. For the BMU-ICI project, UNDP Belarus organised technical activities relating to the development and execution of onsite rewetting measures, e.g. engineering planning, rewetting documentation, and co-operation with executing companies.

**Ministry of Natural Resources and Environmental Protection of the Republic of Belarus**

The Ministry has a strong influence on the overall policy of the Belarus government on peatland management, accepting that peatland rewetting is key to the mitigation of climate change. Being the Party to the UNFCCC and Kyoto Protocol, the Ministry represents the Designated Focal Point for the coordination and approval of Joint Implementation (JI) and Voluntary Emission Reduction (VER) projects.

### 7.1.3 Project Results

Key project outcomes include:
- Rewetting of 14,000 ha peatlands (see chapter 7.2);
- Development of an international peatland carbon standard and an internationally approved baseline and monitoring methodology for peatland rewetting and conservation (see chapter 7.3);
- Assessment of GHG emissions and biodiversity values at project sites before and after rewetting (see chapters 7.3 and 7.4);
- Promotion of the objectives and results of the project at a range of events, including climate change related conferences and climate change talks within UNFCCC (see chapter 7.5) and at the national and local level (see chapter 7.6);
- Demonstration of sustainable management of peatlands by using plant biomass for fuel production (see chapters 6.4, 7.3, and 7.4);
- Capacity building for Belarusian scientists on GHG emissions measurements, analysis of greenhouse gases, and mapping of GESTs to ensure the continuation of the work beyond the project (see chapter 7.7);
- Development and scientific assistance of a twin project in Ukraine (see chapter 7.9).

The project developed and/or applied a set of new methodologies and instruments:
- The greenhouse gas emissions site type (GEST) approach was a newly developed instrument that enabled the assessment of emission reductions from rewetting of different types of degraded peatlands. The approach builds on the fact that GHG emissions are clearly related to annual water levels and the presence or absence of plant species groups that can be assessed by mapping vegetation in the field (see chapter 3.4);
- The new Verified Carbon Standard (VCS) category 'Peatland Rewetting and Conservation (PRC)' provides the eligible activities, requirements for developing projects and methodologies, as well as the requirements for validation, monitoring and verification of peatland rewetting projects (see chapter 5.3);
- The harvesting of biomass from wet and rewetted peatlands for energy production ('paludiculture') which not only provides improved habitat conditions for threatened fen mire biodiversity and additional income for local communities but an additional climate effect through the substitution of fossil fuels.

### 7.1.4 Long-term project impacts

The BMU-ICI project created and tested the essential components for the rewetting of degraded peatlands for the benefit of carbon and biodiversity in Belarus, setting the scene for future replication of the project model in other countries. Activities implemented within the project will lead to changes in four main areas:
- Climate protection and adaptation to climate change: The GHG reduction potential in Belarus with up to 520,000 ha available for rewetting is immense. Additional positive climate effects include the reduction in the number and size of peatland fires that increase the GHG release considerably and prevents the fire-based re-emission of radioactive substances in the Chernobyl region. The development of the VCS-PRC standard and the internationally ap-

proved baseline and monitoring methodology for peatland rewetting and conservation will facilitate further rewetting activities in Belarus. In addition, the appropriate use of biomass from rewetted peatlands can also provide positive climate effects by substituting fossil fuels. Rewetting activities provide additional evapotranspiration cooling, improve regional landscape hydrology and reduce nutrient run-off (e.g. nitrogen) into surface waters.

- Economic effects: With the VCS-PRC standard and subordinate regulations, a new tool for generating income through rewetting activities has been developed that will provide funding for the maintenance of rewetted sites over the longer term and additionally generate funds for further rewetting activities in Belarus (see chapter 5.3). Income-generating opportunities for local communities lie both in the restoration activities as well as in biomass harvesting from rewetted peatlands. The latter combines traditional land use with new ways of processing biomass from peatlands (e.g. pellet and briquette production), thus revitalising rural economies adjacent to peatlands and decreasing energy dependence on neighbouring countries with minimal damage to the environment (see chapter 6.4).
- Conservation of biological diversity: By rewetting degraded peatland and subsequently managing the restored areas, the project benefits both threatened habitats and threatened species of national and international importance. Several project sites have the potential to benefit two globally threatened bird species – the Aquatic Warbler *Acrocephalus paludicola* and the Greater Spotted Eagle *Aquila clanga* (see chapter 4.1 and 7.4).
- Replication: The experiences made by the project can be applied to other drained peatlands in Belarus as well as in Ukraine (see chapter 7.9) and European Russia. The VCS standard developed for peatland restoration in temperate regions will also provide important foundations for the rewetting of peatlands in tropical regions.

## 7.2 Site selection and rewetting actions

Hans Joosten, Annett Thiele & Olga Chabrouskaya

### 7.2.1 Site selection criteria for rewetting

Not all drained peatlands are equally suitable for an emission reduction project aimed at generating carbon credits. The most suitable are those sites where the least technical and financial effort yields the highest emission reductions, or better still, where the greatest number of verifiable emission reductions can be reached. The criteria for identifying such sites relate to the given characteristics of the peatland itself and the chosen accounting rules that may differ from standard to standard. In this chapter we discuss the criteria from the perspective of the Verified Carbon Standard, the most important standard on the voluntary market (see chapter 5.3).

### 1. The peat layer is at least 30 cm thick

This criterion must guarantee that emission reductions can be achieved over the entire project period. Within a crediting period of 20 years, depletion of a layer of peat >30 cm thick will normally not take place (cf. Couwenberg et al. 2010). If the peat layer is thinner, there is a realistic chance that, in the baseline, i.e. without rewetting, all peat would have disappeared before the end of the project period and no peat emissions would take place anymore. Non-compliance with this criterion does not make the area unsuitable for rewetting; it only means that:

a) More detailed peat depth mapping (with e.g. 10 cm depth intervals) is required so that areas can be identified where the peat depletion time is shorter than the project period and reductions are only claimed for those years and areas in which a peat layer would be present in the baseline scenario, or that

b) Areas with less than 30 cm of peat within the project boundaries are conservatively excluded from accounting for the emission reductions.

In addition, areas where a thin (< 30 cm thick) layer of peat is underlain by organic and especially carbonate rich gyttja should for the time being be treated similarly (i.e. the gyttja should not be considered to be peat) as no emission factors for surfacing gyttja are yet available.

**2. The peatland is deeply drained**
This criterion must guarantee that rewetting can achieve sufficient emission reductions, as existing wet areas do not bring significant reductions upon (further) rewetting. The mean annual water level should be at least 20 cm, preferably more than 40 cm under the surface level. This criterion can easily be assessed by field observation of water levels and prevailing vegetation (see chapter 3.4, Couwenberg et al. 2011).

**3. The peatland is rewettable up to the surface**
The largest emission reduction takes place when mean annual water levels are brought to some 10 cm below the surface. To arrive at such a level, winter water levels should be at least up to the surface, probably higher. Raising of the mean water level to less than 20–30 cm below the surface generally does not lead to sufficient emissions reductions to compensate for the rewetting investments made. This criterion requires an in-depth knowledge of local and regional hydrological and hydraulic conditions in order to estimate which average water levels can be reached where. A first and general assessment can often be made on the basis of relief conditions, topography, peat stratigraphy and the degree of peat decomposition, and by assessing the hydro-genetic peatland type.

**4. The area is not and would not have been abandoned**
This criterion is dualistic. On the one hand, abandonment indicates that there will be little resistance against rewetting, because the area is anyhow not used. On the other hand abandonment may lead to spontaneous water level rise in the project area through collapsing ditch banks, vegetation filling ditches, beaver activity and ongoing peat subsidence. Emission reductions resulting from spontaneous rewetting may not be accounted because they are not 'additional' (see below).

**5. The area is and would have been without shrub or forest growth**
Abandoned drained sites have the unfavourable feature of possible tree and shrub encroachment in the baseline. This encroachment and associated carbon sequestration will be stopped or hampered by rewetting, which reduces the amount of realised emission reductions in the project. With substantial forest growth in the baseline (that will die off as a result of rewetting), the possibility exists that, over the (limited) crediting period, the project will generate more emissions than the baseline scenario. To exclude this risk, an accurate estimate of forest encroachment and growth (age, yield classes) and the associated carbon sequestration in the baseline has to be made.

**6. The area must generate little leakage**
Rewetting should not result in a displacement and increase of emission producing activities outside the project area. Leakage occurs, for example, when the project leads to higher water levels and consequent decreased forest growth in areas adjacent to the rewetted project area. Another form of leakage is 'activity shifting', the situation that activities that formerly were practised in the project area (e.g. drainage, peat extraction, fertilisation) move to outside the area and create new emissions there. A classic example of leakage was the plan at Vyhanascanskaje (see chapter 3.5) to use the pump capacity that would become superfluous by rewetting part of the farm to improve drainage in another peatland area. The increased emissions from leakage have to be subtracted from the emission reductions realised in the project area, and thus decrease the total volume of carbon credits that are generated. Leakage is a function of current use. If sites are abandoned or under very low intensity use, no leakage problems are to be expected. If the sites are under high intensity use, the chance of leakage is large and leakage has to be counteracted by providing alternative sources of income.

**7. The rewetting of the site must be 'additional'**
Additionality means that the beneficial effects for the climate would not have happened without activities deliberately implemented for climate change mitigation. Developments that are already happening are not 'additional'. Many peatland sites in Belarus are planned to be rewetted according to government programs and national Belarusian law, e.g. to prevent fires or because they are unsuitable for agriculture and forestry. The project proponent must demonstrate additionality and make plausible that the site would not otherwise be rewetted in absence of the project.

### 7.2.2 General considerations

None of the site selection criteria is exclusive, but the aim should be to try and comply with all of them. The reality in Belarus has, however, shown that currently few sites are made available for rewetting that comply with criteria 4 and 5.

## 7.2 Site selection and rewetting actions

Table 38: Project sites rewetted in the framework of the BMU-ICI project. Land use before rewetting: A = Agriculture, P = Peat extraction, D = Drainage and abandonment. Emission reduction figures are calculated using a forward-looking baseline over a 20 year period and assuming optimal rewetting.

| Name of project site | Peatland type | Area (ha) | Land use before rewetting | Date of rewetting | Costs of measure (USD) | Methods used | Estimated emission reduction (t $CO_2$ ha$^{-1}$ yr$^{-1}$) | Total potential emission reduction (t $CO_2$-eq. yr$^{-1}$) |
|---|---|---|---|---|---|---|---|---|
| Hrycyna-Starobinskaje | Fen | 3,505 | P | June 2009 | 35,759 | Sluices, earth dams | 0.44 | 1533 |
| Zadzienauski Moch | Raised bog | 753 | P | July 2010 | 48,874 | Earth dams | 2.95 | 2063 |
| Scarbinski Moch | Fen with bog lenses | 1,323 | P | June 2010 | 42,362 | Earth dams | 2.28 | 3205 |
| Poplau Moch | Fen | 415 | P | July 2010 | 47,323 | Earth dams | 2.12 | 878 |
| Obal | Fen | 1097 | P | May 2009 | 53,726 | Earth dams, rock filled dam, irrigation trenches | 0.60 | 1061 |
| Dakudauskaje | Raised bog | 1,946 | D | June 2010 | 42,000 | peat dam with 1.5 m deep anti-leakage screen, earth dams | 2.39 | 3820 |
| Zada | Raised bog | 2,471 | D,P | Autumn 2011 | 33,928 | Earth dams | 3.05 | 10298 |
| Dalbeniski | Raised bog | 5,501 | D,P | Summer 2011 | 12,000 | Earth dams | 1.58 | 7539 |
| Horeuskaje | Fen | 191 | D | Autumn 2011 | in planning process | Earth dams | 0.01 | 3 |

For sites that do not fulfil all of the above criteria, preliminary reduction estimates (e.g. based on greenhouse gas emission site types (GESTs) and the use of succession models, cf. Couwenberg et al. 2011, chapter 3.4 and 3.5) are necessary to separate the 'better' sites from the less productive ones.

Sites that show limited overall net-emission reductions (below 2 t $CO_2$-eq ha$^{-1}$ yr$^{-1}$) may still be viable for rewetting (also for emission reduction) but not for creating marketable carbon credits. The large uncertainty in the overall low reduction values prohibits reaching a positive effect when following a conservative approach in the assessment of emission reductions.

### 7.2.3 Steps from site selection to practical rewetting

When implementing peatland rewetting in Belarus, the following procedures have to be followed (see chapter 2.4):
- The development of the 'scientific justification';
- Approval of the site selection process;
- The elaboration of an engineering plan;
- Formulation of an ecological (if necessary) and/or state expertise for the engineering plan;
- The implementation of building works.

According to the experience of the BMU-ICI project, the time schedule was as follows: The preparation of the 'scientific justification' took three to six months. The site selection was approved within one month. The engineering plans were prepared within three to six months. Review by state experts demanded another two to three months, and construction works were implemented within another three to six months.

In the case of a Verified Carbon Standard (VCS) project, the first baseline monitoring can be carried out in parallel to the development of the 'scientific justification'. The collection of data necessary for a Project Document (PD; see chapter 7.3) includes the mapping of vegetation using GESTs and, if necessary, peat depth surveys which takes another five to eight months. The total time required from site selection to finishing the construction works will be between one and two years.

Fig. 66: Location of peatlands in Belarus rewetted within the BMU-ICI project (black dots) or mentioned elsewhere in this book (white dots).

### 7.2.4 Rewetting sites

The sites rewetted in the BMU-ICI project are rather evenly spread across Belarus, with a slight concentration in the northern part (Fig. 66). The areas of the sites varied between 200 and 3,500 ha. In total the nine rewetted sites cover c. 17,200 ha, about one third of them are fens and two thirds raised bog sites. Most of the sites have been drained for peat extraction and some have been affected by drainage and subsequent abandonment before project measures began. Technical rewetting has been realised mostly through the construction of earth dams to block drainage ditches. In Dakudauskaje a 1.3 km long peat dam with a 1.5 m deep anti-leakage screen was constructed to protect the project site against negative effects from the adjacent peat extraction areas. The expected emission reductions calculated using a forward-looking baseline over a 20 year period and assuming optimal rewetting are up to 3 t $CO_2$-eq ha$^{-1}$ year$^{-1}$ (Zada and Zadzienauski Moch), summing up to c. 30,400 t $CO_2$-eq. year$^{-1}$ for all project sites. Corresponding figures are presented in Table 38 and are described in detail in chapters 8–8.6.

## 7.3 Climate actions

Irina Voitekhovitch, Annett Thiele & Merten Minke

### 7.3.1 Introduction

Several years ago it was impossible to perform carbon offsetting projects in peatlands, because there was no method to estimate and monitor emissions from different types of peatlands and, therefore, no possibility to assess emission reductions from rewetting. The recent method developed by Greifswald University (Greenhouse gas Emission Site Types (GEST)) approach; see chapter 3.4 and Couwenberg et al. 2011), enabled the BMU-ICI project to assess the baseline and project emissions in Belarus by calibrating the model to the climate and the natural conditions of Belarus. The GEST model also facilitated the development of project design documentation and the process of approval within national and international carbon procedures for the further determination and validation of emission reductions.

The main areas of climate-related work within the BMU-ICI project were:
- Improvement of the GEST approach and its calibration to Belarusian conditions through greenhouse gas measurements on Belarusian peatlands and research on the relationship between vegetation and water levels in Belarus (see also box 27);
- Development of an international peatland carbon standard (under the Verified Carbon Standard);
- Development of an internationally approved baseline and monitoring methodology for peatland rewetting and conservation;
- Assessment of greenhouse gas (GHG) emissions at the project sites before and after rewetting;
- Development of a strategy to sell carbon credits;
- Demonstration of sustainable biomass use from wet peatlands (paludiculture).

### 7.3.2 Improvement and calibration of the GEST approach

A meta-analysis of long-time observation of gas fluxes in peatlands in Central Europe has shown that $CO_2$ and $CH_4$ emissions from peatlands correlate well with the mean annual water level (see chapter 3.4 and Couwenberg et al. 2011). Vegetation is a good proxy for mean annual water level and can be used to assess the current emissions from peatlands and make a prognosis for 'with rewetting' and 'without rewetting' scenarios (see chapters 3.4 and 3.5).

The approach used by the BMU-ICI project can be applied in all temperate peatlands. When applying the concept, it has to be taken into account that the composition and the indicative value of vegetation may vary between different climatic regions. Measurements of GHG emissions are required to verify and calibrate the already defined GESTs for Belarus (and the Ukraine), to fill gaps for GESTs that lack data on GHG emissions, and to contribute to the GHG emission data set that constitutes the basis for the GEST approach. Such an improvement and regional calibration based on greenhouse gas measurements and vegetation studies in Belarus is currently under development by a research group set up by the BMU-ICI project (see also box 27).

GHG fluxes are measured with a standardised methodology that fits the requirements of the above-mentioned GHG emission data set. Measuring sites include drained, rewetted and natural sites in three Belarusian peatlands – a bog in the north (Raznianskaje in Biarezinski zapaviednik), a fen in central Belarus (Barcianicha), and four fens in the south of the country (located in and around Sporauski biological zakaznik). The research is carried out by Belarusian and German scientists applying a static chamber method (see chapter 3.3) that is especially suited to measure $CO_2$, $CH_4$ and $N_2O$ on different vegetation types. To obtain adequate GHG flux measurements for the GEST approach, at least one year of regular measurements is needed in order to allow for the calculation of an annual GHG balance. To minimize the effect of exceptional yearly weather conditions, measurements over two or more years are recommended. Within the BMU-ICI project

> **Box 27**
>
> **CIM in Belarus**
>
> Merten Minke & Annett Thiele
>
> The German Centre for International Migration and Development (CIM) is a joint operation of GIZ (Deutsche Gesellschaft für Internationale Zusammenarbeit) and the German Federal Employment Agency, which was founded in 1980 to orientate development work towards demands in developing and threshold countries. Companies, governmental and non-governmental organisations and institutions that are in need of expertise that is currently lacking in their country may ask CIM for support. CIM then searches in Germany and the European Union for appropriate candidates, presents them to the employer and provides the selected post holder with organisational and financial support. This human resources placement organisation operates today in 75 countries and has placed 800 European managers and technical experts. In addition, the CIM Returning Experts Programme supports people from developing, emerging and transition countries who have completed training or university studies in Germany to return to their home country and to contribute to their country's development. CIM is largely financed by the German Federal Ministry for Economic Cooperation and Development (BMZ) but also welcomes co-financing.
>
> CIM's strategic activities in Belarus are mainly focussed on development policy areas agreed between Belarus and Germany, which include economic promotion with a special focus on small and medium-sized enterprises, and support for non-governmental organisations. The tasks of the nine CIM Integrated Experts (IE) and three CIM Young Professionals in Belarus are based on economic, educational, social, and environmental subjects. For example, one of the positions is at 'IBB Johannes Rau' (a German-Belarusian education and conference centre, a linchpin of the East European network, and a place for freedom of speech). Another is at the Chernobyl Children's Centre 'Nadezhda'. Also the Head of the German-Belarusian business club (the only organisation representing the interests of German companies in Belarus) is a CIM Integrated Expert. The focus of another CIM post is in the National Academy of Sciences, Belarus, to improve international scientific cooperation. The key area of the CIM expert for science and technology at the International Sakharov Environmental University, Minsk is the development of renewable energy sources in Belarus.
>
> CIM is also active in nature conservation, especially peatland protection. Two young professionals are working at the non-governmental organisation APB (BirdLife Belarus). Both scientists are co-financed by the Royal Society for the Protection of Birds (RSPB), UK, and work in the field of assessing greenhouse gas emission reductions from peatland rewetting in Belarus. They test, calibrate, and adapt the GEST approach (see chapter 3.4) for peatlands in Belarus and provide APB and its partners (National Academy of Sciences of Belarus) with the knowledge of how to apply and improve this approach. To facilitate their work and to build in-country capacity, young Belarusian scientists have been trained within the BMU-ICI project in GHG measurement techniques and gas analytic methods at the Centre for Agricultural Landscape Research (ZALF) in Müncheberg, Germany, and in vegetation ecology at Greifswald University (see chapter 7.7). The Belarusian trainees are part of the GHG research team at APB led by the CIM Young Professionals. Two specialists of the team are supported from CIM's Returning Experts Programme. The work of the team includes year round measurements of GHG exchange and water level fluctuations in selected peatlands at representative vegetation types, the calibration of the GEST approach, and the filling of important gaps (see chapter 3.4). In addition, with the help of the preliminary assessment tool, the GHG emissions from degraded and rewetted peatlands are estimated as the basis to prepare documents to sell carbon credits on the voluntary carbon market.

the majority of GESTs were measured for two years, while site types additionally selected in the course of the research were covered only by one-year of measurements within the project's duration. Ongoing research activities after the end of the project will take place (see chapter 9.1).

### 7.3.3 Development of an international peatland carbon standard (Verified Carbon Standard VCS)

The VCS Standard (formerly Voluntary Carbon Standard, see chapter 5.3) provides a global standard for GHG emission reduction and removal projects. It uses as its core International Organisation for Standardisation (ISO) requirements and the two principal documents of the program are the VCS Program Guide and the VCS Standard (http://www.v-c-s.org/). The VCS Program Guide describes the rules and requirements governing the VCS program, whereas the VCS Standard provides the requirements for developing projects and methodologies, as well as the requirements for validation, monitoring and verification of projects and GHG emission reductions and removals. A VCS Standard is supported by other documents that provide further requirements specific to agriculture, forestry and other land use (AFOLU) projects and methodologies.

Peatlands were recognized as acceptable project category under the VCS in March 2011 by inclusion of a fifth project category – Peatland Rewetting and Conservation (PRC). Until then eligible AFOLU project categories had been Afforestation, Reforestation and Revegetation (ARR), Agricultural Land Management (ALM), Improved Forest Management (IFM) and Reduced Emissions from Deforestation and Degradation (REDD). Within the BMU-ICI project, specific PRC requirements have been elaborated and agreed (see chapter 5.3).

### 7.3.4 Development of an internationally approved baseline and monitoring methodology

Registration under VCS requires a baseline and monitoring methodology. A methodology for the rewetting of peatlands drained for peat extraction, forestry or agriculture was developed by Silvestrum and Greifswald University based on the GEST approach. The methodology is valid for Belarus and other countries with a temperate climate. The methodology is a 'new methodology', eligible under the PRC project category of the VCS AFOLU. It includes procedures that have not been covered in other existing approved methodologies, being the first of its kind under the Rewetting of Drained Peatlands (RDP) sub-category of the PRC project category.

The methodology centres on the use of the ground vegetation composition as a proxy for peatland emissions, known as the 'GEST' approach (see above and chapter 3.4). It outlines transparent and conservative methods to estimate the reduction of net greenhouse gas emissions resulting from project activities implemented to rewet drained peatlands in temperate climatic regions. It allows for the estimation of GHG emissions from drained and rewetted peatlands and also accounts for changes in carbon stocks in selected non-peat carbon pools. The methodology can be used as stand-alone guidance or in conjunction with other methodologies that have provisions to account for the use of biomass from project activities that are proposed to reduce GHG emissions.

### 7.3.5 Assessment of GHG emissions at the project sites before and after rewetting

A monitoring protocol is required to outline the method and procedure for assessing GHG emissions from drained and rewetted peatlands, which is incorporated into the methodology. The protocol is used for the development of a monitoring report for the further verification of emission reductions achieved. To assess the emission reductions generated through the rewetting of a peatland, it is necessary to describe the situation before rewetting as a starting point for the prediction of future baselines, and to monitor the development of the area after rewetting (at intervals described in the monitoring plan of the project document, PD). The first monitoring survey after rewetting must be conducted prior to the verification of the first monitoring report.

As GHG fluxes are assessed using GESTs, the monitoring plan focuses on detailed mapping of vegetation as well as on measuring water levels. Separate chapters are dedicated to the assessment of carbon sequestration and loss in above and below ground living wood biomass, carbon stock in the dead wood and carbon sequestration and loss in litter. The proposed schedule of work and execution periods is also included in the monitoring protocol. The two types of assessment include:

- Before rewetting: map of current vegetation and location of vegetation relevés, map of shunt species cover, database with data of vegetation relevés, table with names and area of vegetation types, map of sample plots with data for tree species, yield classes, age, density; calculated sequestration of living biomass and carbon stock of litter/dead wood, table with water level monitoring data and a map of water level measurement points;
- After rewetting: map of current vegetation and vegetation relevés, map of shunt species cover, database with data of vegetation relevés, table with names and area of vegetation types, map of sample plots with data for tree species, yield classes, age, density; calculated sequestration/loss of living biomass and carbon stock change of litter/dead wood, table with water level monitoring data and a map with additional water level measurement points.

Monitoring of vegetation and water levels was carried out once before (mainly during 2009/2010) and once after rewetting (in 2011) at the BMU-ICI project sites. It was implemented mainly by scientists of the National Academy of Sciences of Belarus with support by a CIM vegetation expert (see box 27). Data from vegetation relevés are also used to assess plant biodiversity before and after rewetting as well as to design sampling schemes for invertebrate monitoring stratified for GESTs.

### 7.3.6 Strategy to sell carbon credits

During the BMU-ICI project, a carbon investment opportunity ('Belarus peatland rewetting project') was elaborated. It provided a brief description of the project proposed, its aims and purposes, including the national and international project cycle, some facts about other rewetting projects already carried out in Belarus and their benefits, estimated emission reductions and feasible voluntary emission reductions. It also outlined the time schedule and options for project implementation, prospective implementation costs and prognosis of advance payments structure, options and guiding features of the carbon transaction structure, as well as project proponent structure.

The aim of the strategy was to inform foreign companies about the core facts of the proposed project and its carbon potential, with a view to investment and carbon credit purchase opportunities for the companies. The project consisted of two groups of sites: new sites with predicted rewetting and emission reduction potential, and existing sites, already rewetted by the GEF and BMU-ICI projects, amounting to 15 sites with a total area of 27,460 ha and an annual emission reduction potential of roughly 60,000 t $CO_2$-eq. within 10 years. The project is planned to be implemented in 2012, with support of the Belarusian Ministries of Forestry and of Environment and based on a newly established Centre of Sustainable Peatland Management (see chapter 7.7).

### 7.3.7 Demonstration of sustainable biomass use from wet peatlands (paludiculture)

The use of biomass from wet peatlands has been tested at Sporava (see box 24 in chapter 6.4) to build up a functioning, self-sustaining and profitable enterprise, demonstrating alternative sustainable land use on wet and rewetted peatlands and its benefits for climate mitigation and biodiversity protection.

The biomass is used to produce fuel briquettes, which can substitute peat briquettes commonly used in rural areas around the peatland. Apart from financing the necessary habitat management, this increases fuel sustainability and income generation within the community and the replacement of fossil fuels additionally contributes to climate protection. This should serve as a demonstration project that could be replicated in other rewetting projects in Belarus and abroad. It is the first time that such an activity has been implemented in Belarus and builds on recent experience from Poland and Germany (see chapter 6.4). The climate mitigation effect of this activity could be estimated when using an approved United Nations Framework Convention on Climate Change – Clean Development Mechanism (UNFCCC CDM) methodology. This will be pursued by an EU-AID project on 'wetland energy', a follow-up project to the BMU-ICI project (see box 23 in chapter 6.4).

## 7.4 Biodiversity actions

Franziska Tanneberger & Rob Field

### 7.4.1 Introduction

The main areas of biodiversity-related work in the BMU-ICI project included:
- Monitoring of flora and fauna before and after rewetting at project sites (including the development of a monitoring protocol);
- Literature studies;
- Preparations for a Climate, Community and Biodiversity Alliance (CCBA) validation;
- A demonstration project on biomass use from wet peatlands.

### 7.4.2 Monitoring

Monitoring of flora and fauna was carried out once before (mainly in 2009/2010) and once after rewetting (mainly in 2011). It comprised monitoring of :
- Vegetation (via greenhouse gas emission site type (GEST) mapping, see chapter 3.4 and 7.3);
- Birds;
- Invertebrates (ground beetles).

At two sites (the bog Dalbeniski and the fen Vyhanascanskaje), the bird communities at non-rewetted control sites were monitored in the baseline situation (S. Levy pers. comm.).

Vegetation monitoring was not conducted on its own, but as part of GESTs monitoring (see chapter 7.3). Plant species number, diversity and the occurrence of target species were recorded from vegetation relevés conducted for the identification of GESTs. A list of potential target species was provided and field workers were asked to pay particular attention to these (partly inconspicuous) species.

Bird monitoring included:
- An early morning transect survey based on standard Breeding Bird Survey methods already used in Belarus (with a transect of at least 2 km length and recording the location of birds in distance bands from the route of the transect; the distance bands were (i) within 25 m of the transect, (ii) between 25 and 100 m of the transect, and (iii) over 100 m from the transect; all birds, including those flying, were recorded);
- A mid-day raptor survey (to monitor, for example, Greater Spotted Eagle *Aquila clanga*);
- An evening survey for crepuscular species (e.g. Corncrake *Crex crex*, Rails *Porzana* spp., Little Bittern *Ixobrychus minutus*, Bittern *Botaurus stellaris,* Great Snipe *Gallinago media*) and possibly Aquatic Warbler *Acrocephalus paludicola.*

The distance of transect walked was adjusted to account for site area. On sites up to 2,500 ha, a single transect of around 2 km was walked. Monitoring at larger sites required additional transects (selected at the same area:length ratio) to be walked in order to achieve a reasonable representation of the site. In such cases, the locations of sampling units were determined using a sub-sampling protocol (i.e. over-laying a 4 km$^2$ grid onto each site and selecting at random, two 4 km$^2$ grid cells to be the sampling units for starting the two transects). For consistency, the grid cells immediately southwest and immediately northeast of the central 4 km$^2$ grid cell were used, unless impossible due to logistical constraints. Transect routes were identified on maps using Global Positioning System (GPS) so that future visits could be made to exactly the same locations. Two visits were performed, the first in the second half of May and the second visit in the first half of June, but at least two weeks apart.

Invertebrate monitoring used a stratified sampling design according to vegetation types mapped in the course of GEST mapping. Monitoring began in 2010 on seven sites and included all major vegetation types recorded before rewetting. Invertebrates were caught in pitfall (Barber) traps with 15 traps per plot, used at least once between May and September for 100 trap-days. A particular focus was given to Carabid beetles, as these species are a good indication of changes in water conditions, and are well-studied in Belarus.

To date, it is only possible to draw preliminary conclusions about the effects of rewetting from the project sites Obal and Hrycyna-Starobinskaje, which were rewetted in 2009 by the Global Environment Facility (GEF) project (see also chapter 2.4 and 4.2). Based on data from the GEF project and after-rewetting data sampled in 2010 within the BMU-ICI project, some initial conclusions can be drawn: Whilst vegetation composition on rewetted areas changed in favour of species indicating wet conditions, it remained relatively stable in non-rewetted areas. On bird monitoring transects on rewetted sites, the number of bird species typical for peatlands and shore and water conditions increased between 2008 and

2010 and those typical of forests and dry open areas decreased. It is expected that a trend towards wetland species will increase in the future. Since such developments take time, it is aimed at continuing the monitoring after the end of the BMU-ICI project.

One of the planned outcomes of the project was that characteristic wetland species, e.g. the globally threatened Aquatic Warbler *Acrocephalus paludicola* (see box 8 in chapter 4.1) and Greater Spotted Eagle *Aquila clanga*, would increase. Due to logistic and legal constraints, it was not possible to rewet a large area of drained fen, and only three of the eleven project sites were fens:
- Horeuskaje (198 ha, overall area 20,000 ha) has a potential for both species since it is located close to Aquatic Warbler breeding sites Dzikoje and the Jasielda floodplain and to Greater Spotted Eagle breeding sites);
- Hrycyna-Starobinskaje (3,505 ha) has a potential for both species since it is located within 30–50 km from the nearest Aquatic Warbler breeding site and as Greater Spotted Eagles could build their nests on the mineral islands, if they are not disturbed, their nearest breeding sites would be 11 km to the West and 25 km to the South; and
- Poplau Moch (415 ha) has no potential for Aquatic Warblers (outside the distribution range in Belarus) and only small potential for Greater Spotted Eagles (and only if wet small-leaved forest occurs and human disturbance is low; the nearest breeding site being only in Biarezinski zapaviednik at > 150 km distance).

While the concrete project outcome for the target species is rather small, there are two further aspects why the BMU-ICI project benefits the target species indirectly:
- Building on the efforts of the GEF project, the precedent of peatland restoration and sustainable management in Belarus has been set (by two new state standards on the ecological restoration of peatlands: the development of methodical guidelines on the ecological restoration of degraded peatlands (Kozulin et al. 2010a); and the development of national documents on the sale of emission reductions from rewetted peatlands on the voluntary carbon market), which will hopefully open the door to the rewetting of a much greater number of fen sites in the future;
- By implementing the biomass pilot activities in Sporava, the project will demonstrate the feasibility of using biomass from wet peatlands, thus aiding the reintroduction of mowing and vegetation management at Aquatic Warbler sites, which are currently declining in suitability due to abandonment and increased eutrophication (see below and chapter 6.5).

Lessons learnt from biodiversity monitoring include that:
- Understanding of the potential of restoration to deliver priority habitats and species (literature review) is crucial for selecting rewetting sites, assessing potential project benefits, and designing a monitoring scheme;
- All biodiversity monitoring activities should aim at spatial coherence between biodiversity monitoring areas and GEST monitoring areas (i.e. locations of hydrological measurement devices and of vegetation assessment);
- On a country-wide scale, non-representative and biased site selection is less likely to benefit target species and habitats.

### 7.4.3 Literature studies

Since rewetting of peatland is a new concept in Belarus, an international literature survey (Tanneberger 2011) was undertaken to describe the long-term effects of rewetting on peatland biodiversity. A database of information on the study region: peatland type and condition before rewetting; treatment; studied taxa; number of years since treatment; and whether control sites have been included in the study design or not, was developed. It contains details of 100 studies, with most (94) focusing on plants, far less on animals (e.g. 10 studies on birds; see also chapter 4.2). The search also yielded a greater number of studies from bogs than from fens. Few studies include control sites, and the bulk of literature describes effects after less than 10 years rewetting. Findings from the literature study were used to derive management recommendations for the project sites (see chapter 8.2–8.7).

### 7.4.4 CCBA validation

The Climate, Community & Biodiversity (CCB) Standards of the Climate, Community & Biodiversity Alliance (CCBA) identify land-based projects that are designed to deliver robust and credible greenhouse gas reductions whilst also delivering net positive benefits to local communities and biodiversity. Use of the CCB Standards requires that independent, accredited auditors determine conformity with CCB Standards at two stages, at validation and at verification. A CCB validation is an assessment of the design of a land-based carbon project against each of the CCB Standards criteria. A CCB verification is an evaluation of a project's delivery of net climate, community, and biodiversity benefits against the project's validated design and monitoring plan. Verification must be performed at least every five years.

Preparations for validation and verification of the project started in 2011. Initial results include:
- Checking of existing successful CCB Project Design Documents (PDDs);
- Short-listing of two sites for which documentation will be prepared (Zada and Dalbeniski raised bogs);
- Forming a project team of project staff and experts from the National Academy of Sciences of Belarus.

Draft PDDs for the two short-listed sites are scheduled for September 2011 (S. Valasiuk pers. comm.).

### 7.4.5 Biomass activities

The use of biomass from wet peatlands is being tested at Sporava (see chapter 6.4, box 27 and also chapter 7.3) to build a functioning, self sustaining and profitable enterprise, demonstrating alternative sustainable land use on wet and rewetted peatlands and its benefits for climate mitigation and biodiversity protection. Sporova is a stronghold of Aquatic Warblers and other internationally important fen-mire biodiversity and urgently requires vegetation management to stop habitat deterioration. Vegetation (planned: 63 ha in 2011 and 325 ha in 2012) will be mown using a 'ratrak', a specialist mowing machine for the removal of biomass, driving on caterpillar tracks, which minimise damage to the delicate peat soil. The machine to be used in Belarus has been purpose-built, based on the experiences of several groups working in wetland management, especially the Polish Society for the Protection of Birds – OTOP BirdLife Poland (Lachmann et al. 2010). The first results from Biebrza National Park in Poland indicate that such management has a positive impact on Aquatic Warbler breeding success (L. Lachmann pers. comm.). The outcome of the biomass pilot project in Belarus will be accessible to all interested parties (e.g. to other key protected areas and rewetting projects with biodiversity objectives in Belarus and elsewhere where similar overgrowth problems exist).

## 7.5 Policy actions

Hans Joosten

As the current structure of the Kyoto Protocol makes accounting for rewetting of peatlands very unattractive (see chapter 5.4), collaborators of the BMU-ICI project supported the delegation of the Republic of Belarus in trying to improve the position of peatland rewetting in the UN Framework Convention on Climate Change (UNFCCC) and its Kyoto Protocol. This lobbying work could build on and add to much of the preparatory and complementary work carried out by other countries and by environmental non-governmental organisations (see chapter 5.2). Project involvement started during the first 2009 UNFCCC meeting in Bonn, when Belarus and Iceland were supported in drafting a joint text for the discussions in the Kyoto Protocol Working Group (AWG-KP): "'Wetland restoration' is a direct human-induced activity to reduce emissions of greenhouse gases (GHG) and thus limiting carbon stock degradation by restoring degraded wetlands. If elected, the activity will include accounting for human-induced drainage of wetlands that results in increased emissions of GHG and reduction of carbon stocks." This text was the starting point for further discussions on the concrete definition and modalities of the proposed activity.

The involvement of project collaborators (predominantly Hans Joosten, but also Irina Voitekhovitch and Viktar Fenchuk) in the Belarus delegation to the UNFCCC in the years to follow allowed direct interventions in the informal and formal discussions. In close cooperation with Wetlands International and its network, an effective exchange of information to the UNFCCC delegates was started (Table 39). This direct participation in the UNFCCC negotiation process offered the unique opportunity to address UNFCCC parties' concerns and misconceptions on peatlands almost 'on demand':
- To address the concern that peatland rewetting could not be monitored, reported and verified, an overview of novel Measurable, Reportable, Verifiable (MRV)-methods was produced (Joosten & Couwenberg 2009, cf. Couwenberg et al. 2011);
- To illustrate the feasibility of wall-to-wall land-based accounting, an analysis of National Inventory Submissions of 10 European countries (including Belarus) was presented (Barthelmes et al. 2009);

Table 39: Peatland lobbying at UNFCCC: Participation of BMU-ICI project collaborators, as delegates of the Republic of Belarus, peatland orientated information booths, expert meetings and side-events, lead organisers and publications launched. GEC: Global Environmental Centre, Kuala Lumpur; WI: Wetlands International, Ede; ECA: Ecosystems Climate Alliance (NGO Coalition); RBY: Republic of Belarus; CIFOR: Center for International Forestry Research, Bogor; IPCC: Intergovernmental Panel for Climate Change.

| Year | Date | Place | Participation | Delegation | Information booth(s) | Expert meeting(s) | Side event(s) | Lead organiser(s) of booth(s)/event(s) | Publications |
|---|---|---|---|---|---|---|---|---|---|
| 2006 | November 6–17 | Nairobi | X | | X | | X | GEC, WI | Hooijer et al. (2006) |
| 2007 | December 3–14 | Bali | | | X | | X | GEC, WI | |
| 2008 | August 21–27 | Accra | | | | | | | |
| 2008 | December 1–12 | Poznan | X | | X | X | X | WI | Kaat & Joosten (2008) |
| 2009 | March 29 – April 8 | Bonn | X | X | | | X | WI, ECA | |
| 2009 | June 1–12 | Bonn | X | X | X | X | X | WI | Couwenberg (2009a), Joosten & Couwenberg (2009) |
| 2009 | August 10–14 | Bonn | X | X | | | | | Barthelmes et al. (2009), Couwenberg (2009b) |
| 2009 | Sept. 28–Oct. 9 | Bangkok | X | X | | | | | |
| 2009 | November 2–6 | Barcelona | X | X | X | | X | WI | Joosten (2009d) |
| 2009 | December 7–19 | Copenhagen | X | X | X | | X | WI, RBY | Pena (2009) |
| 2010 | April 9–11 | Bonn | X | X | | | | | |
| 2010 | May 31–June 11 | Bonn | X | X | | | | | |
| 2010 | August 2–6 | Bonn | X | X | X | | X | WI | |
| 2010 | October 4–8 | Tianjin | X | X | X | | | ECA | WI & University of Greifswald (2010) |
| 2010 | Nov. 29 – Dec. 11 | Cancun | X | X | X | | X | WI, CIFOR, IPCC, RBY | |
| 2011 | April 4–8 | Bangkok | X | X | | | | | |
| 2011 | June 6–17 | Bonn | X | X | | | X | ECA, CIFOR, IPCC | |

- To arrive at better estimates for the GHG emission reduction potential by peatland rewetting, a critical analysis of Intergovernmental Panel for Climate Change (IPCC) emission default values for managed peatlands was made and better values proposed (Couwenberg 2009a, 2011);
- To demonstrate the enormous role drained peatlands play in global GHG emissions, a review paper on Southeast Asian peatland emissions was prepared (Couwenberg et al. 2010);
- To address the concern for methane emissions from rewetted peatlands, a review of state-of-the-art knowledge was prepared, including new default value proposals for tier 1 and tier 2 approaches (Couwenberg 2009b);

- To inform the discussion on national reduction targets, the first ever overview of peat carbon stocks and drainage related $CO_2$ emissions of all countries of the World separately, both for 1990 and 2008, was produced (Joosten 2009d);
- To give insight into the opportunities and obstacles with regard to reporting and accounting for GHG fluxes from terrestrial ecosystems, with special attention to peatland rewetting, a Question & Answer booklet was distributed among negotiators (Wetlands International & Greifswald University 2010).

These publications, the myriads of policy briefs and recommendations issued, especially by Wetlands International, and many face-to-face meetings with negotiators, helped to remove concerns and convinced the UNFCCC delegates of the feasibility to include peatland rewetting under the Kyoto Protocol and under Reducing Emissions from Deforestation and forest Degradation+ (REDD+). Inclusion under these UNFCCC instruments would open up perspectives for financing peatland rewetting through the compliance carbon market.

> On 6 June 2009, during the second UNFCCC meeting of 2009 in Bonn, Vladimir Tarasenko, head of the Belarus delegation, suddenly passed away. Born in 1967, Vladimir Tarasenko was key in the elaboration of documents for the accession of Belarus to the Kyoto Protocol, for the development of the National Kyoto Action Plan, and for building the institutional and legal framework for climate change mitigation in Belarus, including the promotion of peatland issues under the UNFCCC and the Kyoto protocol. Under his leadership Belarus initiated the process to include peatland restoration under a future UNFCCC policy.

## 7.6 Communication and awareness raising

Lydia Pshenitsyna

### 7.6.1 Introduction

Communications in such a large and innovative project like the BMU-ICI project are vital to ensure that a variety of audiences at the local, national and international level understand the benefits of peatland rewetting and its significant positive impacts on climate change mitigation, biodiversity conservation, and local community development. A communication strategy was developed at the beginning of the project to ensure that all key internal and external stakeholders were provided with accurate information about the objectives, structure and activities of the project. The strategy also sought to ensure that key messages and results were conveyed in effective accessible formats and in a timely manner.

### 7.6.2 Inside Belarus

Articles were produced for Belarusian national and local newspapers. Radio interviews were carried out to promote the concept of restoring peatlands in Belarus. Local media coverage of site-based work was widely used to ensure local understanding, and to start building local ownership and support for the project.

The project established cooperation with several schools and other educational institutions in Belarus (see box 29). It was presented at a forum of ecological non-governmental organisations (NGOs) in 2009 and in 2011 in Belarus and received wide interest and support for peatland restoration in Belarus. Photo exhibitions of eye-catching peatland pictures were shown in 2010 and 2011 in Belarus. To complement the pictures, famous musicians were invited to synchronize the exhibitions. A virtual literary photo exhibition was displayed on one of the most popular news pages (www.tut.by) and was dedicated to the international wetlands day on the 18 February 2011. The photo-exhibition stimulated substantial public interest in the project.

In October 2011, a final project conference was organised with the involvement of a wide range of international and local stakeholders from all over the world. Two short video films and the Best Practice Handbook were disseminated at the conference in both the Russian and English language.

### 7.6.3 Outside Belarus

In spring 2010, an international project website (www.restoringpeatlands.org) was launched. The website reflected the three-way partnership and included both the Belarus project and the related Ukrainian peatland rewetting project (see chapter 7.9). The website explained the objectives, structure, activities, results and outputs of the project, conveyed key advocacy messages, and recognized the donor and partner commitments.

Printed materials about the project included leaflets, booklets, and CDs for distribution at various levels. They covered topics such as peatland restoration in Belarus, new international standard for peatland rewetting, the greenhouse gas emission site types (GEST) approach, and paludiculture. Many of these materials can be downloaded from the existing project website. Media coverage was gained outside Belarus e.g. by an article in the German weekly journal 'Der Spiegel' (2008) and by presentations in German radio shows.

The project was presented at international scientific meetings in Germany, UK, Czech Republic, Poland, Turkey, Austria, and Ukraine, including, for example, the 4th Annual Meeting of

**Box 28**

**Speak about peat!**

Andrea Strauss

Used wisely, meetings accompanying large-scale projects can do much more than updating partners on project success and agreeing on next steps. They may help to involve a wider expert audience and other sectors into implementation and the broader application of project approach and best practices. They can serve to build networks, and to engage in public outreach. The special atmosphere of the International Academy for Nature Conservation offers ideal surroundings for intense and effective meetings. Situated on the Isle of Vilm in northeastern Germany, the Academy offers facilities for meetings for up to 55 participants, as well as for excursions in the region. The Academy provides a forum for debate on international nature conservation issues and is open for cooperation.

The BMU-ICI project was accompanied by a workshop series, with three meetings on 'The future of peatlands in Central and Eastern Europe – their ecosystem services and sustainable use in the view of climate change' between 2008 and 2010. The seminars were organised by the German Federal Agency for Nature Conservation (BfN) and its International Academy for Nature Conservation, in cooperation with the Michael Succow Foundation (MSF) and the RSPB. The workshops were mainly funded by the German Federal Ministry for the Environment, Nature Conservation and Nuclear Safety (BMU). Each workshop was attended by about 40 participants, mainly experts, from Belarus, Ukraine and the Russian Federation, but also experts and practitioners with scientific, management, and land use backgrounds from Poland, Germany, the Netherlands, and Turkey.

The meetings facilitated an exchange between the Eastern European countries and a wider expert audience on the lessons learnt from the project and assisted in setting up a similar project in Ukraine. New approaches and methodologies with regard to peatland restoration with climate and biodiversity benefits, and on funding through carbon trading were presented. All meetings dealt with technical and scientific questions of peatland rewetting, sustainable management of rewetted areas and sound scientific methods for measuring and quantifying greenhouse gas emissions avoided by peatland rewetting. During excursions to rewetted peatlands in northern Germany, approaches in emission measurements (Fig. 67 Colour plates III) as well as practical experiences in the management of rewetted peatlands were discussed.

The workshop series reacted to current issues in the project and each workshop built on the previous one. The first meeting focused on the criteria for choosing rewetting sites. Participants also highlighted the need to include peatland restoration into carbon trading. The second workshop in 2009 dealt with questions concerning leakage and with the progress in completing the Verified Carbon Standard Peatland Rewetting and Conservation. Meeting results were used by partners in preparation for the UNFCCC Conference of the Parties (COP) in Copenhagen. A feasibility study on the use of biomass from rewetted peatlands in Belarus was also presented to highlight how the positive climate effects by peatland rewetting may potentially be enhanced by adapted management of the sites. The last workshop dealt with the subsequent monitoring of rewetted sites. It also enabled a debate on potential principles for carbon offsetting of peatlands and a more in-depth discussion on the concept of paludiculture.

The cooperation established will in future benefit peatland conservation and restoration in Belarus, Ukraine, and Russia. Meeting evaluations proved a need for additional training and exchange on scientific and methodological questions of peatland rewetting and sustainable use of rewetted sites, as well as on decision guidance for carbon offsetting from rewetted peatlands.

## 7.6 Communication and awareness raising

> **Box 29**
>
> **Cooperation with educational institutions in Belarus**
>
> Annett Thiele
>
> A synergy between project activities and pupils' understanding of the values of peatlands was achieved by setting up several scientific youth projects with schools and other educational institutions in the Brest and Minsk regions.
>
> A very active school in the village of Zdzitava (Paliessie) was interested in helping the BMU-ICI project from the first day they heard about it during a presentation that was held in Minsk in 2009. The school is situated on the northern border of one of the most famous protected areas in Belarus, the Sporauski zakaznik, which holds the second biggest number of Aquatic Warblers *Acrocephalus paludicola* in Belarus (see chapter 4.1). During the first visit to the school, together with Prof. Michael Succow, children reported about school projects such as cutting willows and reed to keep the mire open and to fuel the local heating facility with harvested material and sang a song they composed themselves about their beautiful village. They live with the peatland throughout the year: In winter while cutting willows, in spring while counting flowering species, and in summer while helping their parents mowing part of the land to have bedding and fodder for their horses. The Zdzitava school children were very happy and proud when a group of German and Belarusian scientists came to ask for their assistance in peatland investigations outside their school door. With support of schoolteachers and pupils, a micrometeorological station (which is needed to model greenhouse gas fluxes) was installed at the school ground, and since 2009, children have been using water level measurement devices for the project (Fig. 68 Colour plates III). With this work they have been the first pupils in Belarus to take part in scientific competitions in the field of peatlands and climate change. They won several prices at the local, regional, and country level by presenting the results of their investigations.
>
> A second successful cooperation was set up with the Ecological Centre for Children and Youth in Barisau, Minsk region. Members of this centre helped to assess greenhouse gas emissions from agriculturally used and rewetted peatlands. They took part in several international conferences for young scientists and won prices in Moscow, St. Petersburg and will soon take part in an international conference in Sweden.

the European Chapter of the Society of Wetlands Scientists (SWS) in Erkner/Germany in 2009, the International Union for Conservation of Nature (IUCN) conference 'Investing in peatlands' in Durham/UK in 2010, and the European Geoscience Union (EGU) General Assembly in Vienna/Austria in 2011. Three workshops, organised through the International Academy for Nature Conservation on the Isle of Vilm, Germany, in cooperation with project partners took place in 2008–2010 and were attended by representatives from NGOs, governments, and scientific institutions from all over Europe (see box 28).

The publication of several scientific papers and abstracts linked to project activities contributed substantially to the ongoing scientific debate on peatland restoration. Examples included articles in journals such as Peatlands International (2009), Journal of Sustainable Energy and Environment (2011), and Hydrobiologia (2011).

The project was also presented at multiple United Nations Framework Convention on Climate Change (UNFCCC) meetings, e.g. in Poznan, Poland (2008) together with the Government of Belarus and with a 'Belarus official side event' in Copenhagen, (2010; see also chapter 7.5). The international broadcaster Deutsche Welle created a seven minutes documentary on the project in 2010, showing the beauty of peatlands, rewetting activities on project territories, the scientific measurements and the first rewetted peatlands from a birds eye view, which was shown in 30 countries. A second documentary created by the Royal Society for the Protection of Birds (RSPB) in 2011 summarises the success of the project on all its diverse fields.

## 7.7 Capacity building

Merten Minke, Franziska Tanneberger, Wendelin Wichtmann

### 7.7.1 Introduction

In the BMU-ICI project, the assessment of greenhouse gas (GHG) emission reductions from peatland rewetting is a central task to realise and quantify 'carbon credits' and to create a long-term income stream to finance the future management of existing sites and rewetting of new sites. The greenhouse gas emission site types (GEST) approach was the key mechanism that was used to assess GHG emissions in the project and the model was tested, adapted, refined, and verified for Belarusian peatlands. This included measuring GHG emissions at field sites and studying the relationship between vegetation and water levels in Belarus (see chapter 3.3 and 3.4). Neither annual GHG emissions nor vegetation forms have been previously studied and published for Belarusian peatlands. In addition, expertise and techniques to assess vegetation forms and measure the GHG exchange in peatlands were lacking in Belarus before the project. Capacity building through the BMU-ICI project focused on providing training in measuring and assessing GHG emissions from peatlands for Belarusian graduates. Trainees were supported by two German experts from the German Centre for International Migration and Development (CIM) working in Belarus throughout the project (see box 27 in chapter 7.3).

Several German students received practical training through the project and one Bachelor thesis was prepared at Greifswald University (Germany) concerning the characteristics of biomass from wet peatlands of Belarus. With regard to peatland rewetting, two seminars for local engineering specialists on 'principles and methods of peatland rewetting' with a special focus on peatland hydrology, were held in autumn 2010 and spring 2011, both in Belarus.

### 7.7.2 Training made through the BMU-ICI project

In order to build up the necessary capacity, a training program for Belarusians was developed at Leibniz Centre for Agricultural Landscape Research (ZALF), Müncheberg and Greifswald University, in Germany. The training did not only seek to transfer knowledge, but to also establish an interdisciplinary workgroup and a gas analytic laboratory in Belarus to complete all the research required during and after the project with respect to assessing GHG emissions from peatlands.

ZALF offered training for Belarusians for:
- Two junior scientists to conduct, analyse and evaluate trace gas measurements (one year);
- One 'electronic' expert to install, operate and repair mobile $CO_2$ measuring sets and weather stations (six months);
- One 'gas analytic' expert for operating and repairing a stationary, automatic, gas chromatographic system for the analysis of $CH_4$ and $N_2O$ (three months).

Greifswald University offered a Master course in landscape ecology and nature conservation (two years) and a short-term training in skills related to landscape ecology (one month) for Belarusian graduates.

The training at ZALF was crucial for the successful implementation of the project since the CIM GHG expert (see chapter 7.3) alone was not able to carry out all of the GHG measurements at a sufficient number of locations. During the project the trained young scientists increased their skills and had the opportunity to improve their qualification through a PhD course, so that Belarus would gain long-term capacity for the study of ecosystem GHG exchange. Another young scientist was invited to directly assist the CIM expert in all aspects of the GHG research, from site selection and preparation to measuring and was thus trained directly 'on the job'.

The Master course at Greifswald University aimed to support the CIM vegetation expert in developing vegetation forms for Belarusian peatlands and to build long-term capacity in peatland ecology and vegetation science. However, because the Master course ran for two years, the trained Belarusian scientists were not able to assist during the first two years of the project. Therefore another young scientist was employed directly to assist the CIM expert and was trained 'on the job' by the CIM expert in all aspects of assessing vegetation types and applying the GEST concept.

### 7.7.3 Selection and posting of trainees

Finding appropriate candidates for the ZALF training took some time. Announcements in Russian and English language were made in autumn 2008 in Belarusian universities and institutes. In December 2008, the first interview was organised with the participation of APB-BirdLife Belarus (Viktar Fenchuk, Merten Minke) and ZALF (Jürgen Augustin) at the Institute for Nature Management (National Academy of Sciences (NAS) of Belarus) where the GHG assistant and the future expert for gas analytics were selected. In March 2009 both future junior GHG scientists were identified but they could not start their training at ZALF until August 2010, because they had to graduate first. An expert in electronics with experience in setting up meteorological stations and carbon dioxide measuring techniques was recruited via the APB office at the Scientific-Practical Centre for Bioresources (SPC; NAS of Belarus). The future expert for gas analytics was sent to ZALF between April – June 2009 and the future junior GHG scientists between August 2009 – April 2010. A suitable candidate for the Master course at Greifswald University was found by the Belarusian project partner APB just when the project started. However, leaving Belarus for two years was quite difficult because of differences in flexibility between the German and the Belarusian educational systems. This made some ingenuity necessary to make a successful start of the course possible.

### 7.7.4 Content of the trainings

The expert for gas analytics was mainly trained to use the gas chromatograph (set up, calibration, trace gas analyses, transfer processing of data, service and maintenance), but was also introduced to GHG field measurements and GHG exchange calculations. The two junior GHG scientists were trained in GHG field measurements with the chamber method (see chapter 3.3) and GHG exchange calculations, as well as in related tasks like site selection and site preparation, sampling of site parameters, air analyses using the gas chromatograph, and analysing meteorological data. They studied and compared different methods to model the annual balances of $CO_2$, $N_2O$, and $CH_4$ and wrote their Master thesis on that topic.

The 2-year master course in Landscape Ecology and Nature Protection at Greifswald University was attended by international postgraduates and experienced graduates whilst undertaking practical work experience. It is designed to enable students from all over the world to assess landscapes and their functioning, to identify and evaluate changes to and the potentials of ecosystems, and, finally, to develop sustainable land use concepts. Based on multidisciplinary and individual mentoring, students are prepared for careers in international administration in non-governmental organisations as well as in research institutes and private companies. The multidisciplinary curriculum imparts theoretical knowledge and practical experience in ecology, biology, geosciences, wetland ecology and restoration, environmental ethics, communication, and economy. It emphasises all aspects of mire ecology e.g. in the following modules: Mire Ecology, Restoration Ecology (both are obligatory in this project), Palaeoecology, Global Change, Aquatic Ecology and Vegetation Ecology, which were attended by the Belarusian student. The one-month training course by Michael Succow Foundation (MSF) and Greifswald University for the 'assistant' to the CIM vegetation expert in early 2010 included courses in Geographic information system (GIS) & Remote Sensing, GESTs, project management, soil sample (C/N) analysis, and statistics.

### 7.7.5 After the training

The expert for gas analytics returned to Belarus at the end of June 2009 and performed her first gas chromatographic analyses in Minsk in July 2010, when the gas chromatograph was installed (which was one year later than planned due to difficulties in importing it). During the first year she was mainly helping with site selection and preparing documents for the import of measuring techniques to Belarus. The junior GHG scientists returned to Minsk in April 2010, when all measuring equipment (apart from the gas chromatograph) had been imported. They helped in preparing the measuring chambers and setting up the research sites. Both were lucky to be included in the CIM program as 'returning experts' that encourages foreigners who have secured additional qualifications in Germany to return to their home country through an addition payment to cover staffing costs for two years.

After the successful completion of the MSc thesis about the 'Relationship between vegetation and water level in Jelnia bog (Belarus) and comparison of water level measurement approaches' (Broska 2010), the graduate returned to Belarus. She now works at the Institute of Experimental Botany (Laboratory of Geobotany and Vegetation Mapping) of the National Academy of Sciences of Belarus and finished her last year of the postgraduate course Botany at the Belarusian State University, with a special focus on peatlands.

All trained young scientists are interested in completing a PhD on different aspects of the GEST research. One will study the importance of peat and plant characteristics for $CH_4$ emissions immediately after rewetting, others will try to improve or develop new models for $CO_2$, $CH_4$, $N_2O$, and the vegetation specialists will study the vegetation succession after rewetting or abandonment of peatlands to develop a tool for predicting future GHG emissions. ZALF and Greifswald University will support their studies after the end of the project.

### 7.7.6 Beyond the end of the project

It was vital to ensure that capacity was maintained after the end of the project. To keep the personnel and technical capacity, all project partners agreed to establish an International Centre for Sustainable Peatland Management at the SPC at the National Academy of Sciences of Belarus (see Fig. 69). This will be the future key scientific entity for quantifying GHG emission reductions from peatland restoration, testing sustainable peatland management systems, and preparing $CO_2$ certificates from peatland restoration.

In parallel, the SPC plans to create a 'laboratory for the gas dynamics of ecosystems'. The laboratory will integrate the group and the gas analytical equipment. It will be linked to the International Centre for Sustainable Peatland Management and will carry out research into the consequences of peatland management on carbon, nitrogen, phosphorous, and greenhouse gas ($CO_2$, $CH_4$, $N_2O$) budgets, but it will also be open to work in other ecosystems types, such as in forest.

## 7.8 Lessons learnt

Jack Foxall

### 7.8.1 Introduction

The BMU-ICI project in Belarus has been, worldwide, the first peatland-based climate mitigation and biodiversity restoration project that has been designed, implemented and brought to the market. In order to use the voluntary carbon market, the project had to develop and employ new scientific, technical and financial expertise. The experimental and innovative project provided new insights into peatland rewetting, assessing greenhouse gas emissions, and eventually selling carbon credits. Knowledge and understanding

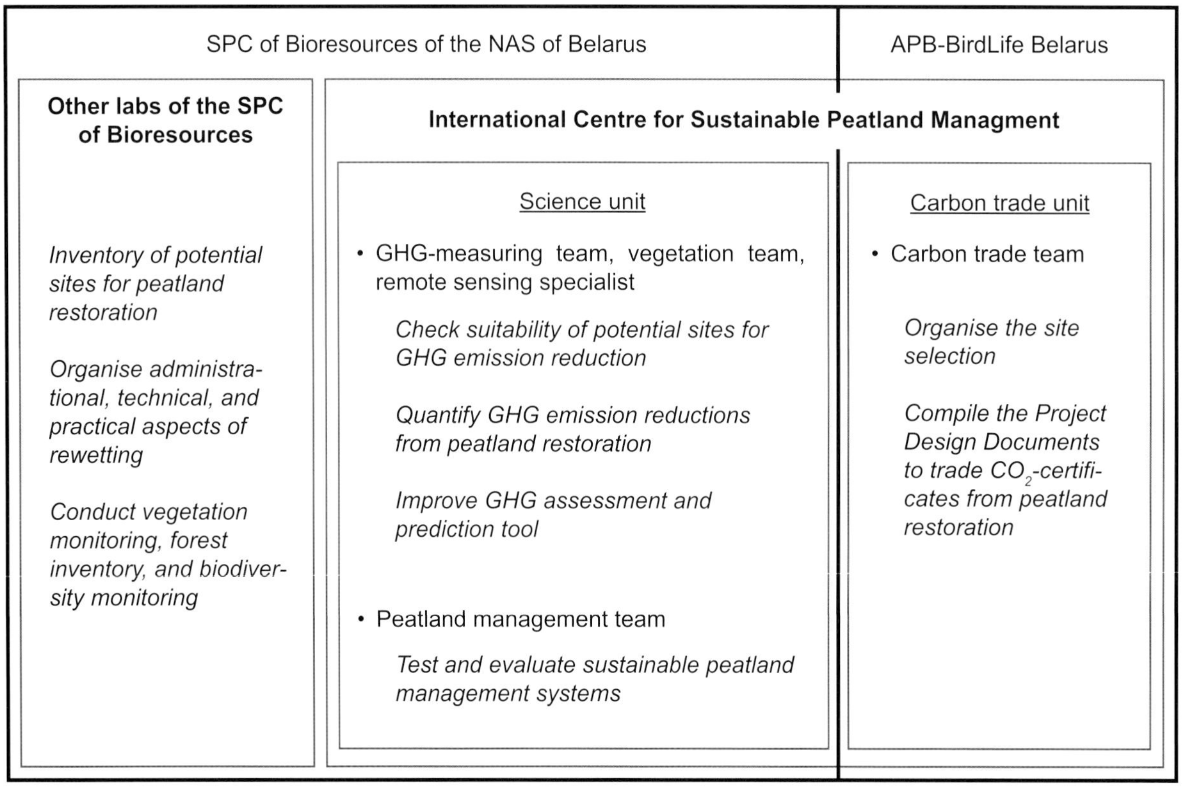

Fig. 69: Responsibilities of the planned International Centre for Sustainable Peatland Management after the end of the project.

## 7.8 Lessons learnt

has been furthered at all levels, from local to international, and new documentation has been produced which will not only serve Belarus beyond the end of the project, but also other countries who wish to implement a similar approach. This chapter looks at some of the key lessons learnt from the project. It first deals with project successes and highlights the components that made the project work well. It then lists the challenges faced by the project, all of which should be considered in the development of future projects of this kind.

### 7.8.2 Project successes: lessons learnt during project implementation

- A wide range of expertise was required to implement the project, drawing not only on the skills of project staff, but also of international experts and consultants. Key areas of essential expertise and scientific competence included: carbon and peatlands, biodiversity, accessing carbon markets, communications, community development, engineering and construction, economics, business strategy and finance;
- Local capacity building was imperative in fields as diverse as greenhouse gas (GHG) measurement and carbon policy, to communications and project management. Training was required to enable Belarusian experts to manage future carbon and biodiversity projects. As the project started from zero, building up capacity in GHG measurements was only possible by support provided by the Centre for International Migration and Development (CIM). On the other hand the project benefited enormously from existing Belarusian expertise in vegetation and peatland sciences and ecology;
- In-country partners were essential to carry out the majority of project implementation. Local organisations with a vested interest in ensuring the long-term success of rewetting are furthermore crucial to sustain and progress the achievements beyond the end of the project;
- Flexibility was key to operating such an innovative project, particularly in terms of the allocation of staff resources and finances. Some activities finished more quickly than expected, others took considerably longer. It was essential to earmark contingency funding throughout the project in order to meet demands unforeseen at the beginning of the project;
- Strong written and verbal communications were essential both between partners and sub-contractors (particularly when conversing in multiple languages), as well as externally to public officials, non-governmental organisations (NGOs), and potential buyers of carbon credits. Breaking new ground in an unknown field required very careful communications planning, education, and public awareness raising;
- Strong project management at all levels was needed to manage the many facets of the project involving a multi-disciplinary team, and the delegation of clear tasks between partners, groups and individuals;
- A visionary attitude by all people involved was required, to consider at all times the long-term impacts of rewetting on the ecosystem, on local populations, and on GHG emissions. Similarly external influences must be anticipated at the political level and on the international carbon credits market. All of these factors are in constant flux;
- Climate project thinking has until now largely been dominated by forest. Peatland restoration for reducing GHG emissions is at an early stage of academic rigour. As a consequence, many new concepts had to be developed and existing concepts adapted to peatland reality. Special attention was required for issues like peat depletion time, forward looking baselines and additionality (e.g. spontaneous filling ditches, beavers). At the start of the project, it was impossible to perform carbon-offsetting projects in peatlands, because there was no standard to judge projects against and no methodologies to assess and monitor emissions from different types of peatlands. A practical methodology is now completed and an international peatland carbon standard under the Verified Carbon Standard (VCS) now exists;
- Biomass from rewetted peatlands can be sustainably managed whilst providing continued GHG emissions reductions and biodiversity benefits. Biomass harvesting and production of biofuels can help to create jobs, contribute to the protection of threatened bird species, provide a sustainable fuel for heating, reduce the pressure on peatlands from peat extraction, and on the long term protect further peatlands from peat excavation;
- Spatial coherence between monitoring activities, i.e. the same monitoring localities for biodiversity and GHG emission assessment, (e.g. hydrology and vegetation as proxies) should be pursued to increase efficiency and facilitate synergies.

### 7.8.3 Project challenges: lessons learnt for future replication

- It appeared to be necessary to plan for and anticipate set-backs and to factor in that breaking new ground can take considerable time. For example, measuring GHG emissons from peatlands required importing equipment that had to pass through strict import controls requiring lengthy paperwork; developing a methodology for assessing emissions using vegetation and water level as a proxy required considerable fieldwork (over at least two years to account for inter-annual variability), substantial study of the scientific literature, and discussion; the new carbon standard required extensive multi-stakeholder and peer review; building capacity and providing training in a new field was time consuming, whereas personnel with the right balance of skills in an emerging field were often hard to source;
- We learned that rewetting sites need to be very carefully chosen for a project of this calibre and scale to achieve the highest possible emission reductions (see also chapter 7.2). In the project proposal stage the volume of achievable emission reductions was calculated as a proportion of the total area of peatlands that could be rewetted. During site selection, it emerged that much less agricultural land was available than originally foreseen and that mainly areas were presented for rewetting that yielded lower than average emission reductions. Sites that show limited overall net emission reductions may still be interesting for rewetting for biodiversity, for providing livelihood opportunities for local people, and even for emission reduction, but not for creating marketable carbon credits;
- Local, regional, and national governmental support proved to be essential to ensure long-term access to peatlands for rewetting, and support to legislative change in favour of peatland rewetting. It is very hard to predict what could change in terms of government legislation but official support should be gained wherever possible;
- At the beginning of the project, an insufficient number of control sites to measure and monitor

---

**Box 30**

**Is peat extraction compatible with peatland rewetting?**

Hans Joosten

Belarus is globally leading in reducing emissions by rewetting drained peatland. On the other hand the country continues peat extraction, uses (and exports) peat on a large scale for fuel and even plans to expand its fuel peat extraction considerably. In judging this paradox the following issues have to be considered:
- Peat is a fossil fuel, similar to coal or oil. Its emission factor is 106 t $CO_2$/TJ against 98.3 for coal (anthracite), 73.3 for fuel oil and 52.2 for natural gas. Replacing other fossil fuels by peat thus increases greenhouse gas emissions;
- Belarus has in the past drained 1.5 million ha of peatlands for peat extraction, forestry and agriculture. Therefore for continued peat extraction no pristine peatlands have to be drained. 36,800 ha of peatlands are currently used for peat extraction;
- Some 800,000 ha drained peatlands are considered to be degraded and largely available for rewetting. Rewetting decreases greenhouse gas emissions. To what extent this rewetting compensates for increased emissions from peat burning depends on the areas and volumes involved;
- Parts of some peatlands are impossible to rewet because of inadequate relief and water conditions. If peat extraction is necessary, it should be concentrated on these areas;
- Harvesting biomass from rewetted peatlands (paludiculture) may provide an economically feasible alternative to peat as a fuel. Biofuels from rewetted peatland share with peat the benefits of energy diversification, national energy independence, and sustainable rural employment, but have a much better greenhouse gas and biodiversity profile.

Trading carbon from peatland rewetting on the voluntary market is, most of all, a matter of long-term reliability. This implies that all plans related to peatland rewetting and peat extraction should be part of a transparent, integral and consistent long term peatland policy.

Fig. 78: The eastern lake in Zada (2010; photo: Semion Levy).

Fig. 79: The former lake in the central part of Zada (2010; photo: Semion Levy).

Fig. 80: The former extraction site before rewetting of Zada (2010; photo: Semion Levy).

Fig. 83: Dry grassland with *Calamagrostis epigeios* before rewetting of Hrycyna-Starobinskaje (photo: Annett Thiele).

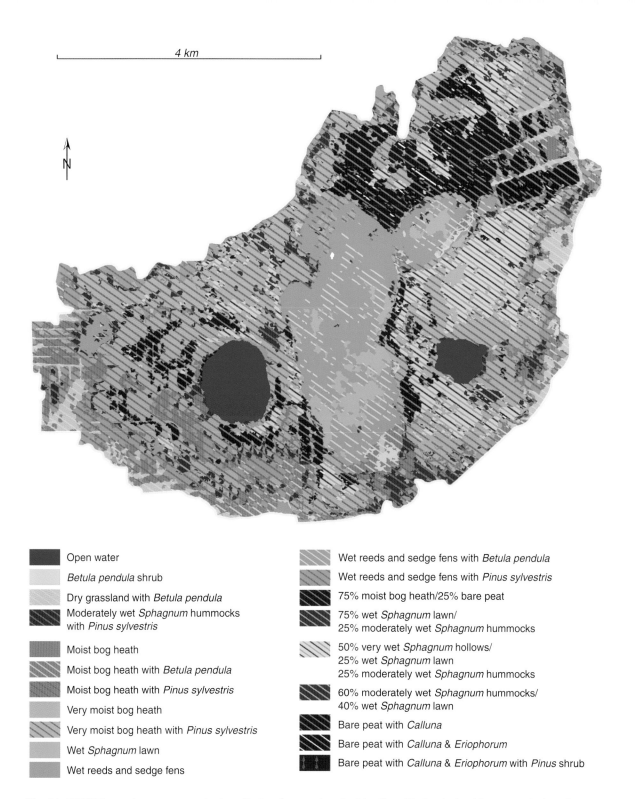

Fig. 81: GEST (greenhouse gas emission site type) map of project territory Zada, derived from spectral angle mapper (supervised classification, based on 48 training sites) and knowledge-based summarization of classes (map: Irina Kokhan & Annett Thiele).

Fig. 84: GEST (greenhouse gas emission site type) map of project territory Hrycyna-Starobinskaje, derived from minimum distance classifier (supervised classification, based on 45 training sites) and knowledge-based summarization of classes (map: Alexei Kozulin & Annett Thiele).

Fig. 85: Dry grassland with *Calamagrostis epigeios* after rewetting of Hrycyna-Starobinskaje (photo: Annett Thiele).

Fig. 86: Hrycyna-Starobinskaje after rewetting (April 2010; photo: Annett Thiele).

Fig. 88: The depleted part before rewetting of Scarbinski Moch (June 2009; photo: Annett Thiele).

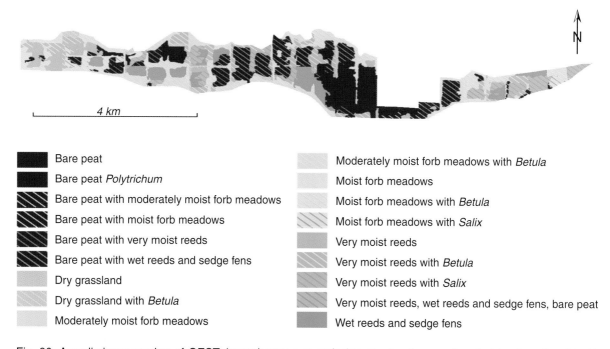

Fig. 90: A preliminary version of GEST (greenhouse gas emission site type) map of project territory Scarbinski Moch, derived from on-site mapping (map: Sergei Tsai & Annett Thiele).

## Colour plates, Part IV

Fig. 89: The former bog lens in the year of rewetting of Scarbinski Moch (2010; photo: Annett Thiele).

Fig. 93: A part of Dakudauskaje overgrown with *Calluna vulgaris* and sparse trees before rewetting, with the peat extraction field in the background (photo: Nina Tanovitskaya).

Fig. 94: Dwarf shrub vegetation of *Ledum palustre* and *Vaccinium uliginosum* with young birches before rewetting of Dakudauskaje (2009; photo: Annett Thiele).

Fig. 95: Newly created pools close to the dam after rewetting of Dakudauskaje (May 2010; photo: Annett Thiele).

Fig. 98: Dammed-up channel near Belaje Lake at Jelnia (photo: Annett Thiele).

Fig. 100: Surroundings of the closed channel at Jelnia, the limited rewetting influence is clearly visible by the strip of cotton grass, fading out in moist bog heath (photo: Annett Thiele).

Fig. 97: Jelnia – one of the country's largest and most beautiful bogs in the northern peatland region (photo: Sergey Plytkevitch).

Fig. 99: Positive development at the fringe of Jelnia through dams installed in 1999–2000. The formerly burnt site with moist bog heath has developed to cotton grass/peat moss-lawns and peat moss hollows (photo: Annett Thiele).

biodiversity were sought. It is imperative for ensuring stringent monitoring that targets and protocols are in place at the very start of the project;
- Planning for the long-term is essential to ensure a sustainable income stream for peatland rewetting and management beyond the end of the grant-financing period. This was not an easy task at the beginning of the project when many factors, e.g. national policies, carbon market opportunities, and emission reductions potential three years later were difficult to foresee. The structures and procedures for the long-term management of a rewetting project for carbon credit generation need to be drafted early in the project and amended as the project develops. At least a year and a half before the end of grant funding a strategy (and action plan) needs to be drafted that provides detail on the exit strategy and thus on the permanence of management, control and verification.

### 7.8.4 Summary of lessons for the implementation of future peatland rewetting projects

The project has demonstrated how it is possible to carry out peatland rewetting for GHG emission reductions and how these reductions can be accounted for on the international carbon market. Future work in this field will benefit from the increased worldwide attention to peatlands, from the new global peatland carbon standard Verified Carbon Standard – Peatland Rewetting and Conservation (VCS-PRC), from the VCS baseline and monitoring methodology for the rewetting of drained peatlands, and from the much strengthened greenhouse gas emission site type (GEST) approach, all facilitated and developed by the project. Belarus now has trained human resource potential to continue and expand peatland rewetting beyond the lifetime of this project as well as to apply market-based mechanisms for climate mitigation and biodiversity conservation. A commodity has been created with quality standards, verification methodologies, and a price, bringing together both the producer and the consumer. So what are the major lessons that can be drawn from this experimental phase?

One key success factor was that a national-based non-governmental Organisation (NGO) performed the majority of the project implementation. This has helped to ensure long-term ownership of the process and a vested interest in the longer-term vision for rewetting. Another key success factor was the ability for a wide range of staff and experts to cooperate in an experimental field, being open to new ideas, and willing to be flexible as the project developed.

Key challenges for the sustainability and expansion of the project in Belarus and for its replication in other countries include gaining ongoing political support for rewetting, and ensuring that the right expertise and knowledge is available in-country. Additionally, thorough project planning and implementation is required, with clear objectives for both the initial developmental phase (preparing for carbon sales), as well as for the longer term (sale of carbon emission reductions and further rewetting).

The experiences made by the project can be applied to other drained peatlands in Belarus as well as in Ukraine (see chapter 7.9) and European Russia. However, it should be noted that some experiences are specific to Belarus. For example, land access was a political, not a financial issue, and therefore economic experiences cannot unequivocally be extrapolated to other countries. Nevertheless, it is hoped that future projects in this field will build on the first stage of successes and challenges in order to expand the whole concept to other countries where peatlands are a threatened ecosystem.

## 7.9 The BMU-ICI twin project in Ukraine

Susanne Bärisch, Galyna Poshtarenko, Zbigniew Karpowicz & Jack Foxall

### 7.9.1 Background

Peatlands cover c. 1.4 million ha, i.e. 2.3%, of the Ukrainian land surface (Truskavetskiy 2010). Most of them can be found in the large lowland areas of the Ukrainian Paliessie, in the north of the country, bordering with the southern part of Belarus. Laying at the southern margin of the mixed forest zone and being characterized by an increasingly continental climate to the east, the area hosts many peatland plant species at their southern and eastern distributional limits. Several protected areas and Ramsar sites underline the importance of the Paliessie lowlands for ecosystem and species conservation, for example for the Aquatic Warbler *Acrocephalus paludicola*.

Ukraine hosts the second largest breeding population of this globally threatened bird.

However, about 50% of the total peatland area of Ukraine has been destroyed or severely degraded. Large-scale drainage of peatlands started in 1873 for peat extraction and agricultural use (Minayeva et al. 2009). Although peat extraction in Ukraine declined and large-scale agricultural practices were abandoned after 1990, the exploitation of peatlands continued. Small-scale agriculture by local communities is still practiced today on the former cooperative farms (kolkhoz) peatlands and here, even when sites are abandoned, it is unlikely that they become rewetted owing to the lack of clear and prescriptive legislation. International climate mitigation activities, however, now offer new opportunities for the restoration of degraded peatlands in Ukraine. Such ecosystem-based climate mitigation projects could produce both carbon credits for sale as well as new forms of sustainable land use, with additional benefits for both biodiversity and local communities.

### 7.9.2 Project details

Since December 2009, the International Climate Initiative (ICI) of the German Ministry for the Environment, Nature Conservation and Nuclear Safety (BMU), working through the Kreditanstalt für Wiederaufbau Entwicklungsbank (KfW – Development Bank), has financed the project 'Avoidance of greenhouse gas emissions by restoration and sustainable management of peatlands in the Ukraine', which aims to address these opportunities. The project is a partnership between The Royal Society for the Protection of Birds (RSPB), UK, the Ukrainian Society for the Protection of Birds (USPB), Ukraine, and the Michael Succow Foundation (MSF), Germany, in cooperation with the Government of Ukraine.

Within the project 20,000 ha of degraded peatlands are due to be rewetted by November 2012. To support further peatland restoration, an additional area of 10,000 ha will be made ready for rewetting after the project ends. Rewetting sites have been selected based on a scoring system taking into account the size of the peatland, its emission reduction potential, its ownership structure, and other parameters. The selected sites are situated in the three most peatland-rich oblasts lying in the Ukrainian Paliessie: Volyn, Rivne, and Chernihiv (holding 21%, 21% and 14% of the total Ukrainian peatland area, respectively (Thiele et al. 2008). The project sites are fen peatlands that have previously been transformed into hay meadows, pastures, and crop fields (Fig. 70 Colour plates III). However, many of the sites were abandoned over the last decade or so (Fig. 71 Colour plates III). They are characterized by low and fluctuating water levels and a strongly degraded upper peat layer. Typical peatland vegetation is either absent or only present in drainage ditches while the dry peat on abandoned sites is colonized by grasses, ruderal communities and an increasing cover of shrubs and trees. Anticipated emission reductions from rewetting of 20,000 ha are in the range of 200,000 t of $CO_2$-eq. per year, between 2012 and 2020.

Similar to the BMU-ICI project in Belarus, emission reductions are assessed based on the greenhouse gas emission site type (GEST) approach (Couwenberg et al. 2011). The project promotes further research in this field, including additional GHG measurements in Belarus on site types that occur in Ukraine but are not yet covered. In addition research is being carried out on how to carry out large-scale mapping of GESTs using remote sensing, on the amount of biomass available on peatlands, and on GHG emissions as a result of peatland fires. The GEST approach is also being adapted and applied for use by a National Inventory System for GHG emissions from drained peatlands in order to fulfil reporting requirements to the United Nations Framework Convention on Climate Change (UNFCCC).

### 7.9.3 Implications of land ownership

In contrast to Belarus where project sites are only state-owned, drained agricultural land in Ukraine is 80% owned by private persons and local cooperatives. The single land parcels are small (2–3 ha) and highly fragmented. Key to the success of this project is to identify innovative mechanisms for land access that allow long-term peatland restoration to be implemented on a large scale. The development and realization of regional communication and partnership strategies as well as cooperation models, and local level participatory planning has been imperative. The first key solution identified by the project to access privately owned land was to develop and gain official approval for changes to the Oblast Land Use Plans in each of the three project regions. These plans serve as legal documents, which define the land management regime of rewetted lands and provide legal restrictions for rewetted land use change in the future. A second key solution identified by the project was to enhance the establishment of village cooperatives to consolidate small private land-owners for easy-access to land for rewetting. Both of these approaches have enabled access to private lands. Communities have welcomed opportunities for alternative income-generation, and for solutions to reduce the likelihood of peatland fires.

In order to facilitate the sale of carbon emission reductions from peatland sites in Ukraine, an entity was required to consolidate the numerous small land-owners and village cooperatives in order to generate a single pool of carbon emission reduction units for carbon markets. The entity (or Carbon Fund) is in the early stages of being formed by the project. It will be used as the vehicle to channel funds for the long-term management of peatland areas for the benefit of climate, biodiversity and local communities from 2012. Contracts between the entity and land owners/ village cooperatives will be made in order to sell predominantly carbon credits, but also biomass for sustainable fuel production or electricity generation, as well as medical herbs.

Further measures to ensure long-term sustainability of the rewetted sites include the incorporation of project sites into the protected area system of Ukraine. The project has prepared scientific justifications for the extension of nature protection territories in the Chernihiv region and submitted these for approval to the Ukrainian Ministry of Environment and Natural Resources. In total, about 16,000 ha of lands, including peatlands restored in the framework of the project, are planned for incorporation into the Ukrainian fund of protected areas. The promotion of peatland-protected areas will be part of the commitment of the country to create new protected areas and to improve the management of existing protected areas under the Life Web Initiative of the Convention on Biological Diversity.

# 8 Practical rewetting examples

## 8.1 Introduction

Franziska Tanneberger & Annett Thiele

The BMU-ICI project comprises nine project sites (see Table 38 and Fig. 66 in chapter 7.2). Five of them are described in detail in chapters 8.2–8.6: The bogs Dalbeniski, Zada, and Daukudauskaje and the fens Hrycyna-Starobinskaje and Scarbinski Moch. Dalbeniski and Zada are large bogs in the north of Belarus that have been rewetted using soil dams. The rewetting approach of Dakudauskaje, located in central Belarus, includes the construction of an anti-leakage dam between the rewetting site and the active peat extraction site. The two fens, one located in the south and one in the north, feature soil dams and modified sluices and soil dams only for rewetting, respectively. As a sixth practical rewetting example, the famous Jelnia mire (chapter 8.7) is included and related rewetting activities are presented.

The four project sites not described in detail are Zadzienauski Moch, Poplau Moch (see also box 14 in chapter 4.4), Obal, and Horeuskaje. These sites are briefly described here:

The project site Zadzenauski Moch is located in the Poozerie, a huge ground moraine region with thousands of lakes and peatlands in the north of Belarus, in Viciebsk oblast. It is a partly forested, partly open raised bog of 753.3 ha that was drained in the 1950s and was subject to peat milling until the 1990s. Due to that the northern part of the territory is severely degraded. Areas with bare peat or sparse moss-vegetation cover have lain abandoned for 20 years. The surrounding not-extracted parts show intensified pine and birch growth. Nevertheless the former hummock-hollow-lawn complex structures have remained but species composition has shifted to drier types, than is typical for such structures. The southern part of the peatland, reaching into the territory of the Russian Federation, has remained wet and open. Rewetting was performed by closing the channels with soil dams in the centre of the extraction site. Several bigger soil dams were installed in the surrounding channels to collect the discharging rainwater. The success of rewetting is only visible after some years and strongly dependent on the amount of precipitation in the following seasons.

Poplau Moch is a fen peatland, with some buried bog peat, located in the same region as Zadzienauski Moch. It was subject to peat extraction from the 1950s until the 1990s. The entire site covers 414.6 ha, of which 350 ha have been exploited by peat extraction (milling). It lay abandoned after the 1990s and was prone to fire until the rewetting activity started. During the ongoing succession, areas became overgrown by birches. Approximately 78% of the peatland area is now covered by young birch forest. Already before rewetting, the drainage system partly collapsed or the beaver helped to keep part of the area wet. After rewetting numerous soil dams blocking the side channels hinder run off to the main channel and maintain wet conditions in large parts of the site.

Obal is a bog peatland that is also located in the Poozerie in Viciebsk oblast. The territory covers 2,555 ha in total and 1,097 ha of it have been rewetted in the frame of the project. The peat extraction activities started in 1922 and are still continuing on 122 ha outside of the project territory. The extraction was performed by milling and excavating, which left a very inhomogeneous surface. Since the rewetting in winter 2008/2009 via numerous soil dams and one stone-gravel overflow dam a diverse mosaic of birch forests, pine forest, mixed forest, flooded bare peat with emerging reeds and regenerating hummock-hollow complexes on the excavated sites are developing.

The fen peatland Horeuskaje is located in the valley mire regions of the Jasielda River, close to the famous conservation site Sporauski zakaznik and covers 190.51 ha. It underwent drainage in the 1970s which was followed by peat extraction. As there remained a substantial peat layer the area was given over to the local kolkhoz. Today it is an abandoned agricultural site with a sparse cover of young birch forest, bare peat and plants indicating dry conditions. Rewetting will be done by closing the channels with soil dams in the surrounding area.

## 8.2 Dalbeniski

Irina Voitekhovitch, Olga Chabrouskaya, Annett Thiele & Franziska Tanneberger

### 8.2.1 Site characteristics

| Location | Viciebsk region, Sarkauscyna district (Fig. 72). |
|---|---|
| Coordinates | 55°18'905"N, 27°11'812"E. |
| Total area | 5,501 ha. |
| Area of restoration site | 5,501 ha. |
| Peatland type | Bog (with a steep slope of c. 9 m height difference). |
| Peat depth | Average: 4.2 m; maximum: 8.0 m. |
| Water supply and discharge | Precipitation; drainage system includes a main channel (2.5 m deep), levee channels (7–8 m wide, 2–2.5 m deep), and ditches (every 20 m, up to 1.5 deep) in the extraction site and drains to rivers Dzisna and Janka. |
| Land use history | In the early 1990s (until 1994) peat extraction in the central part of the bog for Sarkauscyna agro-chemical factory, about 0.5 m of peat was removed over 180 ha and another 200 ha were prepared for peat extraction (drained, trees removed); surrounding lands are used for cattle grazing, hay-making and crop farming. |
| Land use before rewetting | Belonged to the State Forest Enterprise 'Pastavi Forestry', main land use is forestry, hunting, and recreational fishing; a blueberry plantation is run by the Forest Enterprise on the former peat extraction site. |
| Vegetation before rewetting | Severely influenced by fire a huge mono-dominant 'desert' of heather (Fig. 75 Colour plates III) had developed in the centre, around the lake (Fig. 73 Colour plates III), and at the higher parts of the raised bog. The single standing pine trees have been burnt and single standing birch trees have colonized the area. The northern part was partly spared from the fires and hummock-hollow-lawn-complexes remained. In the southern part the fires were very severe. These areas are overgrown by poplar and birch thickets, with the moss *Polytrichum* spec. in the understorey over large parts of the area that have been flooded by beaver. Due to beaver activities and other regeneration processes (see also Fig. 74 Colour plates III) peat moss and sedge reeds have developed. |
| Fire hazard before rewetting | Fire hazard before rewetting. |

### 8.2.2 Restoration approach

Restoration measures aim at decreasing discharge of precipitation water from the peatland. This discharge occurs by channels and ditches that drain the former peat extraction site and also Lake Asviata (which had already reduced one third in size), and from forestry ditches in the marginal zone of the peatland (Kozulin et al. 2010c). The rewetting activities will concentrate on the former extraction site. According to the engineering plan, the construction of 30 dams in the ditches and in the collector channel around the peatland is projected. Sufficient dams have to be installed to guarantee that water level differences over the dams do not exceed 30–40 cm. The initial dams must be 50–70 cm higher than the upper edges of the ditch to ensure that water bypassing the dams is distributed widely over the bog surface. The position of the dams has been planned in the process of development of engineering documentation on the basis of a linear survey of the drainage network along the main channels with cross-section measurements every 200 meters.

## 8.2 Dalbeniski

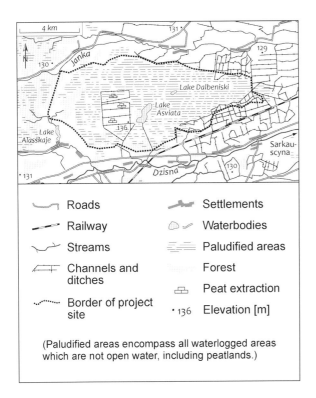

Fig. 72: Location of the rewetting site Dalbeniski (preliminary border or project site; map: Stephan Busse).

### 8.2.3 Greenhouse gas emission site types (GESTs) and expected climate benefits

Before rewetting, the dominant greenhouse gas emission site types (GESTs) are 'moist bog heath' (c. 70% of the area) and 'very moist bog heath', with 12.5 t $CO_2$-eq. ha$^{-1}$yr$^{-1}$ and 10 t $CO_2$-eq. ha$^{-1}$yr$^{-1}$, respectively. The latter type (Fig. 76 Colour plates III) is characterised by the presence of Sphagnum. The parts that have not been affected severely by fires remained wet and are characterized by a mosaic of GESTs composed of predominantly 'very wet Sphagnum hollows' (12.5 t $CO_2$-eq. ha$^{-1}$yr$^{-1}$), and 'wet Sphagnum lawn' (5 t $CO_2$-eq. ha$^{-1}$yr$^{-1}$) and to a lesser extent of 'moderately wet Sphagnum hummocks' (0.5 t $CO_2$-eq. ha$^{-1}$yr$^{-1}$). The parts severely affected by fires along the southern, eastern, and western borders of the peatland have been overgrown by Betula pendula and Populus tremula thickets. Parts rewetted by beaver have formed 'wet Polytrichum lawns' (2 t $CO_2$-eq. ha$^{-1}$yr$^{-1}$), whereas less severely burnt areas rewetted by beaver have developed to 'wet reed and sedge fens' with Carex lasiocarpa, Phragmites australis, and Sphagnum angustifolium (8.5 t $CO_2$-eq. ha$^{-1}$yr$^{-1}$). The former peat extraction area is covered with huge areas of 'bare peat with Polytrichum' (7.5 t $CO_2$-eq. ha$^{-1}$yr$^{-1}$) or just 'bare peat' (7.5 t $CO_2$-eq. ha$^{-1}$yr$^{-1}$), partly overgrown by Betula pendula (Fig. 74 Colour plates III).

It is preliminarily estimated that rewetting will reduce greenhouse gas emissions by 1.6 t $CO_2$-eq. ha$^{-1}$yr$^{-1}$ calculated over a 20-year project period using a forward-looking baseline (see chapter 3.5). Probably detailed spatial analysis will lead to a lower result as the main emission reductions are assumed to take place through water level rise in the centre of the peatland, where 'moist bog heath' could develop to 'very moist bog heath'. Such development is, however, not likely since in the short term the effective rewetting of bogs is hardly feasible. On the former extraction area, the scenario assumes a substantial rise of the water level over two thirds of the area (one third is a blueberry plantation) leading to development of a mosaic of 'wet Sphagnum lawn' with Eriophorum vaginatum.

### 8.2.4 Expected biodiversity benefits

Before the 2002 fires, five plant species of the 2005 Belarusian Red Data Book typical for large, undisturbed bogs have been recorded in Dalbeniski: Carex paupercula, Salix myrtilloides, Rubus chamaemorus, Huperzia selago, and Corallorhiza trifida. These as well as other rare species found in the peatland (Salix lapponica, Betula humilis, Dacthylorhiza maculata, Plathanthera bifolia, Glyceria lithuanica, and Carex disperma) are expected to expand after restoration. Dalbeniski is also part of the Important Bird Area 'Jianka floodplain'. It is expected that some of the 146 breeding bird species (38 Species of European Conservation Concern SPEC 1–3) will benefit from the restoration measures (Kozulin et al. 2010c).

### 8.2.5 Preliminary results and lessons learnt

Restoration works for the rewetting of Dalbeniski peatland have started in late 2010 and completion of the construction works is expected by July 2011. This includes 30 soil dams and combined soil and wooden dams.

### 8.2.6 Management recommendations

A key issue for future management is cranberry (Vaccinium oxycoccos) picking, which is an important income source for local people. The rewetting activities will probably enhance cranberry growth and thus gain support among local communities. The road built for the extraction machines running from the border of the peatland to the centre facilitates berry picking. Maintenance of this road – if not counteracting rewetting – would raise support for the measures by berry-pickers.

## 8.3 Zada

Irina Voitekhovitch, Olga Chabrouskaya, Annett Thiele & Franziska Tanneberger

### 8.3.1 Site characteristics

| | |
|---|---|
| Location | Viciebsk region, Mijory and Sarkauscyna districts (Fig. 77). |
| Coordinates | 55°25'879"N, 28°01'125"E. |
| Total area | 5,383 ha. |
| Area of restoration site | 3,380 ha. |
| Peatland type | Bog (with a slope of up to 5 m height difference). |
| Peat depth | Average: 3.0 m; maximum: 6.0 m. |
| Water supply and discharge | Precipitation; located on a watershed, the underlying fen has been supplied by waters from the Dzisna (the left tributary of West Dzvina) river basin; before drainage rivers flew out only from the boundary zone of the bog; since the 1970s an extensive network of ditches drains the western and southern parts of the bog (for forestry) and the north-eastern part (for peat extraction); the rivers are canalised. |
| Land use history | Drainage started in the 19th/early 20th century, in the early 1970s a forestry reclamation system was built in the western part of the peatland and a substantial part of the site (185 ha) was allocated to milled peat extraction, which started in 1975; in 1999 the entire territory was transferred to forestry without any restoration measure. |
| Land use before rewetting | The area belonged to the State Forest Department 'Dzisna forestry'; main forms of land use were forest management and hunting. |
| Vegetation before rewetting | The non-extracted parts that surround the lakes (Fig. 78 Colour plates IV) consisted of degraded heather communities with partly sparse peat moss and an open pine cover. The central part was a lake that fell dry and is now covered with moss-cotton grass-birch communities (Fig. 79 Colour plates IV), the border areas were quite wet with cotton grass-peat moss communities, the former peat extraction areas werenearly completely overgrown by heather and single standing birch and pine trees (Fig. 80 Colour plates IV), small areas with bare peat and heather or cotton grass occurred (see also Fig. 81 Colour plates IV). |
| Fire hazard before rewetting | Very high due to recreational use by local people (e.g. mushroom and berry picking) and due to the fact that no water reservoirs or other water sources for fire fighting were available within the very fire-prone parts. |

### 8.3.2 Restoration approach

Studies have shown that the drainage system negatively affects the entire peatland. Effects include the expansion of heather and trees at the expense of typical bog species such as *Sphagnum* mosses and the overgrowing of Strecna and Ilava lakes by aquatic vegetation (the latter almost completely in 2010). To ensure the restoration of the hydrological regime it is planned to raise the water level to around the soil surface by building cascades of dams with c. 30–40 cm water level

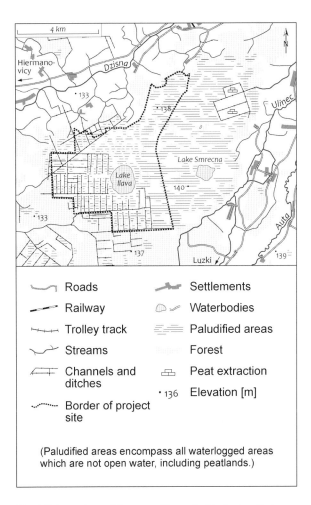

Fig. 77: Location of the rewetting site Zada (preliminary border or project site; map: Stephan Busse).

difference over the dams (Maksimenkov et al. 2007a). According to the most recent engineering plans, in total 50–60 dams will be constructed.

### 8.3.3 Greenhouse gas emission site types (GESTs) and expected climate benefits

Before the start of rewetting activities, the spatially important GESTs are 'moist bog heath' and 'very moist bog heath' together with mosaics of degraded hummock-hollow-lawn structures and 'wet reeds and sedge fens'. 'Moist bog heath' covers the highest and therefore driest, degraded non-extracted parts, but also big parts of the former extraction site. Whereas *Calluna vulgaris* is the characteristic species for both GESTs, the distinction between the two types lies in the cover of *Sphagnum*. 'Moist bog heath' has very low to zero *Sphagnum* cover, as the water level is too low, and *Calluna* is accompanied by other dwarf shrubs like *Ledum palustre* and *Vaccinium uliginosum*, and by *Polytrichum strictum*. The emission value is 12.5 t $CO_2$-eq. $ha^{-1}yr^{-1}$, caused by $CO_2$ resulting from the permanent peat decomposition. The 'very moist bog heath' has less dwarf shrubs and a higher cover of *Sphagnum* species, mainly *Sphagnum magellanicum* on the hummocks and *Sphagnum angustifolium* in the lawn level. The frequent occurrence of *Vaccinium oxycoccos* and *Andromeda polifolia* shows that the water levels are higher resulting in somewhat lower emissions of 10 t $CO_2$-eq. $ha^{-1}yr^{-1}$. Hummock-hollow-lawn complexes have remained and are represented by a mosaic of GESTs of predominantly 'very wet *Sphagnum* hollow' (12.5 t $CO_2$-eq. $ha^{-1}yr^{-1}$), and smaller proportions of 'wet *Sphagnum* lawn' (5 t $CO_2$-eq. $ha^{-1}yr^{-1}$) and 'moderately wet *Sphagnum* hummock' (0.5 t $CO_2$-eq. $ha^{-1}yr^{-1}$), respectively. The dried-out hollows of the former hummock-hollow-lawn-complexes are predominantly occupied by 'moderately wet *Sphagnum* hummocks' and to a lesser extent by 'wet *Sphagnum* lawn'. In the dry lake, 'wet reeds and sedge fens' (8.5 t $CO_2$-eq. $ha^{-1}yr^{-1}$) with *Carex nigra*, *Phragmites australis*, *Sphagnum angustifolium*, and *Calla palustris* cover substantial areas of the central part. In the eastern part, tree growth was substantially enhanced by drainage (Fig. 81 Colour plates IV; A. Skuratovich, D. Dubovik, and S. Levy pers. comm.).

The changes in vegetation composition after rewetting will be minor since only half of the peatland is allowed to be rewetted (the eastern part) and large parts of the site are already quite wet (e.g. former lake and hummock-hollow-complexes). In case of optimal rewetting, the 'moist bog heaths' will develop to 'very moist bog heaths', leading to a decreased growth of the trees (and consequent less $CO_2$ sequestration). Detailed spatial analysis will enable more detailed estimation of the emission reduction in future.

### 8.3.4 Expected biodiversity benefits

It is expected that the restoration measures will favour hummock-hollow complexes and weaken pines, birches, and heather. This will benefit five species of the 2005 Red Data Book of Belarus (*Huperzia selago*, *Linnaea borealis*, *Salix myrtilloides*, *Rubus chamaemorus*, and *Vaccinium microcarpum*) as well as other threatened species such as *Salix lapponica*, *Betula humilis*, and *Utricularia intermedia*. It is also expected that eleven Red Data Book bird species recorded in

Zada before 2006, but not or only in small number after 2006 (Greater Spotted Eagle *Aquila clanga*, Black Stork *Ciconia nigra*, Osprey *Pandion haliaetus*, Lesser Spotted Eagle *Aquila pomarina*, White Grouse *Lagopus lagopus*, Crane *Grus grus*, Golden Plover *Pluvialis apricaria*, Eagle-Owl *Bubo bubo*, Short-eared Owl *Asio flammeus*, Curlew *Numenius arquata*) will benefit from the restoration measures (Maksimenkov et al. 2007a).

### 8.3.5 Preliminary results and lessons learnt

The design documents have already been developed and approved and completion of the construction works is expected by autumn 2011.

### 8.3.6 Management recommendations

Stronger rewetting results are likely to be reached by rewetting the entire peatland, rather than only part of it. The increase in cranberry growth due to the rising water levels will enable local communities to get an extra income from the seasonal collecting of berries (see box 18 and box 19).

In addition, fishing and hunting will probably benefit: The stabilisation of the hydrological conditions of the lakes will lead to an increase in fish production and will have a positive impact on populations of game and thus on hunting, both inside and outside the peatland. Such activities should, as long as they do not counteract restoration, be supported.

## 8.4 Hrycyna-Starobinskaje

Irina Voitekhovitch, Olga Chabrouskaya, Annett Thiele, Wendelin Wichtmann & Franziska Tanneberger

### 8.4.1 Site characteristics

| Location | Minsk region, Saligorsk district (Fig. 82). |
|---|---|
| Coordinates | 52°42'092"N, 27°19'296"E. |
| Total area | Overall area c. 30,000 ha, depleted area 3,860 ha. |
| Area of restoration site | 3,505 ha. |
| Peatland type | Fen. |
| Peat depth | Average: 0.1–0.3 m. |
| Water supply and discharge | Fed by floods, groundwater, precipitation and surface discharge; located in the floodplain of Morac river (part of Paliessie lowland), only a small part (Krivichi) is located on the first river terrace of Sluch river; catchment area is to 50% covered by mixed forests, Morac river is channelled and separated from the extracted part of the peatland by dams; before rewetting the straightened Morac was connected to the extracted part only by channels and had (due to backwater at a nearby sluice in the river) a water level of 0.5 m higher than the surface of the extracted part. |
| Land use history | Peat extraction by milling for the peat briquetting factory Starobinski from 1960 to 1997; peat extraction was performed in a spatially complex manner depending on peat depth which resulted in large differences in terrain relief after abandonment; first rewetting activities started in 1995 to prevent fires. |
| Land use before rewetting | Land belonged to the Starobinski State Forest Enterprise; main forms of land use were forestry, fishing, and hunting. |
| Vegetation before rewetting | Temporary open water areas were surrounded by cattail reeds and common reed (23%), areas with willow shrubs (23%), birch forest (9%), dry grassland (4%; Fig. 83 Colour plates IV), young forest with shrubs (21%), bare land (0.2%), and moderately wet areas (18%); see also Fig. 84 Colour plates IV. |

## 8.4 Hrycyna-Starobinskaje

| Fire hazard before rewetting | Within the mosaic of flooded and dry (with water level at 30–100 cm below soil surface) the latter had a very high risk of fire; in very dry years (1997, 1998, 2002) and especially in spring, dry grassland vegetation and open peat used to catch fire easily and fires expanded over hundreds of hectares. |
|---|---|

### 8.4.2 Restoration approach

Measures aimed at raising the water levels implemented by the peat factory prior to the start of the project turned out to be effective in reducing fire hazard, yet they were insufficient to attain the main goal of fen mire restoration. Research conducted in 2007 showed that the water supply to the territory was extremely unstable. In spring, when water was coming from the catchment areas, the lowlands were flooded but as early as to the end of May the water level had fallen to the extent that bare peat was exposed (e.g. in 2007 the area of open water in April and May was 1,500 ha, and in June-October only 300–400 ha). At several sites the groundwater level remained 30–100 cm below the soil surface throughout the year. The surface of the degraded peatland consists of a mixture of peat residues, sand, and gyttja (lake sediments). Sparse birch and willow woods grew at sites with insufficient water supply (Maksimenkov et al. 2007b).

To restore the fen mire, it has been planned to raise the water all over the site to soil surface by closing main channels in a multistage way depending on the results of the levelling survey. Water level difference over the dams has been planned to be limited to c. 30 cm. In restoration planning, the requirement to avoid flooding of adjacent farmlands had to be taken into account.

### 8.4.3 Greenhouse gas emission site types (GESTs) and expected climate benefits

Before rewetting, the project site was dominated by the GESTs 'flooded tall reeds' (1 t $CO_2$-eq. $ha^{-1}yr^{-1}$), 'moderately moist forb (any herbaceous plant other than grass) meadows' (20 t $CO_2$-eq. $ha^{-1}yr^{-1}$), 'drowned grassland' (up to 77 t $CO_2$-eq. $ha^{-1}yr^{-1}$), 'wet reeds and sedge fens' (8.5 t $CO_2$-eq. $ha^{-1}yr^{-1}$), and 'dry grassland' (20 t $CO_2$-eq. $ha^{-1}yr^{-1}$). The intensively extracted areas were flooded in spring and autumn and reeds and sedge fens have developed. Areas used for placing machinery along the channels and close to the central dam were overgrown by 'dry grassland'. Areas that have been extracted less deep are covered with 'moderately moist forb meadows' (with high emissions) and are partly covered with *Betula pendula* thickets or forests (Fig. 84 Colour plates IV).

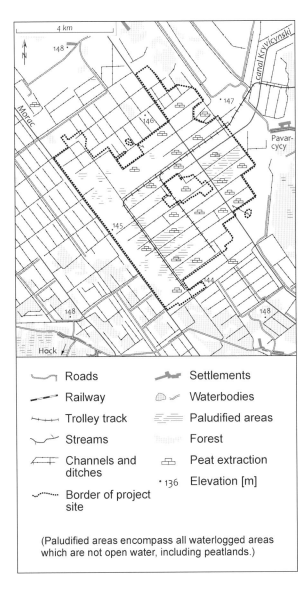

Fig. 82: Location of the rewetting site Hrycyna-Starobinskaje (map: Stephan Busse).

It is estimated that greenhouse gas emissions will be reduced by rewetting by 0.5 t $CO_2$-eq.ha$^{-1}$yr$^{-1}$, calculated over a 20 year period, using a forward-looking baseline (see chapter 3.5). It is expected that mosaics of 'flooded tall reeds' and 'wet reeds and sedge fens' will colonize the areas of open water as the water level fluctuations are getting smaller (Fig. 86 Colour plates IV). The water level rise will affect the higher elevated parts only marginally, but will cause die-off of birch trees on the intermediately elevated areas and along the channels.

### 8.4.4 Expected biodiversity benefits

It is expected that typical fen sedge vegetation (sedge communities, orchids) will benefit as well as several threatened animals such as Bittern *Botaurus stellaris*, Great White Egret *Egretta alba*, Black Stork *Ciconia nigra*, Corncrake *Crex crex*, Kingfisher *Alcedo atthis*, various waders, and possibly also Aquatic Warbler *Acrocephalus paludicola* (Maksimenkov et al. 2007b). Habitat management for the latter, globally threatened species would require vegetation management, most likely mowing and biomass removal.

### 8.4.5 Preliminary results and lessons learnt

The flatness of the site and the presence of sluices on the major channels made the rise of water levels in 2009 easily possible by closing two of these facilities (Fig. 18 Colour plates I and Fig. 85 Colour plates IV). The extracted areas on the left bank of the river were drained by two pumping stations, located at the mouths of the main channels. After their dismantling the groundwater level rose to 0.5–0.7 m below the peat surface over most of the area, meaning that the elimination of water pumping did not lead to the restoration of the former natural water level. Therefore, during rewetting activities a number of soil dams and a metal-concrete flume were built. This resulted in a very good water level and no fires in 2009.

However, the new water levels were so high that they influenced the adjacent agricultural land. Intervention by the respective stakeholders let to a reduction of the height of the stoplog gates at the sluices and, consequently, to significantly lower water levels again. In order to correct this situation, in 2010 one more very large soil dam was constructed that again led to rise of water level.

All facilities involved in this scheme are in working order according to a survey in May 2011. It was, however, observed that the water table difference between the dammed main channel and the adjacent smaller channel that controls the water tables in the north-eastern part and in the agricultural area was more than 2 m. This water table should be regularly checked during the year.

Data from a vegetation monitoring plot illustrate the development at the site with and without rewetting (Table 40) from 2008 (before rewetting) to 2010 (after rewetting). Whereas in the rewetted part, vegetation composition changed in favour of species indicating wet conditions, it remained relatively stable in the non-rewetted part (Pugachevsky et al. 2010).

Along the bird monitoring transect, the number of bird species typical for peatlands increased from 10% to 15% in the overall species composition and that of 'shore and water conditions' from 22% to 30%, respectively. The number of 'forest' species decreased from 33% to 26%, as did species of 'dry open areas' (from 8% to 5%). The monitoring of amphibian and reptile communities reflects also rather substantial changes: The abundance of species typical for hydrophilous conditions and for forests and peatlands increased (e.g. in case of the Moor Frog *Rana arvalis* from 5 individuals/ha in 2008 to 12 individuals/ha in 2010). In the overall species composition, hydrophilous species increased from 23% to 35% and species typical for forests and peatlands from 31% to 43%, respectively.

### 8.4.6 Management recommendations

There are large areas with grassland and weed vegetation that could not be rewetted because they are adjacent to agricultural lands or are located higher than the extracted areas. Further possibilities for raising water tables should be checked here. Generally, it should be considered to put such areas into regular agricultural use e.g. for hay or silage production to minimise losses of nutrients that are anyway set free by mineralization of the peat due to insufficient water levels. This applies also to extended areas (in total >200 ha) adjacent to the main channels and the railway line passing the project area from southwest to northeast and along the main drainage channels. In practice, this would be difficult to realise since the project site belongs fully to the forest enterprise which is not dealing with agriculture.

Another management option concerns the reeds that have developed following rewetting. If they in time develop into more productive stands, biomass harvesting for the production of fuels (bio-

# 8.5 Scarbinski Moch

Table 40: Ground water level and cover of selected plant species at the rewetted and non-rewetted part of the vegetation monitoring plot 1 at Hrycyna-Starobinskaje before (2008) and after (2009, 2010) rewetting (from Pugachevsky et al. 2010).

| Plot | Species | 2008 | 2009 | 2010 |
|---|---|---|---|---|
| KU-1 | (summer ground water level, in cm) | -0.7 | -0.3 – -0.4 | -0.25 – -0.35 |
| (rewetted) | Salix cinerea | 34 | 20 | 20 |
| | Agrostis stolonifera | 60 | 1 | 0 |
| | Carex rostrata | 1 | 1 | 0 |
| | Carex lasiocarpa | 1 | 0 | 0 |
| | Phragmites australis | 6 | 14 | 15 |
| | Lemna minor | 0 | 50 | 30 |
| | Spirodela polyrhiza | 0 | 5 | 30 |
| | Bryum pseudotriquetrum | 70 | 0 | 0 |
| | Drepanocladus aduncus | 3 | 0 | 0 |
| KU-1* | (summer ground water level, in cm) | > -1.0 | > -1.0 | > -1.0 |
| (not | Betula pubescens | 70 | 75 | 70 |
| rewetted) | Achillea millefolium | 2 | 1 | 5 |
| | Cerastium holosteoides | 1 | 1 | 2 |
| | Hieracium umbellatum | 3 | 1 | 1 |
| | Hypericum perforatum | 2 | 1 | 1 |
| | Phragmites australis | 0 | 1 | 0 |

mass briquettes, pellets etc.) should be considered. This may also benefit target species such as Aquatic Warbler. Areas covered by forest should be harvested for production of wood chops as a heating fuel in order to abate water consumption by transpiration of trees. This would also avoid that vast areas with dead trees appear, which often leads to local population opposing rewetting measures. Open channels and deeply inundated areas are already used by fishermen and this kind of utilization could be stimulated in future. The area is really very popular among fishermen.

## 8.5 Scarbinski Moch

Irina Voitekhovitch, Olga Chabrouskaya, Annett Thiele & Franziska Tanneberger

### 8.5.1 Site characteristics

| Location | Viciebsk region, Dubrovno district, 28 km northwest of Dubrovno and c. 1 km north of the Brest-Moscow highway (Fig. 87), close to the Russian border. |
|---|---|
| Coordinates | 54°42'191"N, 30°51'238"E. |
| Total area | 1,427 ha. |
| Area of restoration site | 1,322 ha. |
| Peatland type | Fen with two bog lenses (Fig. 88 Colour plates IV). |
| Peat depth | Average: 3.4 m; maximum: 7.0 m. |

| Water supply and discharge | Fed by precipitation and groundwater; discharging in two directions since it is located on a watershed between the Baltic and Black Sea basins; in spring extensive parts of the peatland are inundated by melt water (1–2 months flooding). |
|---|---|
| Land use history | Drainage started in 1918, peat extraction by Osintorf Peat Extraction Enterprise in 1927. In 1957 hydrotorf extraction (i.e. with dredger under wet conditions) was replaced by milling. A peat briquette factory using peat from Scarbinski Moch and the near-by Osinovskoe peatland has been operating since 1970. In the mid 1980s 852 ha of Scarbinski Moch were under extraction or depleted. The bog lenses in the central part (c. 15% of the site) were only slightly extracted since the peat (with low degree of decomposition and low bulk density) was not suitable for briquetting. Since 1972 depleted parts of the peatland were transferred to the Orsa State Forest Enterprise for so-called 're-naturalisation' and forestry, with the drainage network still active or gradually overgrowing. |
| Land use before rewetting | Orsa State Forest Enterprise, the land owner of the restoration site. |
| Vegetation before rewetting | 60% of the site was covered by shrubby willow (at moist sites) or birch (at dry sites) thickets on the eastern and western edge, along channels, and on a site abandoned for a long time. Reeds and forbs constituted c. 6% of the site, 32% were bare peat, partly with pioneer vegetation of e.g. *Eriophorum* spec. (Fig. 89 Colour plates IV; see also Fig. 90 Colour plates IV). |
| Fire hazard before rewetting | Fires were registered almost annually and devastated large territories more than once. They are mainly caused by careless handling of fire by local population (81% of all registered fires) and associated to spring grass burning. Most fires occurred in 1996, in 2002 (75% of the site affected), and in 2006, usually in May. In May fires, litter from the previous season promotes fire, but the wet peat layer usually does not catch fire. In dry years, the upper peat layer in the depleted part of the peatland is very fire prone since the groundwater level falls to 1.0–1.5 m below soil surface. Therefore the Forest Enterprise and Osintorf Peat Enterprise implemented a number of measures that decrease fire hazard: drainage channels were closed with earth dams; fire lines, mineral lines and fire extinction basins were created. |

### 8.5.2 Restoration approach

Investigations by the National Academy of Sciences of Belarus (Maksimenkov et al. 2008) revealed that
- The drainage network leads to rapid discharge of precipitation water into the Viazouka (drains into Black Sea basin) and Viarchita (drains into Baltic Sea basin) rivers and thus completely ruined, together with peat extraction, the formerly existing natural mire complex. (Fig. 89 Colour plates IV);
- Peat extraction was followed by frequent fires that damaged the upper soil layer severely and deteriorated driving safety at the nearby Brest-Moscow highway, the largest highway in the Republic of Belarus;
- Succession after abandonment of peat extraction under continued drainage lead to the formation of plant communities of little value for biodiversity as well as for forest production (with forest development frequently interrupted by fires);

The drainage network before rewetting included two runoff ditches (4 m wide), two main channels (6–8 m wide and 0.5–1.5 m deep), and 21 smaller ditches (5 m wide). Most channels are intact; but some are partly closed with sand blocks. The depleted area was partly flooded (20–30 cm above

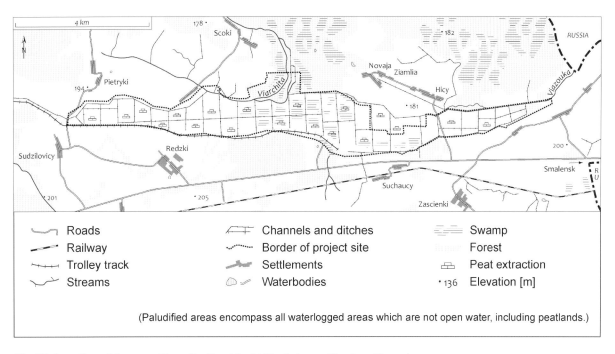

Fig. 87: Location of the rewetting site Scarbinski Moch (map: Stephan Busse).

soil surface) and the groundwater level fell by upto 1–1.5 m below soil surface (Maksimenkov et al. 2008). The restoration plan proposed to close the drainage system with water-retention overflow dams at the northern and southern drainage ditches and the Viazouka river.

### 8.5.3 Greenhouse gas emission site types (GESTs) and expected climate benefits

Before rewetting, the dominant greenhouse gas emission site types (GESTs) at the project site were 'bare peat' (7.5 t $CO_2$-eq. $ha^{-1}yr^{-1}$) and 'very moist reeds' (3.5 t $CO_2$-eq. $ha^{-1}yr^{-1}$), followed by 'moist forb (any herbaceous plant other than grass) meadows' (12.5 t $CO_2$-eq. $ha^{-1}yr^{-1}$) and 'moderately moist forb meadows' (20 t $CO_2$-eq. $ha^{-1}yr^{-1}$; Fig. 88 Colour plates IV). The former bog lenses, situated north, east and west of the watershed, were mainly covered by 'bare peat with *Polytrichum*' (7.5 t $CO_2$-eq. $ha^{-1}yr^{-1}$) or 'moist bog heath'. The central part that had been extracted last, could be characterized as 'bare peat' and on higher elevated areas as 'moderately moist forb meadows'. At the former extraction sites that had been abandoned earlier and had substantially lost height, spontaneous rewetting lead to the development of 'very moist reeds'.

In a prognosis for a 20-year period, it is expected that the former bog lenses (Fig. 88 Colour plates IV) are too high to be influenced by the rising groundwater table. Substantial rewetting effects will only occur when the bog peat has mineralized and thus the surface has lowered to the mean groundwater level. Also the 'very moist reeds' will not get wetter or will get slightly wetter to develop 'wet reeds and sedge fens' (8.5 t $CO_2$-eq. $ha^{-1}yr^{-1}$). On the 'bare peat', 'very moist reeds' will spread or 'bare peat with *Eriophorum*' (7.5 for bare peat, 3.5 for *Eriophorum* t $CO_2$-eq. $ha^{-1}yr^{-1}$) will develop and persist for some time.

It is estimated that by rewetting greenhouse gas emissions will be reduced by 2.3 t $CO_2$-eq. $ha^{-1}yr^{-1}$ calculated over a 20-year project period using a forward-looking baseline (see chapter 3.5). This volume was calculated for optimal rewetting conditions using rough GEST assumptions. First monitoring results will probably decrease this number as it is hard to raise the water level substantially on the higher elevated areas.

### 8.5.4 Expected biodiversity benefits

It is expected that the 2005 Red Data Book of Belarus species occurring currently at the site (Great White Egret *Egretta alba*, Bittern *Botaurus stellaris*, Osprey *Pandion haliaetus*, Short-toed Snake-eagle *Circaetus gallicus*, Corncrake *Crex crex*, and Common Greenshank *Tringa nebularia*) will expand their area of occupancy after rewetting (Maksimenkov et al. 2008). The adjacent not or only slightly drained peatland, Osinovskoe, where rewetting activities have been carried out by the GEF project, could serve as 'donor' for peatland species after restoration.

### 8.5.5 Preliminary results and lessons learnt

The key instrument to rewet the peatland in June 2010 was the construction of water retention overflow dams in the northern and southern drainage ditches and the Viazouka River, as well as renovation of the existing stoplogs at the Viazouka River. Several side channels were closed to stop inflow into the main channel. After the construction work had been finished, water levels in the western and eastern part of the peatland became considerably higher. Large-scale inundations at the minerotrophic part of the peatland showed the functionality of the installed soil dams. The former bog lenses in the central part, however, fall dry for more than 20 days per year (A.A. Vasilievski pers. comm.).

In the year of rewetting (2010), plant communities indicating moister conditions had already increased at the expense of those indicating drier conditions due to spontaneous rewetting (Fig. 91). At a monitoring plot, in 2008 ruderal and wasteland species such as *Cirsium arvense*, *Artemisia vulgaris*, *Chenopodium album*, *Elymus repens*, *Tanacetum vulgare*, *Tussilago farfara* and *Ceratodon purpureus* had 50–80% coverage. In 2010, the cover of species related to peatlands such as *Lycopus europaeus*, *Cirsium palustre*, *Baeothryon alpinum*, *Salix cinerea*, *Calliergonella cuspidata*, and *Betula pubescens* had substantially increased. Formerly bare areas became overgrown with plants, as illustrated at another monitoring plot where the area of bare peat decreased from 30% to 5%. All over the site a change in abundance of *Polytrichum* mosses was observed. *Polytrichum strictum* was almost totally replaced by *Polytrichum commune*. This indicates moister conditions and lower soil acidity (Pugachevsky et al. 2010).

Since the bird monitoring transect became more overgrown by trees and shrubs (mainly birch and willow) and was less affected by spontaneous rewetting in 2008–2010, an increase by 10% in number of birds related to forest was observed. The percentage of bird species associated with trees and shrubs slightly decreased from 34% to 31%. The overgrowing also caused a reduction in the number of water and shore bird species by 3%. The proportion of bog and aquatic bird species remained practically unchanged (Pugachevsky et al. 2010).

### 8.5.6 Management recommendations

To maintain water regulation devices, the state of water-regulating facilities should be monitored and water erosion and destruction by beavers should be prevented. Climate benefits can be maximised on the watershed areas by inundating the areas where possible. Currently, the formerly dry peat has been moistened, which leads to increased microbial activity and possibly to increased greenhouse gas emissions in comparison to the formerly dry peat. Further possibilities to raise the water level and to keep the precipitation water on the watershed should be evaluated. If better rewetting is not possible, topsoil re-

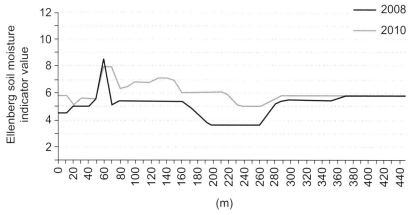

Fig. 91: Mean Ellenberg soil moisture indicator values of plant communities along a monitoring plot located in the lowest part of Scarbinski Moch in 2008 (black) and 2010 (grey). From Pugachevsky et al. (2010).

moval could be considered to level the area and to lower the peat surface. A sensible use for the extracted topsoil should be found, e.g. as fertilizer for agricultural fields. This could possibly reduce pressure on less disturbed peatland soils or on peat deposits that are better suited for rewetting.

Biodiversity would benefit from raising water levels on the watershed (see above) and from the maintenance of large areas of *Phragmites australis* reed bed. Particularly by cutting for biomass, an avifauna typical for this habitat will establish. The complexes of reed bed, other types of swamp, and wet scrub that will develop and be maintained by biomass management following rewetting will support a wide range of passerines and other wildlife typical of reed beds and wet scrub in Belarus. The restored areas of peatland will also form a component of the territories of Lesser Spotted Eagles *Aquila pomarina*. In or-

der for these co-benefits to accrue and be maintained, cutting regimes for both scrub and reeds for biomass should be carefully considered to fall outside the breeding season and be implemented as a multi-year rotation in blocks to ensure suitable habitat is available in every year.

To establish sustainable use of biomass from the rewetted site, it should be investigated to what extent the electric power plant currently provided with loose peat could also be provided with chaffed, loose reed biomass. For limiting evapotranspiration and reducing peat extraction for firing in the electric power plant it is recommended to harvest the birch and willow trees and produce wood chips from them. In the course of forest management fire barriers must be constructed along the perimeter of the project territory and intercepting channels and major channels with water along the site perimeter must be closed.

## 8.6 Dakudauskaje

Irina Voitekhovitch, Olga Chabrouskaya, Annett Thiele & Franziska Tanneberger

### 8.6.1 Site characteristics

| Location | Hrodna region, Lida district (Fig. 92). |
|---|---|
| Coordinates | 53°51'228"N, 25°27'200"E. |
| Total area | Overall peatland area 7,811 ha, depleted area 3,583 ha. |
| Area of restoration site | 1,945 ha. |
| Peatland type | Largely a fen; the restoration site includes a bog (with a slope and c. 2.0 m height difference). |
| Peat depth | Average: 2.8 m, maximum: 5 m. |
| Water supply and discharge | Groundwater and precipitation; the underlying fen is drained by rivers Narva and Lidzjejka; the water level in the drainage network is 0.4–1.5 m below soil surface; the site is surrounded by peat extraction sites in the north and west, forest in the south, and agricultural land in the east. To some extent the entire project area is affected by drainage in the peat mining area. |
| Land use history | Protected area (biological zakaznik; 1,985 ha); a small area of the restoration site was affected by peat mining for agrochemical purposes until c.1980–1990. |
| Land use before rewetting | The restoration site belonged to 'Lida Forest Enterprise' (parts of the surrounding area belonged to Lida Peat Factory). |
| Vegetation before rewetting | Adjacent to the border between the reserve and the extraction site mono-dominant heather without peat mosses but with single standing birch trees had developed (Fig. 93 Colour plates IV). A substantial part of the peatland was subject to fires, thus dwarf shrub rich hummock communities (Fig. 94 Colour plates IV) with substantial pine growth had |

| Vegetation before rewetting (cont.) | developed. A former lake that fell dry hosts reed-peat moss communities. Close to the centrally situated lake widespread peat moss/cotton grass communities indicate that peat formation may have taken place. On the small former peat extraction area a peat moss/cotton grass community had already developed (see also Fig. 95 Colour plates IV). |
|---|---|
| Fire hazard | Very high due to the low water levels. |

### 8.6.2 Restoration approach

The northern and western parts of the project site are directly adjacent to operational peat extraction of the Lida peat factory resulting in progressive degradation of the hydrological regime of the restoration site. The extraction site lies approximately 4 m lower than the surface of the natural remnant and the influence of drainage was huge. This was especially apparent after the construction of a drainage pumping station to complete peat extraction (Kozulin et al. 2008).

Rewetting measures were planned to stop the draining effect of the channel systems on the peatland by setting soil dams in their beds. To prevent discharge to the neighbouring operational extraction an artificial dam of 1.3 km length was constructed. The dam was made out of peat and features a 1.5 m deep polyethylene anti-percolation film. A number of soil dams has been installed in the surrounding of the reserve as well as on the small former extraction area.

### 8.6.3 Greenhouse gas emission site types (GESTs) and expected climate benefits

Before rewetting by the project the dominant GESTs were typical for degraded peatlands – 'moist bog heath' (12.5 t $CO_2$-eq. ha$^{-1}$yr$^{-1}$) and 'moderately wet *Sphagnum* hummocks' (0.5 t $CO_2$-eq. ha$^{-1}$yr$^{-1}$). The first was found in a 200 m broad band adjacent to the border between the extraction area and the reserve. The latter was widespread and dominant on the burnt areas. Dwarf shrubs like *Ledum palustre*, *Vaccinium uliginosum* and peat mosses like *Sphagnum magellanicum* are dominating the area and are overgrown by a substantial layer of young *Pinus sylvestris*. The former peat extraction site had already been rewetted by beaver before project start and was covered with 'very wet *Sphagnum* hollows' (12.5 t $CO_2$-eq. ha$^{-1}$yr$^{-1}$) with *Eriophorum vaginatum* and *Sphagnum cuspidatum*. The dry lake was covered with 'wet reeds and sedge fens' (8.5 t $CO_2$-eq. ha$^{-1}$yr$^{-1}$). Closer to the centrally situated lake, the vegetation cover was similar to undisturbed raised bogs, with mosaics of 'wet *Sphagnum* lawn' (5 t $CO_2$-eq. ha$^{-1}$yr$^{-1}$) and 'moderately wet *Sphagnum* hummocks' (Fig. 95 Colour plates IV).

It is estimated that greenhouse gas emissions will be reduced by rewetting by 2.4 t $CO_2$-eq. ha$^{-1}$yr$^{-1}$ calculated over a 20-year project period using a forward-looking baseline (see chapter 3.5). Remarkable changes in vegetation and emissions are expected to happen directly adjacent to the peat dam with the anti-percolation screen, where 'wet *Sphagnum* lawns' and 'very moist bog heath' are expected to replace 'moist bog heath'. The influence on the entire territory will be small since the slope of the raised bog is the limiting factor.

### 8.6.4 Expected biodiversity benefits

It is expected that natural bog flora (e.g. *Sphagnum* mosses, *Eriophorum* spec., *Drosera* spec., and several wetland orchids) will benefit from the restoration measures. The two 2005 Red Data Book species occurring in the zakaznik (*Vaccinium microcarpum* and *Salix myrtilloides*) are expected to expand.

Although the fauna of the zakaznik was rather rich before rewetting (30 mammal and 104 bird species), typical bog birds such as Crane (*Grus grus*) and Curlew (*Numenius arquata*) breed only in small numbers. Along with vegetation succession towards more natural bog conditions, several Red Data Book species that occur already now in the zakaznik will probably benefit: Crane (*Grus grus*), Black Stork (*Ciconia nigra*), Bittern (*Botaurus stellaris*), Lesser Spotted Eagle (*Aquila pomarina*), Kestrel (*Falco tinnunculus*), Hobby (*Falco subbuteo*), the beetles *Carabus cancellatus* and *Carabus violaceus*, and the Bumblebees *Bombus muscorum* and *Bombus schrenkii* (Kozulin et al. 2008).

## 8.6 Dakudauskaje

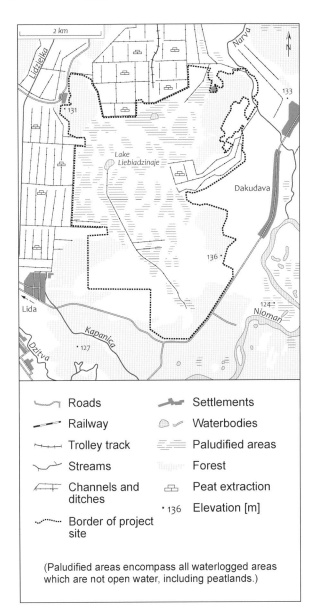

Fig. 92: Location of the rewetting site Dakudauskaje (map: Stephan Busse).

### 8.6.5 Preliminary results and lessons learnt

Construction works have been carried out in March – April 2010. In general, the soil infillings have lead to an increase in water levels (with some differences between north and south) and to the flooding of areas close to the dams. Future dysfunction of the dams by mineralization of the dam material is minimized because the dams have been constructed 20–50 cm higher than the surrounding soil surface, which guarantees long lasting functionality. Especially the U-shaped dams (soil infillings) are a sustainable measure since surplus water from spring floods must bypass the dams towards the rewetting site. Thus, the pressure from high waters on the dams is minimised. The long dam with the incorporated polyethylene screen built in the north of the project area prevents direct surface water outflow from the project site to the drainage channel in the north.

The difference in water tables between the upper (western) and lower (eastern) pools excavated for getting material for dam construction was in May 2011 about 1.5 m in total. This results in draining of the area west (with water tables about 35 cm below surface) and total inundation of the area east of the pool. This is the same for each of the pools and shows that peat for dam construction should be extracted parallel to the contour lines to prevent such effect.

Since the relief declines to the east, extra water from spring flows goes via an overflow in a wooden wall in the east to the drainage system in order to avoid over flooding and damage of constructed dams. Thus, there is no total retention of spring water flow. In other periods all the water is kept on the peatland. Since the effects of rewetting were only visible in short distance from the dam, a larger flooded area at the margins could be useful to get a larger effect on the water level in the entire peatland.

With regard to the sustainability of the rewetting measure, the dam was slightly compacted in spring 2011. Experience of functioning of a similar dam at a rewetting site of the GEF project (Marocna) shows that grass appears at the dam, and the roots of the grass and other vegetation make the dam very solid. But further mineralization of the peat can be expected.

### 8.6.6 Management recommendations

As in other sites, a visual inspection of all dams should be carried out on a quarterly basis, and in case of damage timely repair is required. Special attention should be given to the permanence of the dam separating the restoration site from the operational peat extraction site.

Any effects of subsurface water losses should be monitored on the longterm. High open water tables in the pools will probably lead to increased subsurface water losses from the site to the northern main drain, because the total head differences are very high (>5 m difference in hydraulic head to the bottom of the main ditch).

## 8.7 Jelnia

Irina Voitekhovitch, Tatsiana Broska, Annett Thiele & Franziska Tanneberger

### 8.7.1 Site characteristics

| Location | Viciebsk region, Mijory and Sarkauscyna districts (Fig. 96). |
|---|---|
| Coordinates | 55°33'005"N, 27°50'720"E. |
| Total area | 19,984 ha (reserve: 25,758 ha). |
| Area of restoration site | Restoration measures aimed at improving the condition of the entire area. |
| Peatland type | Bog (one of the largest bog complexes in Central Europe; elliptical / almost quadrangular shape). |
| Peat depth | Average: 3.8 m; maximum 8.3 m. |
| Water supply and discharge | Precipitation (and some discharge of artesian waters from the underlaying bedrock); network of drainage channels and one channelized river. |
| Land use history | In the 1920s/30s several channels were dug from the bog centre to its fringe, linking the Belaje Lake and other smaller lakes which caused a decrease in groundwater level. Later, in the second half of the 20th century, additional channels were dug to increase discharge from other lakes. These activities have significantly changed the macrorelief of the bog. Further substantial changes in the hydrographic network of the mire took place when peat extraction sites were established in adjacent bogs and surrounding mineral areas were drained for agricultural purposes. In 1957–1959, i.e. before the establishment of a reserve, a network of drainage channels and ditches was constructed in the northeastern part of the mire to facilitate peat extraction. Although no peat extraction was carried out, the drainage network exerted and still exerts a significant negative influence on the site. Canalization and regulation of rivers resulted in larger water discharge in spring and an overall drop in the groundwater table in big parts of the mire. |
| Land use before rewetting/ current land use | Forestry (Dzisna forestry enterprise); State nature reserve (1968, enlarged in 1981), various international designations, e.g. Ramsar site (2002) and Important Bird Area (2006); berry-picking, hunting, and fishing; mowing and grazing takes place in some peripheral parts of the reserve. |
| Vegetation before rewetting | Raised bog vegetation with pine and shrub-moss communities was the dominant vegetation type (75% of the total area, Fig. 2 Colour plates I, Fig. 97 Colour plates IV). Forests covered 21% and open tree and shrub formations 2%. Another 2% was occupied by lakes. |
| Fire hazard | Before rewetting, prone to fires with various intensities (Grummo et al. 2010). Fires affecting several thousands of hectares took place e.g. in 1993, 1994, 1998, and 2002. Fires are most frequent at the bog crests and at sites where the hydrological regime is affected. The fires of 1998 and especially those of 2002 had a disastrous effect and devastated almost all natural parts of the mire leaving only the southern part unaffected. 52% of the total area of the mire (and 82% of the open raised bog) were burnt during the 2002 fire. |

## 8.7 Jelnia

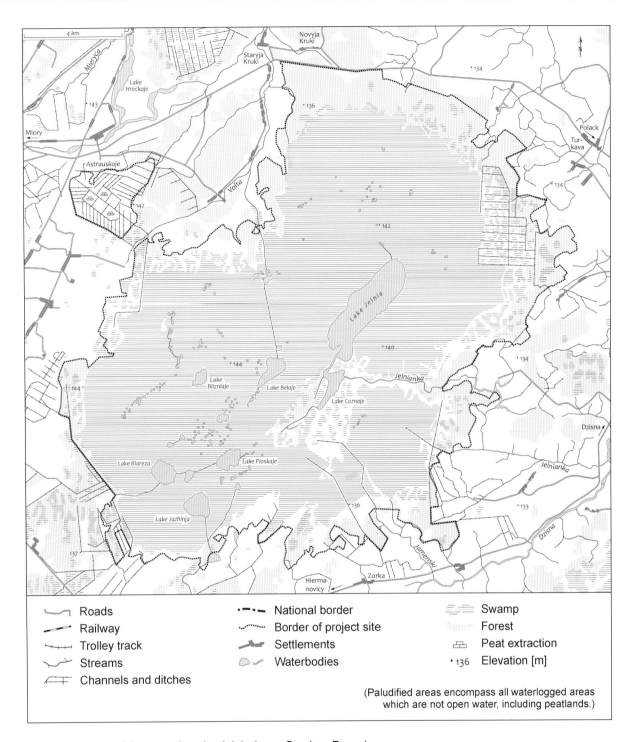

Fig. 96: Location of the rewetting site Jelnia (map: Stephan Busse).

### 8.7.2 Restoration approach

Since 1999 the non-governmental organisation APB-BirdLife Belarus organises works for the improvement of the hydrological regime with the financial support of the Ministry of Natural Resources and Environmental Protection of the Republic of Belarus, of the Royal Society for the Protection of Birds (RSPB) UK, and Wetlands International.

A project in 1999–2002 led to the construction of 17 dams in the peripheral part of the bog closing the channels out flowing from the mire. As a result, the water level in the bog fringes came close to the optimum. Because of the large (natural) height difference from the bog centre to its fringes (about 7 m) these works at the margins did not restore the hydrological regime in the central part of the bog. At present the central raised part of the bog is being drained by the channelized Jelnianka river and a number of man-made channels and regulated rivers. Each of the channels passes through several lakes that significantly increase the drainage influence of each channel. Vegetation and fires show that these channels lower the groundwater table of the bog up to a distance of 500 to 2,000 m.

The rewetting activities in 1999–2002 improved the situation only in the peripheral part of the bog and could not prevent the vast fires in 2002. Therefore, starting from 2006, APB-BirdLife Belarus, with support from Global Environment Facility (GEF) Small Grants Programme and corporate funders (Coca-Cola Beverages Belorussiya, FE) initiated the construction of further 43 dams (Fig. 98 Colour plates IV) in the central part of Jelnia (Belyatskaya et al. 2008). The works started with damming of the biggest channel of Jelnia (4.5 km long) and then the other three main channels and the channelized Jelnianka river. First of all four channels were blocked with overflow dams at the points of their exits from the lakes for rising water level in the lakes. During 2007–2010 more dams were built and enlarged to rewet adjacent, higher areas. More than 20 dams block the oldest channel today. They are continuously checked and, if necessary, repaired by the local population and volunteers. The distance between the dams is c. 300 m (P. Bogoley pers. comm.).

### 8.7.3 Expected climate benefits

The huge areas affected by fire are covered by species-poor moist bog heath vegetation. As these areas emerge over their surroundings (that have been lowered as a result of peat oxidation and shrinkage along water bodies and channels), the rewetting activities have only a small effect, at least on the shortterm. The oldest dam, built in 1999 at the northeastern fringe of the peatland and closing the end of the oldest channel, shows some positive influence in the surrounding of the small lakes (see Fig. 97 and 99 Colour plates IV). The water level of the lakes has increased substantially and an *Eriophorum vaginatum-Sphagnum angustifolium*-lawn has replaced the moist bog heath in the surrounding of this lake complex. This is a first step of the long process to restoring the entire peatland: the lower lying areas get wet first, form new peat and new acrotelm conditions, that can then actively hinder runoff and raise the water level in the central parts. From these restored fringes a re-colonization of the degraded central parts by *Sphagnum* species can be expected. The low-lying channels crossing the areas with moist bog heath may also initiate new *Sphagnum*-mats, but the entire depression must be 'filled' before the *Sphagnum* can spread out to the higher located moist bog heath. The climate benefits of the rewetting activities are very small to zero over a period of 20 years, therefore the gradual rewetting and regeneration will take decades or even up to hundreds of years to be completed.

However, the rewetting of the areas around the channels will lead to decreased risk of peat fires. Mapping of peat fires clearly showed, that regular peat fires of 1993–2002 started from drier areas around the channels and lakes and then spread to as much as 80% of the area of peatland. Taking into account that peat fires caused huge emission of greenhouse gases, improved hydrological situation will help to avoid large-scale emissions.

### 8.7.4 Expected biodiversity benefits

The repeated fires severely decreased the habitat of several peatland bird species, like waders, passerines, and Crane *Grus grus* and resulted in the disappearance of these species from the burned areas. Close to the channels, where dams have been built, a slow regeneration of habitats for birds like Wood Sandpiper *Tringa glareola*, Greenshank *Tringa nebularia*, Whimbrel *Numenius phaeopus*, Curlew *Numenius arquata*, and Crane *Grus grus* is taking place. The gradual return of these species has been observed in the last years (S. Levy pers. comm).

### 8.7.5 Preliminary results and lessons learnt

Thanks to the rewetting activities and the wet years 2009 and 2010, die off of birch trees and heather and an initial colonization of the burnt areas by peat mosses is taking place. Although the retention capacity of the ditches is still limited, the run-off has decreased compared to the situation before rewetting. The fire hazard has reduced severely and the habitat of animals and plants listed in the Red Book of the Republic of Belarus (e.g. *Betula nana*, *Rubus chamaemorus*, *Carex pauciflora*, *Huperzia selago*, *Lobaria pulmonaria*, *Sphagnum molle*) that used to or still inhabit the territory has extended.

Dams constructed in the biggest drainage channel of Jelnia in 2006 show positive results (see Fig 100 Colour plates IV). Vegetation indicative for deeper water levels (e.g. *Calluna vulgaris* heath with *Polytrichum strictum*) is now only found on the higher parts further away from the dams, where rewetting hardly has taken place, because the channel is lying much lower than the surrounding areas as a result of subsidence (Grummo & Thiele 2010, Broska 2010).

The resettlement of the burnt areas with peat moss is much more advanced in formerly forested areas, where the charred trees provide shade, than on open heath land. This is a process of climatic self-regulation: a temporarily increased wood stock can promote regeneration (Edom & Succow 1998, Edom 2001a). Former hollows that are oriented perpendicularly to the ditch today can function as 'irrigation cars' to more distant areas.

### 8.7.6 Management recommendations

To increase water retention in the channels, additional dams along the channel should be built in a way that height differences over the dam are less than 30 cm. Another measure would be to heighten and widen the existing dams in such a way that the water is forced to flow from the dam into the adjacent areas.

The condition of the mire would benefit substantially from a hydrological and climatic buffer zone that encompasses the surrounding land that affects the peatland via evapotranspiration covering the entire surface and subsurface hydrological interaction zones of the mire (Poelman & Joosten 1992, Edom 2001a). As Jelnia is part of a comprehensive region of clustered bogs (see Fig. 66), restoration of nearby degraded peatlands like Dalbeniski and Zada will probably positively influence Jelnia. Similarly the restoration of the directly adjacent bog Astrauskoje, west of Jelnia, would be favourable to stabilize the hydrology of the mire (see chapter 3.5 and Fig. 66).

# 9 Recommended research and monitoring activities in rewetted peatlands

## 9.1 Recommended research activities

Susanne Bärisch & Franziska Tanneberger

### 9.1.1 Introduction

Successful restoration of peatlands relies on the availability of relevant knowledge derived from targeted and effective research and monitoring. Research is done in projects of limited duration at selected study sites, specifically designed to gather information of general importance, be it fundamental facts on greenhouse gas (GHG) emissions, vegetation development, or determining the success of management activities. Research projects are often characterized by innovative approaches either concerning the methods applied or the objects under investigation (e.g. specific species or habitats). Research results can be used to develop model-based (and thus more cost-effective) methods to be applied within monitoring activities (see chapter 9.2) and to guide restoration activities.

Key areas of research related to peatland rewetting for climate and biodiversity benefits are:
- The assessment of GHG emissions from peatlands via proxies;
- The assessment of vegetation development after rewetting;
- The assessment of biodiversity before and after rewetting.

### 9.1.2 Assessment of GHG emissions from peatlands via proxies

Targeted analysis of literature data on GHG emissions from peatlands enabled the development of the Greenhouse gas Emission Site Type (GEST) approach (see chapter 3.4). Although vegetation has been identified to be generally a valuable proxy for assessing GHG emissions from peatlands, other parameters have to be taken into account in emission assessments, especially in case of bare sites, treed sites, and sites experiencing disturbance such as flooding or fires. This chapter focuses on how to further develop the GEST approach and recommends research activities in the following six modules (H. Joosten & J. Couwenberg pers. comm.):
- GEST-HERB: using equilibrium herbaceous vegetation as a proxy for GHG fluxes;
- GEST-WATER: using mean annual water level as a proxy for GHG fluxes;
- GEST-FOREST: using biomass data as a proxy for $CO_2$ sequestration in forests and shrubs;
- GEST-TRANSIENT: using initial biomass quality and quantity as a proxy for GHG emissions in non-equilibrium ('transient') situations after rewetting;
- GEST-FIRE: using fire scars and vegetation as a proxy for GHG emissions from peat fire;
- GEST-PREDICT: using vegetation development models for assessing with-project GHG emission reductions.

The GEST-HERB module can be regarded as basic module of the GEST approach (see chapter 3.4) and is currently improved within the BMU-ICI project and other projects by calibration and complementation ('gap filling') of the existing data on vegetation types and GHG emissions (see box 5 in chapter 3.3). Simultaneous assessment of GHG emission characteristics, vegetation composition, and abiotic parameters from study sites in the project region that are identical or similar to sites for which emissions data are published is necessary to calibrate emission values for given vegetation types. Vegetation forms (see box 6 in chapter 3.4) recently studied in northeastern Germany have been identified as good indicators of site parameters (e.g. water level) and can therefore be used to directly assign GESTs for project sites (Couwenberg et al. 2008). Other types of vegetation classification based on species composition require further elaboration with regard to their indicative value prior to their interpretation in terms of GESTs. This can, for example, be done using Ellenberg moisture indicator values (Ellenberg et

# 9 Recommended research and monitoring activities in rewetted peatlands

Table 41: List of available and required emission values for GESTs (with water level classes, see chapter 3.4) identified for the project region. Emission values are available from literature sources or are currently being measured within the BMU-ICI project in Belarus (see also Table 2 in chapter 3.3). Most of the further required emission values for GEST calibration or gap filling will be measured within the Ukrainian BMU-ICI project (see chapter 7.9).

| List of GESTs for GHG measurements | Available emission data | | GHG measurements BMU-ICI project Belarus | Further required measurements |
|---|---|---|---|---|
| | $CH_4$ | $CO_2$ | $CH_4$ & $CO_2$ | $CH_4$ & $CO_2$ |
| Cultivated peatlands (2-, 3+/2-, 2+) | ✓ | ✓ | ✓ | |
| Cultivated peatlands (3+/2+, 3+) | ✓ | ✓ | ✓ | |
| Cultivated peatlands (4+/3+) | ✓ | ✓ | | Calibration |
| Cultivated peatlands with *Phalaris arundinacea* (4+) | ✓ | ✓ | | Calibration |
| Rewetted cultivated peatlands (5+) | ✓ | ✓ | ✓ | |
| Flooded cultivated peatlands with dying vegetation (6+) | | | ✓ | |
| Moderately Moist Forbs & Meadows (2+) | ✓ | ✓ | | Calibration |
| Moist Forbs & Meadows (3+) | ✓ | ✓ | ✓ | |
| Moist Bog Heath (3+) | ✓ | ✓ | | Calibration |
| Moist Bare Peat, Moist peat with sparse vegetation (3+) | ✓ | ✓ | ✓ | |
| Very Moist Meadows (4+/3+) | ✓ | ✓ | | Calibration |
| Very Moist Meadows, Forbs & Tall Reeds (4+) | ✓ | ✓ | | Calibration |
| Very Moist Bog Heath (4+) | ✓ | ✓ | | Calibration |
| Very Moist Bog Heath W/Mudflats (4+) | ✓ | ✓ | | Calibration |
| Very Moist Tall Sedge Marshes (5+/4+) | ✓ | | | Gap filling |
| Very Moist Tall Sedge Marshes – biomass harvest (5+/4+) | | | | Gap filling |
| Very Moist Peat Moss Lawn (5+/4+) | ✓ | ✓ | | Calibration |
| Wet Tall Sedge Marshes (5+) | ✓ | | ✓ | |
| Wet Tall Sedge Marshes – biomass harvest (5+) | | | | Gap filling |
| Wet Moss Dominated Short Sedge Marshes (5+) | ✓ | | | Gap filling |
| Wet Moss Dominated Short Sedge Marshes – biomass harvest (5+) | | | | Gap filling |
| Wet Short & Tall Sedge Marshes & Reeds w/Moss Layer (5+) | ✓ | | ✓ | |
| Wet Short & Tall Sedge Marshes & Reeds w/Moss Layer – biomass harvest (5+) | | | ✓ | |
| Wet Tall Reeds (5+) | ✓ | | ✓ | |
| Wet Tall Reeds – biomass harvest (5+) | | | | Gap filling |
| Wet Peat Moss Lawn (5+) | ✓ | ✓ | ✓ | |
| Wet Peat Moss Hollow (5+) | ✓ | ✓ | ✓ | |
| Wet *Polytrichum Lawn* (5+) | | | | Gap filling |
| Flooded Tall & Short Reeds, different dominant species (6+) | ✓ | | ✓ | |
| Flooded Tall & Short Reeds, different dominant species – biomass harvest (6+) | | | | Gap filling |
| Wet peat with *Carex* species (5+) | | | ✓ | |
| Wet peat with *Eriophorum angustifolium* (5+) | | | ✓ | |

## 9.1 Recommended research activities

| List of GESTs for GHG measurements | Available emission data | | GHG measurements BMU-ICI project Belarus | Further required measurements |
|---|---|---|---|---|
| | CH₄ | CO₂ | CH₄ & CO₂ | CH₄ & CO₂ |
| *Polytrichum strictum – Ledum palustre* hummocks (3+) | | | ✓ | |
| Crops on drained fens (2+), different crop cultivation | ✓ | ✓ | ✓ | |
| Abandoned fields and fallows (2+) | | | | Gap filling |
| Grassland cultures (degraded type) with *Juncus effusus, Carex ovalis, Potentilla anserina* (3+) | | | | Gap filling |
| Grassland cultures (degraded type) with strongly alternating wetness (4~, 3~) | | | | Gap filling |

al. 1992), combined with regional expert knowledge. Further research on vegetation forms in other regions should combine vegetation relevés with measurement of site conditions. Resultant communities will allow simpler GEST assignment. Site types that have a real relevance for a project country's rewetting sites, but for which insufficient emissions data are available, need to be assessed by direct GHG measurements (see chapter 3.3). Table 41 provides an interim overview of site types for which literature data are already available, of site types that have been measured within the current BMU-ICI project in Belarus, and of site types where GHG measurements remain to be conducted for GEST calibration and complementation in the project region. Special emphasis should be put on sites where biomass is used (see chapters 6.1 to 6.3 for options and chapter 6.4 for benefits). Biomass harvesting is a promising management option for rewetted sites and may become implemented large-scale. GHG emission characteristics for harvested sites are not yet well studied.

Any additional data on GHG emissions of vegetation types (if conducted with reliable methods, see chapter 3.3) will strengthen the GEST approach since they improve the resolution of GEST assignment as well as the reliability of the applied regressions (see chapter 3.4).

The GEST-WATER module is applied at sites without vegetation or with insufficient species cover to clearly indicate water level conditions. Because of small-scale differences in relative water level due to habitat heterogeneity in peatland sites a large number of water level measuring points are required. Thus, further research should focus on developing cost-effective fine-scaled water level monitor approaches, for example by remote sensing or by hydrological modelling.

The GEST-FOREST module is an interim approach using data on wood biomass as a proxy for $CO_2$ sequestration in forests and shrubs (see examples in chapter 3.5). For calculating the $CO_2$ balance of forested peatlands and those with shrub encroachment, the results of GEST-FOREST complement the emission values assessed with the GEST-HERB module.

As soon as sufficient GHG flux data from forested peatlands are available (from eddy covariance measurements from over the tree and shrub canopy, see chapter 3.3), the GEST-FOREST module will become dispensable and the vegetation types identified in the GEST approach (GEST-HERB) will be complemented by forested vegetation types. Until recently the number of GHG measurements from forested peatlands was very limited since suitable equipment for such measurements is expensive and data analysis requires expert knowledge. As shrub and tree encroachment on peatlands is a function of management, and not only of water level, several forested site types need to be measured. Until sufficient data are available it is relevant to improve the GEST-FOREST module by collecting growth and carbon sequestration data of trees and shrubs on peatlands. Especially for willow species on peatlands the knowledge of yearly increments is still limited.

A large field for additional research offers the GEST-TRANSIENT module: Sites with 'transient' conditions may show completely different emission characteristics from those before and long after rewetting. While annual peat $CO_2$ fluxes can be assessed by using the $CO_2$ regression based on mean annual water level (see chapter 3.4) for each year until equilibrium vegetation types are established, $CH_4$ emissions during transient phases require an integrated assessment of water levels and pre-rewetting biomass quantity and quality. The approach is only required for water levels above -20cm since no $CH_4$ emissions occur on sites with lower water levels.

### 9.1.3 Assessment of vegetation development after rewetting

A sound prediction of future vegetation development (the GEST-PREDICT module) relies on long-term studies of vegetation succession (see chapter 3.5). Rewetting projects for the international carbon market usually require an outlook on developments over a future time span of 30 years and more. Several studies from abandoned temperate peatlands can inform such forward-looking baseline scenarios:
- For 50 and more years after abandonment of cutover bogs e.g. Buttler et al. (1996), Davis & Wilkinson (2004), Girard et al. (2002), Joosten (1995), Laggoun-Defarge et al. (2008);
- For 35 years after abandonment of drained (but not extracted) bogs e.g. Frankl (1996);
- For up to 27 years after abandonment of drained (but not extracted) fens Podlaska (2010) and up to 23 years Mälson et al. (2008).

Long-term studies of purposefully rewetted peatlands are less numerous and stretch in many cases over a shorter period of time. Vegetation development at rewetted temperate peatlands has been studied:
- At rewetted cutover bogs by Joosten (1995; 50 years after rewetting), Poschlod (1990, 1992; up to 40 years), Främbs & Nick (2001; 8–12 years), Lavoie et al. (2005a; 6–10 years), Nick & Weber (2001; 5–12 years), Schmatzler & Hoyer (1994; 12 years);
- At rewetted previously drained (but not extracted) bogs by Jauhiainen et al. (2004; 60 years after rewetting);
- At rewetted previously drained (but not extracted) fens by Jansen & Roelofs (1996; up to 15 years after rewetting), Mälson et al. (2008; 4 years), Steffenhagen et al. (2010; 10 years), Toogood & Joyce (2009; 21 years).

Vegetation prediction is an important tool in rewetting projects to define both baseline and project scenarios. Analogue to the GEST approach (GEST-HERB module), prediction needs to be improved by calibration and complementation. Here, calibration stands for assessing vegetation development characteristics on sites in the project region that are identical or similar to sites for which long-term studies are available in the literature. As vegetation development is determined not only by abiotic site conditions, but also by population dynamics, availability of diaspores etc., such factors need to be taken into account. For the complementation of the vegetation prediction tool long-term vegetation developments on yet insufficiently studied peatland types (e.g. fens drained for agriculture) need to be investigated.

### 9.1.4 Assessment of biodiversity before and after rewetting

The special habitat conditions in peatlands often lead to low diversity, but unique and highly adapted species assemblages. Related to this, strong interactions and dependencies can be found in peatland communities and the loss of one species may lead to the extinction of others (Parish et al. 2008). Thus, basic knowledge of these interactions as well as the effects of management is necessary to assess biodiversity impacts of rewetting peatland habitats.

Specific research needs in the area of biodiversity impacts of rewetting depend on the target species concerned and their habitats. In case of well-known 'flagship' species, such as the Aquatic Warbler (see chapter 4.3), there is already detailed knowledge of the species` distribution and habitat requirements at breeding sites. Furthermore, the effects of habitat management are currently being investigated to develop guidelines for management. However, the understanding of population sizes and processes is still weak. Successful conservation (especially of small remnant populations) has to take key population parameters such as productivity, survival, and between-site movements into account. These processes are essential for understanding the dynamics of local breeding populations, and for properly assessing their future fortunes. Restoring former, and creating new, breeding sites has a lower chance of success if these processes are neglected.

In case of less well-known target species such as insects, spiders and amphibians there is a general need for research on changes in species composition, population size and species abundance in drained and rewetted peatlands. Some invertebrate species exclusively feed on peatland dwarf shrubs (Parish et al. 2008) and the effects of rewetting on host and feeding plants need also to be assessed when specific invertebrate species are the object of concern. Peatland habitats host many plant and animal species itemized in Red Data Books and Red Lists. Peatland rewetting can help to restore habitats for these species. Furthermore, some species, (e.g. *Sphagnum* spp.), act as ecosystem engineers and are of central importance to the targeted habitat. Research on available seed banks or potential sources for immigration will help to identify appropriate management actions, such as active reintroduction of target species.

To assess the effects of management activities on peatland species and habitats, including rewetting and vegetation management, research should be setup in a Before-After-Control-Impact (BACI) design. Comparable control sites without rewetting help to separate management effects from other species-specific or site-specific, as well as off-site effects that may influence local and wider population dynamics (for instance habitat conditions on wintering or passage sites for migratory species, or wider population declines of a species due to climate change).

## 9.2 Recommended monitoring activities

Susanne Bärisch, Franziska Tanneberger, Jens Wunderlich & Hans Joosten

### 9.2.1 Introduction

While research is intended to provide knowledge about the scientific aspects of restoration (see chapter 9), monitoring is applied to track the status of peatlands or the response to management activities at all or at a subset of sites using standardised methods over longer periods. Often, simpler and more cost-effective methods are used in monitoring activities than in research projects.

The effects of once off (incl. maintenance) and regularly repeated management activities must be monitored in each specific case in order to improve the management continuously. So-called adaptive management (see e.g. The Conservation Measures Partnership 2007) is now a standard approach in ecosystem management. It strongly depends on monitoring results in order to critically review and adapt activities with respect to the management goal. Therefore, monitoring is important wherever peatlands are subject to regular, and often laborious or expensive management.

Several monitoring types can be distinguished depending on the aim of monitoring. For restoration projects intended to generate climate and biodiversity benefits, three types are important: (1) implementation monitoring, which assesses implementation of management activities; (2) effectiveness monitoring, which evaluates achievement of the objective of a management activity; and (3) verification monitoring, which determines the correctness of assumptions and applied models (Block et al. 2001, Noss & Cooperrider 1994).

Rewetting projects designed for validation under specific standards have to fulfil the monitoring requirements of these standards. This includes the development of a monitoring plan, regular monitoring, and clear reporting after implementation of the rewetting activities. Monitoring reports under the Verified Carbon Standard (VCS; see chapter 7.3) have to quantify greenhouse gas (GHG) emissions in the baseline and in the with-project situation, as well as GHG emissions from leakage (see chapter 5.3). Thus, monitoring must not be restricted to the project site but must include areas that might be affected by project activities. Permanence is a further condition to be monitored. A regular implementation monitoring helps to identify and address problems with regard to permanence. Under Climate, Community & Biodiversity (CCB) Standards (see chapter 5.3 and 7.4) and other standards, additional plans have to be developed for Community and Biodiversity Impact Monitoring.

### 9.2.2 Greenhouse gas monitoring via vegetation and water levels

Changes in greenhouse gas emissions can be assessed by monitoring changes in vegetation and water levels (i.e. via GESTs, see chapter 3.4 and 3.5). This monitoring also allows monitoring the effectiveness of the rewetting measures and verifying and adapting the assumptions and models applied for predicting vegetation developments in the with-project scenario. The monitoring of vegetation and water levels on control sites should further be applied to verify and adapt the assumptions for the baseline scenario.

Monitoring vegetation changes includes: (1) changes in floristic composition and derived vegetation types, (2) changes in the cover of shunt species, and (3) changes in the biomass carbon stock. All monitoring campaigns should use the same standardized form that has been developed for the project-monitoring plan.

In some cases, vegetation composition does not adequately reflect site conditions. On bare peat with sparse vegetation and in transition periods after rewetting, when the vegetation has not yet adapted to the new site conditions (c. 3–5 years after rewetting), a direct monitoring of water levels is needed to better reflect site conditions and assess GHG emissions. For verification monitoring after the transition period (to check whether GESTs have developed as predicted in the project description), monitoring can follow a frequency of once every five years, as is common in carbon projects. To monitor the effectiveness of the measures on a shorter time frame and to be able to adapt rewetting measures, a detailed direct hydrological monitoring of the rewetted sites is, however, recommended. Ginzler & Steiner (1997) compared three methods for water level measurements on peatlands:

1. Observation wells with manual readout of the recent ground water level;
2. Minimum-maximum recorders (Bragg et al. 1994) with manual readout of the recent water level and of the minimum and maximum water levels since the last observation;
3. Automatic water level measuring devices.

The methods differ in initial costs, frequency of field readout as well as accuracy of the water level data achieved. Observation wells have the lowest initial costs but require a high frequency in manual readout (preferably once a week). Water levels can only be compared between observations and the real water level fluctuations are difficult to determine. For effectiveness monitoring, readout should rather take place under dry weather conditions than following rain events. Minimum-maximum recorders ideally provide more defined data on water level fluctuations also for the period between two readouts. With relatively low costs compared to automatic water level measuring devices and a lower required frequency in manual readout compared to observation wells, minimum-maximum recorders are a good trade-off between required resources and data accuracy. The disadvantage of these devices, however, is their susceptibility to damage (e.g. by animals) and the subsequent need for maintenance and repair (Broska 2010). Automatic water level data loggers provide the highest quality of data with the lowest necessary readout frequency. All types of measuring devices are to some extent prone to vandalism and robbery.

Because of the huge potential for peatland rewetting in Central and Eastern Europe and in order to arrive at a more resource-effective monitoring of large-areas, it is necessary to advance the monitoring process, e.g. from GEST mapping in the field to GEST monitoring by remote sensing. Direct monitoring of groundwater levels of drained peatlands by remote sensing is not yet possible, but with the availability of the latest satellite generations, advances have been made in remote sensing of wetland vegetation types and soil humidity. With spatial ground resolutions of less than 1 m, satellite imagery has become more and more interesting for large scale applications (e.g. Frick 2006). Further research in this field is intended to develop and improve remote sensing methodologies for monitoring GESTs.

### 9.2.3 Biodiversity monitoring

The assessment of biodiversity benefits of rewetting requires monitoring of the performance of relevant peatland species at the site before and after rewetting. Demonstration of additional biodiversity benefits of a carbon project allows for a better acceptance of the project and could provide the opportunity to sell carbon credits with a premium price under the CCB Standards (see chapter 7.5).

For continuous assessment of the population status of target species and for the evaluation of the effects of management measures, a regular and standardised monitoring of populations of target and indicator species (see chapter 4.3) and habitats is necessary. Frequency, intensity and scale of the monitoring depend on population and habitat size, and on available financial resources (see chapter 4.3). Key elements to monitor in the species' range are population size, distribution and trend, population state (productivity), habitat size and suitability, and ideally also other demographic parameters. To separate management effects from other site-specific or wider species-specific effects, it is important to monitor also control sites (before-after-control-impact schemes, see also chapter 9.1).

In case of the Aquatic Warbler *Acrocephalus paludicola* (see chapter 4.1), as a minimum the population size must be monitored by regular counts of singing males. These counts should be undertaken synchronously across adjacent breeding sites during the singing activity peaks around sunset. The usual approach includes a first count in late May and a second in late June, matching the peak laying dates for early and late broods, respectively. At smaller sites with up to 50–100 singing males full counts should be feasible. At larger sites, a systematic sampling of randomly or systematically selected representative transects is recommended. Monitoring of breeding performance is recommended for key breeding sites, small-threatened satellite populations (Pomerania, Hungary), and for sites where management measures have taken place. An index based on standardized regular counts of feeding females is recommended for basic monitoring. More detailed studies on nest success with a limited duration are justified and strongly recommended where this index decreases or is generally low. Information on age ratios should already be available from ringing activities during autumn migration. For monitoring purposes, data should be collated and analysed on a regular basis.

Until recently, remote sensing applications for biodiversity assessment at local scale were virtually non-existent. Animal populations and small-scale vegetation communities were not or at least difficult to detect with common remote sensing techniques. This has changed with the introduction of leadoff sub-meter remote sensing satellites and hyperspectral airborne techniques since the beginning of the century. Rapidly developing sensor techniques applied with well performing image processing methods and combined with ancillary geodata offer new opportunities for a biodiversity monitoring (Kerr & Ostrovsky 2003, Wulder et al. 2004, Turner et al. 2003).

### 9.2.4 Monitoring of community benefits

Peatland rewetting can provide environmental, social, and economical benefits for local communities (see e.g. chapter 6.5). Assessing, regular monitoring, and communicating these benefits can enhance the acceptance of the project and related measures. In addition, some carbon standards (e.g. CCB Standards, Gold Standard) require monitoring of the impact of project activities on local communities and sustainable development. The project is eligible under these standards only if a positive impact is verified.

A set of indicators can help to determine and monitor benefits for local communities. For projects registered under the Gold Standard (Ecofys, TÜV-Süd & Field 2009) for example, benefits are assessed and monitored based on specific indicators and potential parameters for each indicator that are expected to best reflect the attributed impact of the project. Air and water quality, biodiversity, and microclimate can be used to indicate environmental benefits for communities. Indicators for social development can include availability of human and institutional capacity, access to clean energy services and recreation possibilities, as well as quality of employment. A third set of indicators addressing the economic and technical benefits can contain quantitative employment, income generation, technology transfer, regional net added value, technological self-reliance, and others.

Under certain conditions negative impacts from rewetting projects can arise. In case of project sites that are still in utilization the opportunity costs of rewetting, i.e. the resigned income from former land use, can be higher than the expected income generated by the project. Additional income generation possibilities, e.g. by biomass use from rewetted peatlands or by tourism (e.g. bird-watching), must be assessed to mitigate negative and eventually enhance positive community effects. A sound cost-benefit analysis of the project activities and a regular verification monitoring of included parameters like biomass yield and visitors is recommended. On abandoned sites with low opportunity costs the potential negative impacts are very low or even absent. However, negative external effects like flooding of adjacent arable lands after damming of drainage ditches can occur and must be monitored and measures adapted.

Effects of project activities on local communities occur on a wider perspective than within the defined project boundaries and on different stakeholder levels. Thus, the scale of monitoring is of importance when designing a community benefits monitoring plan.

### 9.2.5 Data storage and publication

In all kinds of monitoring, systematic and safe storage of data is of utmost importance. Monitoring data can produce much more insights when shared between related studies and over longer period of time. Storage may be in local or preferably in open access international databases, certainly after publication of the data. The publication of monitoring outcomes is of similar importance, not the least in order to prevent 'the wheel is being continuously reinvented'.

# 10 Acknowledgements

This guidebook summarises the outcomes of the project 'Restoring peatlands and applying concepts for sustainable management in Belarus – climate change mitigation with economic and biodiversity benefits'. The project was made possible through funding received from the German Federal Ministry for the Environment, Nature Conservation and Nuclear Safety (BMU) in the framework of its International Climate Initiative (ICI). KfW Entwicklungsbank managed the project on behalf of BMU and supported the project intensively at all stages of its development.

Without the continuous support, interest and progressive attitude of the Ministry of Natural Resources and Environmental Protection of Belarus and of the Ministry of Forestry of the Republic of Belarus the project would not have been possible.

We commend the project partners, the Royal Society for the Protection of Birds, UK, APB-BirdLife Belarus and the Michael Succow Foundation, Germany, for their cooperative efforts to implement this innovative project. Intercultural and interdisciplinary differences were overcome, skills and knowledge were shared and valuable lessons have been learnt as a result. Support by the United Nations Development Programme (UNDP) in Belarus and by the National Academy of Sciences of Belarus was essential for the success of the project.

Like the project, this publication covers a wide range of topics related to peatland rewetting and associated effects, including greenhouse gas emissions and their accounting, biodiversity benefits, and sustainable use of wetland biomass. We owe great respect to the numerous experts who have contributed to the development and finalisation of this book.

Our thanks are extended to the following people for their critical and detailed reviews: Hans Joosten, John Couwenberg, Rob Field, Jenny Schulz, Jack Foxall, Olga Chabrouskaya, Annett Thiele, and Merten Minke – publication would not have been possible without them. Translation, always a challenge with technical, scientific texts, would not have been possible without the support of the following specialists: Alexej Gorin, Tatsiana Broska, Tatsiana Yarmashuk, John Couwenberg, Nadzeya Liashchynskaya, and Iryna Dabravolskaya. René Fronczek and Stephan Busse redrew all the graphs and maps. Proofreading was done by Anne Dowden (English) and Sofya Morozova (Russian) and supervised by Jack Foxall and Olga Chabrouskaya. Invaluable assistance during the final stage was provided by Fanny Mundt. Thanks to all as well as to everybody else who contributed to the preparation of this book. Our final sincere thanks go to E. Schweizerbart Science Publishers, especially to Hilke Bornholdt and Andreas Nägele.

Franziska Tanneberger & Wendelin Wichtmann

# References

Aerts, R., Callaghan, T.V., Dorrepaal, E., van Logtestijn, R.S.P. & Cornelissen, J.H.C. (2009): Seasonal climate manipulations result in species-specific changes in leaf nutrient levels and isotopic composition in a sub-arctic bog. Funct. Ecol. 23: 680–688.

Alm, J., Talanov, A., Saarnio, S., Silvola, J., Ikkonen, E., Aaltonen, H., Nykänen, H. & Martikainen, P.J. (1997): Reconstruction of the carbon balance for microsites in a boreal oligotrophic pine fen, Finland. Oecologia 110: 423–431.

Anderson, P., Tallis, J.H. & Yalden, D.W. (1997): Restoring Moorland. Peak district moorland management project. Phase III report. Unpublished report. Bakewell.

Anderson, P., Artz, R., Bragg, O., Lunt, P., Marrs, R. (2010): Peatland Biodiversity. Draft Scientific Review. IUCN UK Peatland Programme's Commission of Inquiry into Peatland Restoration.

Anderson, R. (2010): Restoring afforested peat bogs: results of current research. Northern Research Station Roslin, FCRN 006, 8 p.

Aquatic Warbler Conservation Team (AWCT) (1999): World population, trends and conservation status of the Aquatic Warbler Acrocephalus paludicola. Vogelwelt 120: 65–85.

Augustin, J. & Chojnicki, B. (2008): Austausch von klimarelevanten Spurengasen, Klimawirkung und Kohlenstoffdynamik in den ersten Jahren nach der Wiedervernässung von degradiertem Niedermoorgrünland. In: Gelbrecht, J., Zak, D. & Augustin, J. (Eds.): Phosphor- und Kohlenstoff-Dynamik und Vegetationsentwicklung in wiedervernässten Mooren des Peenetals in Mecklenburg-Vorpommern – Status, Steuergrößen und Handlungsmöglichkeiten. Berichte des IGB 26: 50–67.

Augustin, J., Merbach, W., Schmidt, W. & Reining, E. (1996): Effect of changing temperature and water table on trace gas emission from minerotrophic mires. Angew. Botanik 70: 45–51.

Baggs, E.M. (2008): A review of stable isotope techniques for $N_2O$ source partitioning in soils: recent progress, remaining challenges and future consideration. Rapid Communications in Mass Spectrometry 22: 1664–1672.

Baker-Blocker, A., Donahue, T.M. & Mancy, K.H. (1977): Methane flux from wetlands areas. Tellus 29: 245–250.

Bakker, J.P. & Olff, H. (1995): Nutrient dynamics during restoration of fen meadows by hay making without fertiliser application. In: Wheeler, B.D., Shaw, S.C, Fojt, W.J. & Robertson, R.A. (eds): Restoration of Temperate Wetlands, p. 143–166. John Wiley, Chichester.

Baldocchi, D.D. (2003): Assessing the eddy covariance technique for evaluating carbon dioxide exchange rates of ecosystems: past, present and future. Global Change Biol. 9: 479–492.

Bambalov, N.N. & Rakovich, V.A. (2005): Rol bolot v biosfere. [Mires in the biosphere] Minsk. [in Russian]

Bambalov, N.N., Belenkiy, S.G., Dubovets, A.G. & Smirnova, V.V. (1981): Genezis i evoliutsiia torfianykh mestorozhdenii Belorussii. [Genesis and evolution of peatlands in Belarus] Torfianaia prom., No. 10. [in Russian]

Dambalov, N.N., Dubovets, A.G. & Belenkiy, S.G. (1990): Bolotoobrazovatelnye protsessy na territorii Belorussii. [Processes of peatland formation in Belarus] Problemy Polesia. Vol.13 p. 75–90. [in Russian]

Bambalov, N.N., Tanovitskiy, I.G. & Belenkiy, S.G. (1992): Razvitie issledovaniy v oblasti genezisa, ispolzovaniia i okhrany torfianykh mestorozhdenii Belarusi. [Research development in formation, utilisation and protection of peatlands in Belarus] Tverdye goriuchie otlozheniia Belarusi i problemy okhrany okruzhaiushchei sredy. Minsk. pp. 27–39. [in Russian]

Barthelmes, A., Couwenberg, J. & Joosten, H. (2009): Peatlands in National Inventory Submissions 2009 – An analysis of 10 European countries. Wetlands International, Ede, 26 p. http://tinyurl.com/meo5hb.

Barz, M., Wichtmann, W. & Ahlhaus, M. (2006): Energetic Utilisation of Common Reed for Combined Heat and Power Generation, Proceedings of the 2nd International Baltic Bioenergy Conference, Stralsund, 02.–04. Nov.

Bellebaum, J. (2004): Feasibility of the Use of Biomass Yielded through Management of Aquatic Warbler Breeding Sites in Pomerania in Caloric or Biogas Power Plants. Desk study prepared for the Royal Society for the Protection of Birds, 31 p.

Belyatskaya, O., Volosuk, S., Grummo, D.G., Tanovitskaya, N.I., Maksimenkov, M., Kozulin, A.V., Ermokhin, M, Skuratovich, A. & Shevtsov, N. (2008): Obosnovanie optimisatsii gidrologicheskovo rezhima tsentralnoy chasti bolota Jelnia, v tselakh predotvrazhenia torfyanykh pozharov, vosstanovlenia narushennykh ekosistem torfyanika i biorasnoobrazia. [Justification of optimization of hydrological regime of central part of Jelnia peatland for prevention of peat fires, restoration of degraded peatland ecosystems and biodiversity]. GEF SGP Report, Minsk. [in Russian]

Benken, T. (1988): Der Einfluß der Renaturierungsmaßnahmen auf die Libellenfauna im NSG "Rotes Moor". Renaturierungsprojekte und Regenerationsprozesse an Mooren in Mittelgebirgslandschaften. Tagung des BMU und der DGMT in der Rhön, September 1988. Kurzfassung der Referate: 10–13.

Benstead, P., Drake, M., José, P., Mountford, O., Newbold, C. & Treweek, J. (1997): The wet grassland guide: Managing floodplain and coastal wet grasslands for wildlife. Edited by Treweek, J., José, P. & Benstead, P.S., Royal Society for the Protection of Birds, English Nature and Institute of Terrestrial Ecology. 254 p.

Best, E.P.H. & Jacobs, F.H.H. (1997): The influence of raised water table on carbon dioxide and methane production in ditch-dissected peat grasslands in the Netherlands. Ecol. Engineering 8: 129–144.

Bibby, C.J. & Lunn, J. (1982): Conservation of reedbeds and their avifauna in England and Wales. Biol. Conserv. 23: 167–186.

Biewer, H., Poschlod, P., Bühler, F., Metzler, S. & Böcker, R. (1994): Wiedervernässung und Wiederherstellung artenreicher Feuchtwiesen im geplanten Naturschutzgebiet Südliches Federseeried (vegetationskundlicher Teil). Ausgangszustand (Vegetation), Versuchsplanung und Methoden. Veröff. PAÖ 8. 289–303.

BirdLife International (2004): Birds in Europe: population estimates, trends and conservation status. Cambridge, UK: BirdLife International. (BirdLife Conservation Series No. 12).

BirdLife International (2009): International Species Action Plan for the Aquatic Warbler Acrocephalus paludicola. BirdLife International

BirdLife International (2011): Important Bird Areas factsheet: Zvanets. Downloaded from http://www.birdlife.org on 23/02/2011.

Blaustein, R.J. (2010): Recarbonizing the Earth. World Watch 23/1: 24–28.

Bliefert, C. (1995): Umweltchemie (1. Nachdruck). VCH Verlagsgesellschaft, Weinheim, 453 p.

Block, W.M., Franklin, A.B., Ward, J.P. Jr., Ganey, J.L. & White, G.C. (2001): Design and Implementation of Monitoring Studies to Evaluate the Success of Ecological Restoration on Wildlife. Restorat. Ecol. 9: 293–303.

Bogdanovskaya-Gienef, I.D. (1953): Tipy vnutrizalezhnoy vody. [Types of subsurface water] Trudy GGI 39(39), Leningrad, p. 81–95. [in Russian]

Bogdanovskaya-Gienef, I.D. (1969): Zakonomernosti formirovania sfagnovykh bolot verkhogo tipa. [Principles of the formation of Sphagnum bogs] Nauka, Leningrad, 186 p. [in Russian]

Bonneville, M.-C., Strachan, I.B., Humphreys, E.R. & Roulet, N.T. (2008): Net ecosystem $CO_2$ exchange in a temperate cattail marsh in relation to biophysical properties. Agric. Forest Meteorol. 148: 69–81.

Bortoluzzi, E., Epron, D., Siegenthaler, A., Gilbert, A. & Butler, A. (2006): Carbon balance of a European mountain bog at contrasting stages of regeneration. New Phytol. 172: 708–718.

Bragg, O.M., Hulme, P.D., Ingram, H.A.P., Johnston, J.P. & Wilson, A.I.A. (1994): A maximum-minimum recorder for shallow water tables, developed for ecohydrological studies on mires. J. Appl. Ecol. 31: 589–592.

Breeuwer, A., Heijmans, M.M.P.D., Gleichman, M., Robroek, B.J.M. & Berendse, F. (2009): Response of Sphagnum species mixtures to increased temperature and nitrogen availability. Plant Ecol. 204: 97–111.

Bretschneider, A. (2010): Moorwald oder Birkenstadium des degenerierten Hochmoores? Über den Umgang mit Birken im Moor. Ann. Sci. Bios. Vosges du Nord-Pfälzerwald 15: 171–178.

Briemle, G. (1978): Pioniergehölze auf Moorbrachen in Abhängigkeit vom Moortyp. Telma 8: 153–169.

Briemle, G. (1980): Untersuchungen zur Verbuschung und Sekundärbewaldung von Moorbrachen im südwestdeutschen Alpenvorland. Dissertationes Botanicae. Vaduz: 286.

Briemle, G. (1990): Natürliche Bewaldungstendenz und Mindestpflege von Moorbiotopen. In: Göttlich, K. H. (Ed.) Moor- und Torfkunde, 3th ed. Schweizerbart, Stuttgart: 496–501.

Brix, H. (1993): Macrophyte-mediated oxygen transfer in wetlands: Transport mechanisms and rates. In: Moshiri, G.A. (ed.): Constructed wetlands for water quality improvement. Lewis Publishers, Boca Raton, p. 391–398.

Brooks, S. & Stoneman, R. (1997): Conserving bogs. The management handbook. Edinburgh: The Stationary Office. 286 p.

Broska, T. (2010): Relationship between vegetation and water level in Yelnia bog (Belarus) and comparison of water level measurement approaches. MSc thesis. Greifswald University. 48 p.

Burvall, J. & Hedman, B. (1998): Perennial rhizomatous grass. The delayed harvest system improves fuel characteristics for reed canary grass. In: El Bassam, N., Behl, R.K. & Prochnow, B. (ed.): Sustainable agriculture for food, energy and industry. James & James Ltd., London. p. 916–918.

Buttler, A., Warner, B.G., Grosvernier, P. & Matthey, Y. (1996): Vertical patterns of testate amoebae (Protozoa: Rhizopoda) and peat-forming vegetation on cutover bogs in Jura, Switzerland. New Phytol. 134: 371–382.

Buttler, A., Grosvernier, P. & Matthey, Y. (1998): Development of Sphagnum fallax diaspores on bare peat with implications for the restoration of cut-over bogs. J. Appl. Ecol. 35: 800–810.

Byrne, K.A., Chojnicki, B., Christensen, T.R., Drösler, M., Freibauer, A., Friborg, T., Frolking, S., Lindroth, A., Mailhammer, J., Malmer, N., Selin, P., Turunen, J., Valentini, R. & Zetterberg, L. (2004): EU peatlands: Current carbon stocks and trace gas fluxes. CarboEurope-GHG Concerted Action – Synthesis of the European Greenhouse Gas Budget, Report 4/2004. 58 p.

Byshnev I.I., Stavrovsky D.D., Pikulik M.M. & Zishachkin A.K. (1996): Atlas nazemnykh pozvonochnykh: Berezinsky biozferny zapovednik. [Atlas of terrestrial vertebrates: Berezinsky Biosphere Reserve] Nauka i Tekhnika, Minsk, 304 p. [in Russian]

Caro, T.M. (2010): Conservation by Proxy: Indicator, Umbrella, Keystone, Flagship, and Other Surrogate Species. Island Press, Washington, DC.

Chanton, J.P. & Whiting, G.J. (1995): Trace gas exchange in freshwater and coastal marine environment: ebullition and transport by plants. In: Matson, P.A. & Harriss, R.C. (eds): Biogenic trace gases: measuring emissions from soil and water. Methods in Ecology, Blackwell Science, Oxford, p. 98–125.

Chapin, F.S. III, Matson, P.A. & Mooney, H.A. (2002): Principles of Terrestrial Ecosystem Ecology. Springer. 436 p.

Chapin, F.S., Woodwell, G.M., Randerson, J.T., Rastetter, E.B., Lovett, G.M., Baldocchi, D.D. et al. (2006): Reconciling carbon-cycle concepts, terminology, and methods. Ecosystems 9: 1041–1050.

Chapman, S., Buttler, A., Francez, A.-J., Laggoun-Défarge, F., Vasander, H., Schloter, M., Combe, J., Grosvernier, P., Harms, H., Epron, D., Gilbert, D. & Mitchell, E. (2003): Exploitation of northern peatlands and biodiversity maintenance: a conflict between economy and ecology. Front. Ecol. Environ. 1: 525–532.

Chojnicki, B.H., Augustin, J. & Olejnik, J. (2007): Impact of reflooding on greenhouse gas exchange of degraded fen peatlands, Proceedings of the 1st International Symposium on Carbon in Peatlands, Wageningen, The Netherlands, 15–18 April 2007.

Clausnitzer, U. & Succow, M. (2001): Vegetationsformen der Gebüsche und Wälder. In: Succow, M. & Joosten, H. (Eds.): Landschaftsökologische Moorkunde. Schweizerbart, Stuttgart: 161–170.

Coppolillo, P., Gomez, H., Maisels, F. & Wallace, R. (2004): Selection criteria for suites of landscape species as a basis for site-based conservation. Biol. Cons. 115: 419–430.

Couwenberg, J. (2007): Biomass energy crops on peatlands: on emissions and perversions. – IMCG-Newsletter 2007/3: 12–14.

Couwenberg, J. (2009a): Emission factors for managed peat soils – An analysis of IPCC default values. Wetlands International, Ede, 16 p. http://tinyurl.com/kmeqet.

Couwenberg, J. (2009b): Methane emissions from peat soils (organic soils, histosols) – Facts, MRV-ability, emission factors. Wetlands International, Ede, 16 p. http://tinyurl.com/lch5jj.

Couwenberg, J. (2011): Greenhouse gas emissions from managed peat soils: is the IPCC reporting guidance realistic? Mires and Peat, Volume 8 (2011), Article 02, 1–10

Couwenberg, J., Augustin, J., Michaelis, D. & Joosten, H. (2008): Emission reductions from rewetting of peatlands. Towards a field guide for the assessment of greenhouse gas emissions from Central European peatlands. Duene/RSPB, Greifswald/Sandy. 27 p.

Couwenberg, J., Dommain, R. & Joosten, H. (2010): Greenhouse gas fluxes from tropical peatlands in south-east Asia. Global Change Biol. 16: 1715–1732. doi: 10.1111/j.1365-2486.2009.02016.x, http://circa.europa.eu/Public/irc/env/biodiversity_climate/library?l=/ghgfluxestropicalpeatlan/_EN_1.0_&a=d.

Couwenberg, J., Thiele, A., Tanneberger, F., Augustin, J., Bärisch, S., Dubovik, D., Liashchynskaya, N., Michaelis, D., Minke, M., Skuratovich, A. & Joosten, H. (2011): Assessing greenhouse gas emissions from peatlands using vegetation as a proxy. – Special Issue Wetland Restoration, Hydrobiologia 674: 67–89

Cowie, N.R., Sutherland, W.J., Ditlhogo, M.K.M. & James, R. (1992): The effects of conservation management of reed beds. I. The flora and litter disappearance. J. Appl. Ecol. 29: 277–284.

Cramp, S. (ed) (1992): The Birds of the Western Palearctic. Vol. 6. Oxford Univ. Press, Oxford.

Crill, P., Hargreaves, K. & Korhola, A. (2000): The role of peat in Finnish greenhouse gas balances. Ministry of Trade and Industry Finland, Helsinki.

Dallimer, M., Marini, L., Skinner, A.M.J., Hanley, N., Armsworth, P.R. & Gaston, K.J. (2010): Agricultural land-use in the surrounding landscape affects moorland bird diversity. Agricult. Ecosys. Environ. 139: 578–583.

Darroch-Thompson, M. D. & Ash, N. (2003): Biofuel Heating at The Farmhouse Old Moor Visitor Centre. A feasibility study for RSPB. Draft. Hunston Egineering Ltd.

Davis, S.R. & Wilkinson, D.M. (2004): The conservation management value of testate amoebae as 'restoration' indicators: speculations based on two damaged raised mires in northwest England. Holocene 14: 135–143.

Denmead, O.T. (2008): Approaches to measuring fluxes of methane and nitrous oxide between landscapes and the atmosphere. Plant Soil 309: 5–24.

Dias, A.T.C., Hoorens, B., van Logtestijn, R.S.P., Vermaat, J.E. & Aerts ,R. (2010): Plant species composition can be used as a proxy to predict methane emissions in peatland ecosystems after land-use changes. Ecosystems 13: 526–538.

Ditlhogo, M.K.M., James, R., Laurence, B.R., Sutherland, W.J. (1992): The effects of conservation management of reed beds. I. The invertebrates. J. Appl. Ecol. 29: 265–276.

Dokuchaev, V.V. (1949): Po voprosu osusheniya bolot voobzhe i v chastnosti ob osushenii Polessya. [About peatland drainage and in particular the drainage of Palessie] Sobr. soch. Vol. 1. Izdatelstvo AN USSR, Moscow, p. 27-65.

Dombrovsky, V. (2010): The diet of the Greater Spotted Eagle (Aquila clanga) in Belarusian Polesie. Slovak Rapt. J. 4: 23–36.

Drobenkov, S.M., Novitsky, R.V., Kosova, L.V., Ryshevich, K.K. & Pikulik, M.M. (2005): The amphibians of Belarus. Advances in amphibian research in the former Soviet Union 10: 1–156.

Drösler, M. (2005): Trace gas exchange and climatic relevance of bog ecosystems, southern Germany. PhD thesis, Technische Universität München. 182 p.

Drösler, M. (2008): Von der Spurengasmessung zur Politikberatung – interdisziplinärer Ansatz und erste Ergebnisse des Verbundprojekts "Klimaschutz – Moornutzungsstrategien". Presentation given at the BfN workshop "Biodiversität und Klimawandel" http://www.bfn.de/4399.html. Accessed 16.11.2010.

Drösler, M., Freibauer, A., Christensen, T.R. & Friborg, T. (2008): Observations and status of peatland greenhouse gas emissions in Europe. In: Dolman,

A.J., Valentini, R. & Freibauer, A. (eds): The Continental-Scale Greenhouse Gas Balance of Europe. Ecological Studies 203: 243–261.

Dubovets, A.G. (1981): O geneticheskikh tipakh torfianykh zalezhey. [On genetic types of peat] Problemy Polesia. Vol.7. p.129–133. [in Russian]

Dyrcz, A. (2010): Polish wetland and marshland birds: looking for undisturbed references. Ardea 98: 301–308

Dyrcz, A., Wink, M., Backhaus, A., Zdunek, W., Leisler, B. & Schulze Hagen, K. (2002):. Correlates of multiple paternity in the Aquatic Warbler (Acrocephalus paludicola). J. Ornithol. 143: 430–439.

Ecofys, TÜV-Süd & Field (2009): Gold Standard Toolkit 2.1. http://www.cdmgoldstandard.org.

Eder, G., Haslinger, W. & Wörgetter, M. (2004): Gutachten energetische Nutzung von Schilfpellets. Im Auftrag des Amtes der Burgenländischen Landesregierung, Abt.9, Wasser- und Abfallwirtschaft, 53 p.

Edom, F. (1991): Untersuchungen zum Wasserhaushalt des Naturschutzgebietes Mothäuser Heide als Beitrag zur Kenntnis gefährdeter Moorökosysteme des Erzgebirges. Diploma-thesis TU Dresden, 87 p., app.

Edom, F. (2001a): Revitalisierung von Regenmooren an ihrer klimatischen Arealgrenze. Kap. 9.3.8. – In: Succow, M. & Joosten, H. (Eds.): Landschaftsökologische Moorkunde: 534–543. Schweizerbart. Stuttgart.

Edom, F. (2001b): Moorlandschaften aus hydrologischer Sicht. Kapitel 5 In: Succow, M. & Joosten, H. (Eds.): Landschaftsökologische Moorkunde. Schweizerbart, Stuttgart: 185–228.

Edom, F. & Golubcov, A.A. (1996): Zum Zusammenhang von Akrotelmeigenschaften und einer potentiell natürlichen Ökotopzonierung in Mittelgebirgsregenmooren. Verhandl. der Gesellsch. f. Ökol. 26, Stuttgart, p. 221–228.

Edom, F. & Succow, M. (1998): Regenmoore. In: Wegener, U. (ed.): Naturschutz in der Kulturlandschaft. Gustav Fischer Jena, Stuttgart, Lübeck, Ulm. p. 150–156.

Edom, F. & Wendel, D. (1998): Grundlagen zu Schutzkonzepten für Hang-Regenmoore des Erzgebirges. – In: Sächsische Landesstiftung Natur und Umwelt (ed.): Ökologie und Schutz der Hochmoore im Erzgebirge. 31–77.

Edom, F., Dittrich, I., Goldacker, S. & Kessler, K. (2007): Die hydromorphologisch begründete Planung der Moorrevitalisierung im Erzgebirge. In: Praktischer Moorschutz im Naturpark Erzgebirge/Vogtland und Beispiele aus anderen Gebirgsregionen. Sächsische Landesstiftung Natur und Umwelt, Akademie, Grillenburg, 19–32.

Edom, F., Stegmann, H., Dittrich, I., Keßler, K. & Schua, K. (2009a): Geoökologische und hydrologische Prinzipien eines möglichen Huminstoffrückhalts in sauren Mooren. Erkenntnisstand und Versuch einer Synthese. Dr. Dittrich & Partner Hydro-Consult GmbH & HYDROTELM – Edom & Stegmann, Bannewitz & Dresden, im Auftrag der Landesdirektion Chemnitz, 39 p. & 5 app.

Edom, F., Keßler, K., Stegmann, H., Wendel, D., Dittrich, I. & Münch, A. (2009b): Hydrologisches und moorkundliches Gutachten zur Konkretisierung von Erhaltungs- und Entwicklungsmaßnahmen für das Moor Stengelhaide im FFH-Gebiet "Mothäuser Heide". Dr. Dittrich & Partner Hydro-Consult GmbH & HYDROTELM – Edom & Stegmann, Bannewitz & Dresden, im Auftrag der LfULG Zwickau, 77 p., 20 app.

Edom, F., Dittrich, I. & Keßler, K. (2010): Hydrogenetische und hydromorphologische Grundlagen der Bewertung von Moor- und Moorwald – Lebensräumen zur Umsetzung der FFH-Richtlinie der EU – Erfahrungen aus dem Erzgebirge. Coll. Tourbières, Ann. Sci. Rés. Bios. Trans. Vosges du Nord-Pfälzerwald – 15 (2009–2010): 230–250.

Efremov, S.P. (1987): Pionernije drevostoi osusennych bolot. [Pioneer shrubs in drained peatlands] Nauka, Novosibirsk, 248 p. [in Russian]

Eggleston, H.S., Srivastava, N., Tanabe, K., Baasansuren, J. & Fukuda, M. (eds) (2011): IPCC Expert Meeting on HWP, Wetlands and Soil $N_2O$. Meeting Report of the IPCC Expert Meeting on HWP, Wetlands and Soil $N_2O$, Geneva, Switzerland, 19–21 October, 2010, Pub. IGES, Japan 2011.

Ellenberg, H., Weber, H. E., Düll, E., Wirth, V. & Werner, W. (1992): Zeigerwerte von Pflanzen in Mitteleuropa. Scripta Geobot. 18: 1–258.

Eugster, W., Kling, G., Jonas, T., McFadden, J.P., Wüest, A., MacIntyre, S. & Chapin, III F.S. (2003): $CO_2$ exchange between air and water in an Arctic Alaskan and midlatitude Swiss lake: Importance of convective mixing. Geophys. Res., 108(D12), 4362, DOI:10.1029/2002JD002653.

Flessa, H., Wild, U., Klemisch, M. & Pfadenhauer, J. (1998): Nitrous oxide and methane fluxes from organic soils under agriculture. Eur. J. Soil Sci. 49: 327–335.

FNR (2006): Handreichung Biogasgewinnung und -nutzung. 3rd edition. Fachagentur Nachwachsende Rohstoffe, Gülzow. 232 p.

Främbs, H. & Nick, K.-J. (2001): Multivariate Analyse der Vegetation im Leegmoor. In: Nick, K.-J., Löpmeier, F.-J., Schiff, H., Blankenburg, J., Gebhardt, H., Knabke, C., Weber, H. E., Främbs, H. & Mossakowski, D.: Moorregeneration im Leegmoor/Emsland nach Schwarztorfabbau und Wiedervernässung. Angewandte Landschaftsökologie 38. p. 91–106.

Francez, A.-J., Gogo, S. & Josselina, N. (2000): Distribution of potential $CO_2$ and $CH_4$ productions, denitrification and microbial biomass C and N in the profile of a restored peatland in Brittany (France). Eur. J. Soil Biol. 36: 161–168.

Frankl, R. (1996): Zur Vegetationsentwicklung in den Rottauer Filzen (südliche Chiemseemoore) im Zeitraum von 1957–1992. Bayreuther Forum Ökologie 37.

Frey, W. & Lösch, R. (1998): Lehrbuch der Geobotanik. Gustav Fischer, Stuttgart.

Frick, A. (2006): Beiträge höchstauflösender Satellitenfernerkundung zum FFH-Monitoring – Entwicklung eines wissensbasierten Klassifikationsverfahrens und Anwendung in Brandenburg. Dissertation. Technische Universität Berlin. Berlin.

Fritz, C., Pancotto, V.A., Elzenga, J.T.M., Visser, E.J.W., Grootjans, A.P., Pol, A., Iturraspe, R., Roelofs, J.G.M. & Smolders, A.J.P. (2011) : Zero methane emission bogs: extreme rhizosphere oxygenation by cushion plants in Patagonia. New Phytologist 190: 398–408.

Frolking, S.E., Roulet, N. & Fuglestvedt, J. (2006): How northern peatlands influence the Earth's radiative budget: Sustained methane emission versus sustained carbon sequestration. J. Geophys. Res. (G), 111, D08S03, doi: 10.1029/2005JG000091.

Gierk, M. & Kalbe, L. (2001): Ökologische Bewertung von Wiedervernässungsgebieten in Brandenburg – dargestellt am Beispiel der Nuthe-Nieplitz-Niederung. Naturschutz und Landschaftspflege in Brandenburg 10: 52–61.

Ginzler, C. & Steiner, G.M. (1997): Moor-Monitoring. In: Traxler, A. (1997): Handbuch des vegetationsökologischen Monitorings – Methoden, Praxis, angewandte Projekte. Teil A: Methoden. Monographien Band 89A. Umweltbundesamt Austria. Wien.

Girard, M., Lavoie, C. & Thériault, M. (2002): The regeneration of a highly disturbed ecosystem: A mined peatland in southern Quebec. Ecosystems 5: 274–288.

Gorham, E. (1991): Northern peatlands – role in the carbon-cycle and probable responses to climatic warming. Ecol. Appl. 1: 182–195.

Gremer, D. & Edom, F. (1994): Entwicklungskonzept Anklamer Stadtbruch. Landschaftsökologische Problemanalyse. Bericht des Botanischen Institutes der Universität Greifswald im Auftrag des Umweltministeriums Mecklenburg-Vorpommern in Schwerin, 59 p. & 13 app.

Gremer, D. & Poschlod, P. (1991): Vegetationsentwicklung im Torfstichgebiet des Haidgauer Rieds (Wurzacher Ried) in Abhängigkeit von Abbauweise und Standort nach dem Abbau. Verhandlungen der Gesellschaft für Ökologie 20: 315–324.

Grummo, D. & Thiele, A. (2010): Der einzigartige Wald-Moorkomplex Jel'nia – Belarus. Aktueller Zustand, Restauration und Monitoring. TELMA 40. p. 195–210.

Grummo, D.G., Sozinov, O.V., Zelenkevich, N.A., Ilyuchik, M.A., Tanovitskaya, N.I., Puchilo, A.V., Grechko, A.M., Skuratovich, A.N., Dubovik, D.V., Vlasov, B.P., Shevzov, N.V., Kuzmicheva, N.A. & Broska, T.V. (2010): Flora i rastitelnost respublikansogo landshaftnogo zakaznika Yelnia. [Flora and vegetation of landscape zakaznik Elnia] National Academy of Sciences of the Republic of Belarus, Institute of Experimental Botany. Minsktipproekt, Minsk, 200 p. [in Russian]

Gryseels, M. (1989): Nature management experiments in a derelict reedmarsh. I: Effects of winter cutting. Biol. Conserv. 47: 171–193.

Guimbaud, C., Catoire, V., Gogo, S., Robert, C., Chartier, M., Laggoun-Défarge, F., Grossel, A., Albéric, P., Pomathiod, L., Nicoullaud, B. & Richard, G. (2011): A portable infrared laser spectrometer for flux measurements of trace gases at the geosphere–atmosphere interface. Meas. Sci. Technol. 22: 1–17. DOI : 10.1088/0957-0233/22/7/075601.

Güsewell, S. & Le Nédic, C. (2004): Effects of winter mowing on vegetation succession in a lakeshore fen. Appl. Veg. Sci. 7: 41–48.

Hahn-Schöfl, M., Zak, D., Minke, M., Gelbrecht, J., Augustin, J. & Freibauer, A. (2011): Organic sediment formed during inundation of a degraded fen grassland emits large fluxes of $CH_4$ and $CO_2$. Biogeosciences 8: 1539–1550.

Häkkinen, J. (2007): Traditional use of reed. In: Ikonen, I. & Hagelberg, E. (ed.): Read up on reed! Part IV, touch and thatch. Southwest Finland regional environment centre. Turku 2007. pp 62–72.

Hamilton, K., Sjardin, M., Shapiro, A. & Marcello, T. (2009): Fortifying the Foundation: State of the Voluntary Carbon Markets 2009, Ecosystem Marketplace & New Carbon Finance.

Hamilton, K., Sjardin, M., Peters-Stanley, M. & Marcello, T. (2010): Building Bridges: State of the Voluntary Carbon Markets 2010. Ecosystem Marketplace, Washington and Bloomberg New Energy Finance, New York, 129 p.

Hartmann, H. (2001): Die energetische Nutzung von Stroh und strohähnlichen Brennstoffen in Kleinanlagen. Gülzower Fachgespräche: Energetische Nutzung von Stroh, Ganzpflanzengetreide und weiterer halmgutartiger Biomasse. FNR. Gülzow. Tangram documents, Rostock p. 62-84.

Hartmann, H. & Hering, T. (2004): Eigenschaften biogener Festbrennstoffe. In: Härdtlein, M., Eltrop, L. & D. Thrän (Eds.): Voraussetzungen zur Standardisierung biogener Festbrennstoffe. Landwirtschaftsverlag Münster. p. 29–47.

Hartmann, H. Thuneke, K., Höldrich, A. & Rossmann, P. (2003): Handbuch Bioenergie-Kleinanlagen. FNR Gülzow. Tangram documents, Bentwisch, 184 p.

Hawke, C.J. & José, P.V. (1996): Reedbed management for commercial and wildlife interests. The Royal Society for the Protection of Birds, Sandy.

Heath, M.F. & Evans, M.I. (2000): Important bird areas in Europe: Priority sites for conservation. 2 vols. Birdlife International; Cambridge.

Heijmans, M.M.P.D., Klees, H., de Visser, W. & Berendse, F. (2002): Response of a Sphagnum bog plant community to elevated $CO_2$ and N supply. Plant Ecol. 162: 123–134.

Heise, G. (1974): Der Seggenrohrsänger – eine vom Aussterben bedrohte Art. Falke 21: 6–10.

Hendriks, D.M.D., van Huissteden, J., Dolma, A.J. & van der Molen, M.K. (2007): The full greenhouse gas balance of an abandoned peat meadow. Biogeosciences 4: 411–424.

Herold, B. (2011): Neues Leben in alten Mooren – Brutvögel in wiedervernässten Flusstalmooren. Hauptverlag.

Hesse, E. (1910): Beobachtungen und Aufzeichnungen während des Jahres 1909. J. Orn. 58: 489–519.

Hodgson, J.G., Grime, J.P., Wilson P.J., Thompson, K. & Band, R.S. (2005): The impacts of agricultural change (1963–2003) on the grassland flora of Central England: processes and prospects. Basic Appl. Ecol. 6: 107–118.

Hofbauer, H., Linsmeyer, T. & Steurer, C. (2001): Machbarkeitsstudie Schilfverwertungsanlage. Endbericht – Kurzfassung, 29.11.2001. Amt der Burgenländischen Landesregierung. Abt. 9, Wasser- und Abfallwirtschaft. Landeswasserbaubezirksamt Schützen/Geb.

Hohmann, T. (1995): Energetische Nutzung von Miscanthus. Schriftenreihe Nachwachsende Rohstoffe. FNR. Band 4. Landwirtschaftsverlag Münster, p.143–158.

Holden, J., Chapman, P.J. & Labadz, J.C (2004): Artificial drainage of peatlands: hydrological and hydrochemical process and wetland restoration. Progress in Physical Geography 28: 95–123.

Holsten, B., Neumann, H., Wiebe, C. & Wriedt, S. (2001): Die Wiedervernässung der Pohnsdorfer Stauung – eine Zwischenbilanz unter Berücksichtigung der Auswirkungen auf die Vegetation sowie die Amphibien- und Brutvogelbestände. Die Heimat 11(12): 195–205.

Hooijer, A., Silvius, M., Wösten, H. & Page, S. (2006): PEAT-$CO_2$, Assessment of $CO_2$ emissions from drained peatlands in SE Asia. Delft Hydraulics report Q3943. http://tinyurl.com/ymlq3r .

Höper, H., Augustin, J., Cagampan, J.P., Drösler, M., Lundin, L., Moors, E., Vasander, H., Waddington, J.M. & Wilson, D. (2008): Restoration of peatlands and greenhouse gas balances. In: Strack, M. (ed.): Peatlands and Climate Change. International Peat Society, Jyväskylä, p. 182–210.

Houghton, J.T., Meira Filho, L.G., Callander, B.A., Harris, N., Kattenberg, A. & Maskell, K. (eds) (1996): Climate change 1995: The science of climate change. Contribution of working group I to the second assessment report of the IPCC. Cambridge University Press, Cambridge, 878 p.

Hübner, E. (1908): Avifauna von Vorpommern und Rügen. Theodor Oswald Weigel, Leipzig.

Hundt, R. & Succow, M. (1984): Vegetationsformen des Graslandes der DDR. Wissenschaftliche Mitteilungen des Institutes für Geographie und Geoökologie der Akademie der Wissenschaften der DDR 14: 61–104.

Hyvönen, N.P., Huttunen, J.T., Shurpali, N.J., Tavi, N.M., Repo, M.E. & Martikainen, P.J. (2009): Fluxes of nitrous oxide and methane on an abandoned peat extraction site: Effect of reed canary grass cultivation. Bioresource Technol. 100: 4723–4730.

IPCC (1997): Revised 1996 IPCC Guidelines for National Greenhouse Inventories. In: Houghton, J.T., Meira Filho, L.G., Lim, B., Tréanton, K., Mamaty, I., Bonduki, Y., Griggs, D.J. & Callander, B.A. (eds): Intergovernmental Panel on Climate Change (IPCC), IPCC/OECD/IEA, Paris.

IPCC (2003): Good Practice Guidance for Land Use, Land-Use Change and Forestry. In: Penman, J., Gytarsky, M., Hiraishi, T., Krug, T., Kruger, D., Pipatti, R., Buendia, L., Miwa, K., Ngara, T., Tanabe, K. & Wagner, F. (eds): Intergovernmental Panel on Climate Change (IPCC), IPCC/IGES, Hayama.

IPCC (2006): 2006 IPCC Guidelines for National Greenhouse Gas Inventories, Prepared by the National Greenhouse Gas Inventories Programme. In: Eggleston, H.S., Buendia, L., Miwa, K., Ngara. T. & Tanabe, K. (eds): IGES, Japan. http://www.ipcc-nggip.iges.or.jp/public/2006gl/vol2.htm.

IPCC (2007): Climate Change 2007: The Physical Science Basis. Contribution of Working Group I to the Fourth Assessment Report of the Intergovernmental Panel on Climate Change. Cambridge University Press, Cambridge. 18 p.

Ivanov, K.E. (1975): Vodoobmen v bolotnych landsaftach. [Water exchange in peatland landscapes] Gidrometeoizdat, Leningrad, 279 p. [in Russian]

Ivchenko, T.G. (2009): Khorologia bolotnych kompleksov Ilmenskogo Zapovednika I ee otobrazenie na geobotanicheskikh kartakh. [Chorology of mire complexes of the Ilemnsky Zapovednik and their depiction on geobotanical maps] Russian Academy of Sciences, Ural Branch & Celyabinsk State University, Entsiklopedia Celyabinsk, 142 p. [in Russian]

Jacobs, C.M.J., Moors, E.J. & van der Bolt, F.J.E. (2003): Invloed van waterbeheer op gekoppelde broeikasgasemissies in het veenweidegebied by ROC Zegveld. Alterra-rapport 840. Alterra, Wageningen.

Jansen, A.J.M. & Roelofs, J.G.M. (1996): Restoration of Cirsio-Molinietum wet meadows by sod cutting. Ecol. Engineering 7: 279–298.

Jansen, A.J.M., Fresco, L.F.M., Grootjans, A.P. & Janlink, M.H. (2004): Effects of Restoration Measures on Plant Communities of Wet Heathland Ecosystems. Appl. Veg. Sci. 7: 243–252.

Jauhiainen, S., Laiho, R. & Vasander, H. (2002): Ecohydrological and vegetational changes in a restored bog and fen. Annales Botanici Fennici 39(3): 185–199.

Jauhiainen, S., Pitkainen, A. & Vasander, H. (2004): Chemostratigraphy and vegetation in two boreal mires during the Holocene. Holocene 14: 769–779.

Joosten, H. (2009a): Peatlands at the UNFCCC in Bonn on their way to Copenhagen. IMCG Newsletter 2009-1: 14–15. http://www.imcg.net/imcgnl/pdf/nl0901.pdf.

Joosten, H. (2009b): The long and winding peatland road to Copenhagen, stage Bonn III. IMCG Newsletter 2009-2: 20–23. http://www.imcg.net/imcgnl/pdf/nl0902.pdf.

Joosten, H. (2009c): Getting peatlands under Kyoto: The long and winding road to Copenhagen, stage Bangkok/Barcelona. IMCG Newsletter 2009-3: 16–23. http://www.imcg.net/imcgnl/pdf/nl0903.pdf.

Joosten, H. (2009d): The Global Peatland $CO_2$ Picture. Peatland status and emissions in all countries of the World. Wetlands International, Ede, 10 p., tables. http://tinyurl.com/yaqn5ya.

Joosten, H. (2010a): Getting peatlands under Kyoto: Arriving in Copenhagen – and now what? IMCG Newsletter 2010-1: 8–11. http://www.imcg.net/imcgnl/pdf/nl1001.pdf.

Joosten, H. (2010b): Getting peatlands under Kyoto: arriving at a moving target in Cancun. IMCG Newsletter 2010/3–4: 13–19.

Joosten, H. & Clarke, D. (2002): Wise use of mires and peatlands – Background and principles including a framework for decision-making. International Mire Conservation Group/International Peat Society, 304 p.

Joosten, H. & Couwenberg, J. (2009): Are emission reductions from peatlands MRV-able? Wetlands International, Ede. 16 p. http://tinyurl.com/mud9a9.

Joosten, J.H.J. (1995): Time to regenerate: Long-term perspectives of raised bog regeneration with special emphasis on palaeoecological studies. In: Wheeler, B.D., Shaw, S.C., Fojt, W.J. & Robertson, R.A. (eds): Restoration of temperate wetlands. John Wiley & Sons, Chichester, New York, Brisbane, Toronto, Singapore. p. 471–494.

Julve, P. (2011): Provisory Holarctic Mire Plant Species List. http://philippe.julve.pagespersoorange. fr/imcg-proj.htm#mireplantlist. Accessed 28/02/2011.

Jungkunst, H.F. & Fiedler, S. (2007): Latitudinal differentiated water table control of carbon dioxide, methane and nitrous oxide fluxes from hydromorphic soils: feedbacks to climate change. Global Change Biol. 13: 2668–2683.

Jungkunst, H.F., Freibauer, A., Neufeldt, H. & Bareth, G. (2006): Nitrous oxide emissions from agricultural land use in Germany – a synthesis of available annual field data. J. Pl. Nutr. Soil Sci. 169: 341–351.

Jurzyk, S. (2004): Protection of Myrica gale L. brushwood in a peatbog under agricultural management. In: Wołejko, L. & Jasnowska J. (eds): The future of Polish mires. Agricultural University of Szczecin. p. 263–270.

Kaat, A. & Joosten, H. (2008): Fact book for UNFCCC policies on peat carbon emissions. Wetlands International, Ede, 26 p.

Käding, H., Weise, G., Knabe, W., Robowsky, K.-D. & Schuppenies, R. (1990): Wert der Quecke (*Agropyron repens* L.) auf Graslandstandorten. Archiv für Acker-, Pflanzenbau und Bodenkunde 34: 723–728.

Kalbitz, K. & Geyer, S. (2002): Different effects of peat degradation on dissolved organic carbon and nitrogen. Organic Geochemistry 33: 319–326.

Kask, Ü., Kask, L. & Paist, A. (2007): Reed as a energy source in Estonia. Tallinn University of Technology (TUT), Thermal engineering Department. In: Ikonen, I. & Hagelberg, E. (ed.): Read up on Reed. Southwest Finland Regional Environment Centre. Turku. p. 102 – 124.

Kastberg, S. & Burvall, J. (1998): Perennial rhizomatous grass – Reed canary grass as an upgraded biofuel: experiences from combustion tests in Sweden. In: El Bassam, N., Behl, R.K. & Prochnow, B. (ed.): Sustainable agriculture for food, energy and industry. James & James Ltd., London. p. 932–937.

Kästner, M. & Flössner, W. (1933): Die Pflanzengesellschaften der erzgebirgischen Moore. II. Veröff. Landesverein Sächs. Heimatschutz, Dresden, 206 p.

Keddy, P. (1989): Effects of competition from shrubs on herbaceous wetland plants: a 4-year field experiment. Canad. J. Bot. 67: 708–716.

Keddy, P., Twolan-Strutt, L. & Shipley, B. (1997): Experimental evidence that interspecific competitive asymmetry increases with soil productivity. Oikos 80: 253–256.

Kerr, J.T. & Ostrovsky, M. (2003): From space to species: Ecological applications for remote sensing. Trends in Ecology and Evolution. 18 (6): 299–305.

Kersten, U., Lindner, H., Melzer, R., Rehberg, U., Staak, R. & Werner, W. (1999): Ergebnisse des Projektes "Regeneration und alternative Nutzung von Niedermoorflächen im Landkreis Ostvorpommern". Stiftung Odermündung, Regionalverband für dauerhafte Entwicklung e.V., Anklam, 57 S.

Kettunen, A., Kaitala, V., Alm J., Silvola, J., Nykänen, H. & Martikainen, P.J. (2000): Predicting variations in methane emissions from boreal peatlands through regression models. Boreal Environm. Res. 5: 115–131.

Klimkowska, A., Kotowski, W., van Diggelen, R., Grootjans, A.P., Dzierza, P. & Brzezinska, K. (2010): Vegetation Re-development After Fen Meadow Restoration by Topsoil Removal and Hay Transfer. Restorat. Ecol. doi 10.1111/j.1526-100X.2009.00554.x.

Kloskowski, J. & Krogulec, J. (1999): Habitat selection of Aquatic Warbler Acrocephalus paludicola in Poland: consequences for conservation of the breeding areas. Vogelwelt 120: 113–120.

Kool, D.M., Dolfing J., Wrage, N. & van Groenigen, J.W. (2011): Nitrifier denitrification as a distinct and significant source of nitrous oxide from soil. Soil Biol. Biochem. 43: 174–178.

Koska, I. (2007): Weiterentwicklung des Vegetationsformenkonzeptes. Ausbau einer Methode für die vegetationskundliche und bioindikative Landschaftsanalyse, dargestellt am Beispiel der Feuchtgebietsvegetation Nordostdeutschlands. PhD thesis, Greifswald University, Greifswald.

Koska, I., Succow, M., Clausnitzer, U., Timmermann, T. & Roth, S. (2001): Vegetationskundliche Kennzeichnung von Mooren (topische Betrachtung). In: Succow, M. & Joosten, H. (Eds.): Landschaftsökologische Moorkunde. Schweizerbart, Stuttgart: 112–184.

Kossoy, A. & Ambrosi, P. (2010): State and Trends of the Carbon Market 2010. Worldbank, Washington, 89 p.

Kotowski, W. (2002): Fen communities. Ecological mechanisms and conservation strategies. Groningen University, PhD thesis.

Kotowski, W., van Diggelen, R. & Kleinke, J. (1998): Behaviour of wetland plant species along a moisture gradient in two geographically distant areas. Acta Bot. Neerl. 47: 337–349.

Kozulin, A.V. (2010): Obosnovanie povtornogo zabolachivania vyrabotannogo mestorozhdenia Astrauskoje. [Scientific Justification of rewetting of peatland Astrauskaje]. Minsk. The scientific and practical centre for bioresources, 37 p. [in Russian]

Kozulin, A.V. & Flade, M. (1999): Breeding habitat, abundance and conservation status of the Aquatic

Warbler Acrocephalus paludicola in Belarus. Vogelwelt 120: 97–112.

Kozulin, A.V. & Tanovitskaya, N.I. (2010): First phase of an inventory of the natural peatlands of Belarus based on satellite images. Unpublished report.

Kozulin, A.V., Maksimenkov, M.V., Grumo, D.G., Shevtsov, N.V., Tanovitskaya, N.I. (2008): Nauchnoe obosnovanie vosstanovlenia narushennogo (dobychey torfa) estestvennogo hydrologicheskaya rezhima zakaznika Dokudauskaje dlya predotvrazheniya pozharov na torfyanikakh i dlya okhrany bioraznoobrazia samogo bolshogo verkhogo bolota v Grodnenskoy oblasti. [Scientific justification of restoration of disturbed (by peat extraction) natural hydrological regime of Dakudauskaje zakaznik for prevention of peat fires and conservation of biodiversity of largest bog in Hrodna oblast.] Scientific and Practical Center on Biological Resources of the National Academy of Sciences of Belarus, 50 p. [in Russian]

Kozulin, A.V., Tanovitskaya, N.I. & Vershitskaya, I.N. (2010a): Methodical Recommendations for ecological rehabilitation of damaged mires and prevention of disturbances to the hydrological regime of mire ecosystems in the process of drainage. Scientific and Practical Center for Bio Resources, Institute for Nature Management of the National Academy of Sciences of Belarus, 40 p.

Kozulin, A.V., Tanovitskaya, N. & Vershitskaya, I. (2010b): Rekommendazii po ekologicheskoy reabilitazii narushennykh bolot i po predotvrazheniyu narushenii gidrologicheskogo rezhima bolotnykh ekosistem pri osushitelnykh rabotakh. [Methodical Recommendations for ecological rehabilitation of damaged mires and prevention of disturbances to the hydrological regime of mire ecosystems in the process of drainage] UNDP, Minsk. 51 p. [in Russian].

Kozulin, A.V., Maksimenkov, M.V, Zhuravlev, D.V., Korzun, E.V., Grechanik, L.N., Skuratovich, A.N., Dubovik, D.V. & Tanovitskaya, N.I. (2010c): Obosnovanie vtorichnogo zabolachivania narushennogo bolota Dalbeniski. [Justification of rewetting of the extracted peatland Dolbeniski]. Scientific and Practical Center on Biological Resources of the National Academy of Sciences of Belarus, 47 p. [in Russian].

Kratz, R. (1994): The conception of goal species in fen meadowland. In: Jankowska-Hufleijt, H. & Golubiewska, E. (eds) Proceedings of the international symposium "Conservation and management of fens", Warsaw-Biebrza, Poland, 6–10 June 1994. Institute for Land Reclamation and Grassland Farming, Falenty. p. 365–370.

Kratz, R. & Pfadenhauer, J. (Eds.) (2001): Ökosystemmanagement für Niedermoore. Strategien und Verfahren zur Renaturierung. Ulmer Verlag, Stuttgart, 317 p.

Kroon, P.S., Schrier-Uijl, A.P., Hensen, A., Veenendaal, E.M. & Jonker, H.J.J. (2010): Annual balances of CH4 and N2O from a managed fen meadow using eddy covariance flux measurements. Special Issue: Nitrogen and greenhouse gas exchange. Europ. J. Soil Science. 61 (5): 773–784.

Kube, J. & Probst, S. (1999): Bestandsabnahme bei schilfbewohnenden Vogelarten an der südlichen Ostseeküste: Welchen Einfluss hat die Schilfmahd auf die Brutvogeldichte? Vogelwelt 120:27–38.

Kulczynski, S. (1949): Torfowiska Polesia. Peat bogs of Polesie. Mem. Acad. Pol. Sc. et Lettres. Sc. Mat. et Nat. Serie B: Sc. nat., 15:, Kraków, p. 1–359.

Kutzbach, L., Schneider, J., Sachs, T., Giebels, M., Nykänen, H., Shurpali, N.J., Martikainen, P.J., Alm, J. & Wilmking, M. (2007): $CO_2$ flux determination by closed-chamber methods can be seriously biased by inappropriate application of linear regression. Biogeosciences 4: 1005–1025.

Kuzyakov, Y. (2006): Sources of $CO_2$ efflux from soil and review of partitioning methods. Soil Biol Biochem 38: 425–448.

Lachmann, L., Marczakiewicz, P. & Grzywaczewski, G. (2010): Protecting Aquatic Warblers (Acrocephalus paludicola) through a landscape-scale solution for the management of fen peat meadows in Poland. Grassland science in Europe 15: 711–713.

Laggoun-Defarge, F., Mitchell, E., Gilbert, D., Disnar, J. R., Comont, L., Warner, B. G. & Buttler, A. (2008). Cut-over peatland regeneration assessment using organic matter and microbial indicators (bacteria and testate amoebae). J. Appl. Ecol. 45: 716–727.

Lai, D.Y.F. (2009): Methane dynamics in northern peatlands: a review. Pedosphere 19: 409–421.

Lamers, M., Ingwersen, J. & Streck, T. (2007): Modelling nitrous oxide emission from water-logged soils of a spruce forest ecosystem using the biogeochemical model Wetland-DNDC. Biogeochemistry 86: 287–299.

Lashof, D.A. (2000): The use of global warming potentials in the Kyoto Protocol. Climatic Change 44: 423–425.

Lautkankare, R. (2007): Reed construction in the Baltic Sea region. In: Ikonen, I. & Hagelberg, E. (ed.) (2007): Reed up on reed. Southwest Finland Regional Environment Centre, Turku, pp 73–80 (www.ymparisto.fi/julkaisut).

Lavoie, C., Marcoux, K., Saint-Louis, A. & Price, J.S. (2005a): The dynamics of a cotton-grass (Eriophorum vaginatum L.) cover expansion in a vacuum-mined peatland, southern Québec, Canada. Wetlands 25: 64–75.

Lavoie, C., Saint-Louis, A. & Lachance, D. (2005b): Vegetation Dynamics on an Abandoned Vacuum-Mined Peatland: 5 Years of Monitoring. Wetlands Ecol. Managem. 13: 621–633.

Leisler, B. & Catchpole, C.K. (1992): The evolution of polygamy in European reed warblers of the genus Acrocephalus: a comparative approach. Ethol. Ecol. Evol. 4: 225–243.

Lenschow, D.H. (1995): Micrometeorological techniques for measuring biosphere-atmosphere trace gas exchange. In: Matson, P.A. & Harriss, R.C. (eds): Biogenic Trace Gases: Measuring Emissions from Soil and Water. Blackwell Science Ltd. Oxford. UK p. 126–163.

Lesnichiy, Iu. D., Verovskaya, T.A., Komarovskaya, E.V. & Kuzmic, S.A. (2011): Prodostavlenie gidrometeorologicheskoy informatsii po AS Sharkovshchina, MS Verchnedvinsk i gidrologicheskomu postu r. Disna – p.g.t. Sharkovshchina, osrednennyj za period 1980–2009 gg. [Hydrometeorological information for AS Sharkovshchina, MS Verchnedvinsk and the hydrological station at Disna river for the period 1980–2009] Respublikanskij Gidrometcentr Minsk, in Order of APB, 12 p. [in Russian]

Lieckweg, T. & Niedringhaus, R. (2010): Auswirkungen von Rückdeichungen an der Nordseeküste auf die Arthropodenfauna am Beispiel einer Maßnahme auf der Insel Langeoog. Naturschutz und Biologische Vielfalt 91: 111–122.

Linacre, N., Kossoy, A. & Ambrosi, P. (2011): State and Trends of the Carbon Market 2011, Washington DC, World Bank.

Lindsay, R. (2010): Peatbogs and Carbon: A Critical synthesis. RSPB Scotland, Edinburgh.

Livingston, G.P. & Hutchinson, G.L. (1995): Enclosure-based measurement of trace gas exchange: applications and sources of error. In: Matson, P.A. & Harriss, R.C. (eds): Biogenic Trace Gases: Measuring Emissions from Soil and Water. Blackwell Science Ltd. Oxford. UK p. 14–51.

Livingston, G.P., Hutchinson, G.L. & Spartalian, K. (2006): Trace gas emission in chambers: a non-steady-state diffusion model. Soil. Sci. Soc. Am. J. 70: 1459–1469.

Lloyd, J. & Taylor, J.A. (1994): On the temperature dependence of soil respiration. Funct. Ecol. 8: 315–323.

Lohila, A., Laurila, T., Aro, L., Aurela, M., Tuovinen, J.-P., Laine, J., Kolari, P. & Minkkinen, K. (2007): Carbon dioxide exchange above a 30-year-old Scots pine plantation established on organic-soil cropland. Boreal Environment Research 12: 141–157.

Loock Biogassysteme GmbH Hamburg (2009): Verfahrensbeschreibung Trockenvergärungsanlage für NaWaRo und Bioabfälle nach dem Loock TNS Verfahren, 11 p. (www.loock-biogassysteme.de).

Lütt, S. (1992): Produktionsbiologische Untersuchungen zur Sukzession der Torfstichvegetation in Schleswig-Holstein. Mitteilungen der AG Geobotanik in Schleswig-Holstein und Hamburg 43.

Maksimenkov, M.V., Novitsky, R.V., Zhuravlev, D.V., Korzun, E.V., Grechanik, L.N., Sudnik, A.V., Verzhitskaya, I.N., Skuratovich, A.N., Shevzov, N.V., Rakovich, V.A & Tanovitskaya, N.I. (2007a): Obosnovanie vosstanovlenia gidrologicheskogo rezhima narushennogo verkhovogo bolota Zada. [Justification of rewetting of the extracted peatland Zada]. Scientific and Practical Center on Biological Resources of the National Academy of Sciences of Belarus, 38 p. [in Russian]

Maksimenkov, M.V., Novitsky, R.V., Zhuravlev, D.V., Korzun, E.V., Grechanik, L.N., Sudnik, A.V., Verzitskaya, I.N., Skuratovich, A.N., Shevzov, N.B., Rakovich, V.A. & Tanovitskaya, N.I. (2007b): Obosnovanie povtornogo zabolachivania nakhodivshegosya pod torforazrabotkoy bolota Grichino-Starobinskoe. [Justification of rewetting of the extracted peatland Grichino-Starobinskoe]. Scientific and Practical Center on Biological Resources of the National Academy of Sciences of Belarus, 32 p. [in Russian]

Maksimenkov, M.V., Novitsky, R.V., Zhuravlev, D.V., Korzun, E.V., Grechanik, L.N., Sudnik, A.V., Verzhitskaya, I.N., Skuratovich, A.N., Shevzov, N.V., Rakovich, V.A,, Tanovitskaya, N.I. (2008): Obosnovanie povtornogo zabolachivania narushennogo bolota Scarbinski Mokh. [Justification of rewetting of the extracted peatland Scarbinski Mokh]. Scientific and Practical Center on Biological Resources of the National Academy of Sciences of Belarus, 47 p. [in Russian]

Maljanen, M., Oskarsson, H., Sigurdsson, B.D., Gudmundsson, J., Huttunen, J.T. & Martikainen, P.J. (2010): Land-use and greenhouse gas balances of peatlands in the Nordic countries – Present knowledge and gaps. Biogeosciences 7: 2711–2738.

Mälson, K., Backeus, I. & Rydin, H. (2008): Long-term effects of drainage and initial effects of hydrological restoration on rich fen vegetation. Appl. Veg. Sci. 11: 99–106.

Margoczi, K., Aradi, E., Takacz, G. & Batori, Z. (2007): Small scale and large scale monitoring of vegetation changes in a restored wetland. In: Okruszko, T., Maltby, E., Szatylowicz, J., Swiatek, D. & Kotowski, W. (eds): Wetlands: Monitoring, modelling and management. Taylor & Francis, London. p. 55–60.

Markowska, A. & Zylicz, T. (1999): Costing an international public good: The case of the Baltic Sea. Ecol. Economics 30: 301–316.

Matthias, A.D., Douglas, N.Y. & Weinbeck, R.S. (1978): A numerical evaluation of chamber methods for determining gas fluxes. Geophysical Research Letters 5: 765–768.

Matthias, A.D., Blackmer, A.M. & Bremner, J.M. (1980): A simple chamber technique for field measurement of emissions of nitrous oxide from soils. J. Environm. Qual. 9: 251–256.

Mauersberger, R., Gunnemann, H., Rowinsky, V. & Bukowsky, N. (2010): Das Mellenmoor bei Lychen – ein erfolgreich revitalisiertes Braunmoosmoor im Naturpark Uckermärkische Seen. Naturschutz und Landschaftspflege in Brandenburg 19: 182–186.

Mazerolle, M. J., Poulin, M., Lavoie, C., Rochefort, L., Desrochers, A. & Drolet, B. (2006): Animal and vegetation patterns in natural and man-made bog pools: implications for restoration. Freshwater Biol. 51: 333–350.

McBride, A., Diack, I., Droy, N., Hamill, B., Jones, P., Schutten, J., Skinner, A. & Street, M. (eds) (2010): The Fen Management Handbook. Scottish Natural Heritage, Perth.

Megonigal, J.P., Hines, M.E. & Visscher, P.T. (2004): Anaerobic metabolism: Linkages to trace gases and aerobic processes. In: Schlesinger, W.H. (ed.) Biogeochemistry, Elsevier-Pergamon, Oxford, UK., p. 317–424.

Meyer, K. (1999): Die Flüsse der klimarelevanten Gase $CO_2$, $CH_4$ und $N_2O$ eines norddeutschen Niedermoores unter dem Einfluß der Wiedervernässung. Göttinger Bodenkundliche Berichte 111: 651–664.

# References

Michaelis, L. & Menten, M.L. (1913): Die Kinetik der Invertinwirkung. Biochemische Zeitschrift 49: 333p.

Middleton, B., Holsten, B. & van Diggelen, R. (2006): Biodiversity management of fens and fen meadows by grazing, cutting and burning. Appl. Veg. Sci. 9: 307–316.

Minayeva, T., Sirin, A. & Bragg, O. (eds) (2009): A Quick Scan of Peatlands in Central and Eastern Europe. Wetlands International, Wageningen, The Netherlands. 132 pp, tabl. 6, fig. 17.

Minke, M., Thiele, A., Augustin, J., Couwenberg, J., Fenchuk, V., Yarmashuk, T., Liashchynskaya, N., Ryzhikov, V. & Joosten, H. (2009): Greenhouse gas emission reduction from peatland restoration in Belarus: testing and adapting a rapid assessment tool. In: Peatlands in the Global Carbon Cycle. Proceedings of the 2nd International Symposium on Carbon in Peatlands, Prague, Czech Republic, 25–30 September 2009. http://www.peatnet.siu.edu/Assets/M.pdf. Accessed 16.11.2010.

Minkkinen, K., Byrne, K.A. & Trettin, C. (2008): Climate impacts of peatland forestry. In: Strack, M. (ed.) Peatlands and Climate Change. International Peat Society, Jyväskylä: 98–122.

Mitchell, C.C. & William, A.N. (1993): Vegetation change in a topogenic bog following beaver flooding. Bulletin of the Torrey Botanical Club 120: 136–147.

Moffat, A.M., Papale, D., Reichstein, M., Hollinger, D.Y., Richardson, A.D., Barr, A.G., Beckstein, C., Braswell, B.H., Churkina, G., Desai, A.R., Falge, E., Gove, J.H., Heimann, M., Hui, D., Jarvis, A.J., Kattge, J., Noormets, A. & Stauch, V.J. (2007): Comprehensive comparison of gap-filling techniques for eddy covariance net carbon fluxes. Agric. Forest Meteorol. 147: 209–232.

Moncrieff, J., Valentini, R., Greco, S., Seufert, G., Paolo, Ciccioli P. (1997): Trace gas exchange over terrestrial ecosystems: methods and perspectives in micrometeorology. Journal of Experimental Botany 48: 1133–1142.

Müller, F. (1988): Die Auswirkungen der Renaturierungsmaßnahmen im NSG "Rotes Moor" auf die Vogelwelt, insbesondere Wiesenbrüter, und deren Eignung als Biotop-Indikatoren. Renaturierungsprojekte und Regenerationsprozesse an Mooren in Mittelgebirgslandschaften. Tagung des BMU und der DGMT in der Rhön, September 1988. Kurzfassung der Referate: 16–17.

Müller-Motzfeld, G. (1997): Renaturierung eines Überflutungssalzgrünlandes an der Ostseeküste. – Schr.-R. f. Landschaftspfl. U. Natursch. 54, 239–263.

Mundel, G. (1976): Untersuchungen zur Torfmineralisation in Niedermooren. Archiv für Acker- und Pflanzenbau und Bodenkunde 20: 669–679.

Neeff, T., Ashford, L., Davey, C., Durbin, J., Fehse, J., Hedges, A., Herrera, T., Janson-Smith, T., Moore, C., Mountain, R., Panfil, S., Tuite, C. & Wheeland, M. (2010): The forest carbon offsetting report 2010, EcoSecurities.

Nick, K.-J. & Weber, H. E. (2001): Entwicklung der Vegetation auf dem wiedervernässten leegmoor in den Jahren 1989–1996. In: Nick, K.-J., Löpmeier, F.-J., Schiff, H., Blankenburg, J., Gebhardt, H., Knabke, C., Weber, H. E., Främbs, H. & Mossakowski, D. (Eds.): Moorregeneration im Leegmoor/Emsland nach Schwarztorfabbau und Wiedervernässung. Angewandte Landschaftsökologie 38. p. 75–90.

Nikiforov, M. E., Tishechkin, A. K., Samusenko, I. E. & Pareyko, O. A. (1995): Formirovanie struktury ornitokompleksov i populyazii modelnykh vidov ptiz. [Formation of avian complexes and population of model species] In: Sushchenya, L. M., Pikulik, M. M. & Plenin, A. E. (eds): Zhivotny mir v zone avarii chernobylskoy AES [Fauna of the zone of the Chernobyl disaster] Academy of Sciences Belarus, Institute of Zoology. Minsk, Navuka i Tekhnika, p. 158–174.

Nilsson, M., Sagerfors, J., Buffam, I., Laudon, H., Eriksson, T., Grelle, A., Klemedtsson, L., Weslien, P. & Lindroth, A. (2008): Contemporary carbon accumulation in a boreal oligotrophic minerogenic mire – a significant sink after accounting for all C-fluxes. Glob. Change Biol. 14: 2317–2332.

NOAA (2010): Annual Greenhouse Gas Index (AGGI) URL: http://www.esrl.noaa.gov/gmd/aggi

Noss, R.F. & Cooperrrider, A.Y. (1994): Saving nature's legacy – protecting and restoring biodiversity. Island Press, Washington.

Okruszko, H. (1956): Zjawisko degradacji torfu na tle rozwoju torfowiska [The influence of peat degradation on peatland development] In: Zagadnienia degradacji i regeneracji gleb lakowych torfowych. Zeszyty problemowe postepow nauk rolniczych 2: 69–112. [in Polish]

Paist, A., Kask, Ü. & Kask, L. (2007): Composition of reed mineral matter and its behavior at combustion. Proceedings of the 15th European biomass conference & exhibition, Berlin, Germany. p. 1666–1669 (digital suppl.)

Page, S.E., Siegert, F., Rieley, J.O., Boehm, H.D.V., Jaya, A. & Limin, S. (2002): The amount of carbon released from peat and forest fires in Indonesia during 1997. Nature 420 (6911): 61–65.

Palkin, G. (2008): Biogas popolnyaet energoresursy. [Biogas as energy resources] Novoe selskoe khozyastvo 2: 138–139. [in Russian]

Pape Moller, M. & Mousseau, T.A. (2009): Reduced abundance of insects and spiders linked to radiation at Chernobyl 20 years after the accident. Biol Lett. 23: 356–359.

Parish, F. & Canadell, P. (2006): Vulnerabilities of the carbon-climate system: Carbon pools in wetlands/peatlands as positive feedbacks to global warming. Asia-Pacific Network for Global Change Research, 66 p.

Parish, F., Sirin, A., Charman, D., Joosten, H., Minayeva, T. & Silvius, M. (eds) (2007): Assessment on Peatlands, Biodiversity and Climate Change: Executive Summary. Global Environment Centre, Kuala Lumpur and Wetlands International, Wageningen, 18 p.

Parish, F., Sirin, A., Charman, D., Joosten, H., Minaeva, T. & Silvius, M. (eds) (2008): Assessment on peatlands, biodiversity and climate change. Global Environment Centre, Kuala Lumpur and Wetlands International Wageningen, 179 p.

Pedersen, A.R., Petersen, S.O. & Schelde, K. (2010): A comprehensive approach to soil-atmosphere trace-gas flux estimation with static chambers. Eur. J. Soil Sci. 61: 888–902.

Pena, N. (2009): Including peatlands in post-2012 climate agreements: Options and rationales. Joanneum Research, Institute of Energy Research, Graz, Report commisioned by Wetlands International, 26 p.

Peters-Stanley, M., Hamilton, K., Marcello, T. & Sjardin, M. (2011): Back to the Future; State of the Voluntary Carbon Markets 2011 Ecosystem Marketplace & Bloomberg New Energy Finance.

Piavchenko, N.I. (1963): Lesnoje Bolotovedenie. [Ecology of forested peatlands] Isdat. Akademii Nauk SSSR, Moskva, 185 p. [in Russian]

Pidoplichko, A.P. (1961): Torfyanye mestrorozhdeniya Belorussii. [Peatlands of Belarus] Akademiya Nauk BSSR, Minsk, 190 p. [in Russian]

Pihlatie, M., Christiansen, J.-R., Aaltonen, H., Korhonen, J.F.J., Nordbo, A., Rasilo, T., Benanti, G., Giebels, M., Helmy, M., Hirvensalo, J., Jones, S., Juszczak, R., Klefoth, R., Lobo do Vale, R., Rosa, A.P., Schreiber, P., Serça, D., Vicca, S., Wolf, B. & Pumpanen, J. (in prep.): Comparison of static chambers to measure $CH_4$ fluxes from soils.

Plachter, H. & Hampicke, U. (eds) (2010): Large-scale livestock grazing. A management tool for nature conservation. Springer, Berlin. 400 p.

Podlaska, M. (2010): Sukzession ungenutzer Moorwiesen in Dolny Slask. Telma 40: 105–118.

Poelman, A. & Joosten, H. (1992): On the identification of hydrological interaction zones for bog reserves. In: Bragg, O.M., Hulme, P.D., Ingram, H.A.P., Robertson, R.A. (eds): Peatland ecosystems and man: an impact assesment. Department of Biological Sciences, Univ. of Dundee, S. 141–148.

Poschlod, P. (1990): Vegetationsentwicklung in abgetorften Hochmooren des bayrischen Alpenvorlandes unter besonderer Berücksichtigung standortskundlicher und populationsbiologischer Faktoren. PhD thesis, Technical University Munich. 331 p.

Poschlod, P. (1992): Development of vegetation in peat-mined areas in some bogs in the foothills of the Alps. In: Bragg, O.M., Hume, P.D., Ingram, H.A.P., Robertson, R.A. (eds): Peatland ecosystems and Man: An impact assessment. Department of Biological Sciences, University of Dundee, and International Peat Society. p. 287–290.

Poschlod, P., Meindl, C., Sliva, J., Herkommer, U., Jäger, M., Schuckert, U., Seemann, A., Ullmann, A. & Wallner, T. (2007): Natural Revegetation and Restoration of Drained and Cut-over Raised Bogs in Southern Germany — a Comparative Analysis of Four Long-term Monitoring Studies. Global Environmentall Research 11: 206–216.

Poschlod, P., Herkommer, U., Meindl, C., Schuckert, U., Seemann, A., Ullmann, A. & Wallner, T. (2009): Langzeitbeobachtung und Erfolgskontrolle in Regenmooren des Alpenvorlandes nach Torfabbau und Widervernässung. In: Schuster, U. (ed.): Vegetationsmanagement und Renaturierung. Festschrift zum 65. Geburtstag von Prof. Dr. Jörg Pfadenhauer. Laufener Spezialbeiträge 2/09: 46–59.

Poulin, M., Rochefort, L., Quinty, F. & Lavoie, C. (2005): Spontaneous revegetation of mined peatlands in eastern Canada. Canadian Journal of Botany 83: 539–557.

Price, J. S. & Whitehead, G. S. (2001): Developing hydrologic thresholds for Sphagnum recolonization on an abandoned cutover bog. Wetlands 21: 32–40.

Prochnow, A. & S. Kraschinski, S. (2001): Angepasstes Befahren von Niedermoorgrünland. Merkblatt 323, Deutsche Landwirtschaftsgesellschaft. 16 S.

Pugachevsky, A.V., Sudnik A.V., Skuratovich A.N., Verzhitskaya I.N., Ermolenkova G.V., Novitsky R.V., Maksimenkov M.V. & Zhuravlev D.V. (2009): Sozdanie integrirovannoy sistemy monitoringa ekosistem (v chasti zhivotnogo i rastitelnogo mira) vosstanavlivameykh torfyannykh bolot... [Design of an integrated monitoring scheme (flora i fauna) for peatlands during restoration...] Report – Part 5. National Academy of Sciences Belarus. 207 p.

Pugachevsky, A.V., Sudnik, A.V., Skuratovich, A.N., Verzhitskaya, I.N., Ermolenkova, G.V., Novitsky, R.V. & Zhuravlev, D.V. (2010): Provedenie monitoringa ekosistem (v chasti rastitelnogo i zhivotnogo mira) vosstanavlivaemykh torfyannykh bolot... [Monitoring of ecosystems (flora and fauna) of restored peatlands for assessment of measures on retoration in the framework of establishment an integrated system of complex monitoring of ecosystems (flora and fauna) of restored peatlands for informational support of management decisions on restoration of degraded peatlands] Institute of Experimental Botany V.M. Kuprevich and Scientific and Practical Center on Biological Resources of the National Academy of Sciences of Belarus. 74 p. [in Russian]

Pumpanen, J., Kolari P., Ilvesniemi, H., Minkkinen, K., Vesala, T., Niinistö, S., Lohila, A., Larmola, T., Morero, M., Pihlatie, M., Janssens, I., Yuste, J.C., Grünzweig, J.M., Reth, S., Subke, J.A., Savage, K., Kutsch, W., Østreng, G., Ziegler, W., Anthoni, P., Lindroth, A. & Hari, P. (2004): Comparison of different chamber techniques for measuring soil $CO_2$ efflux. Agric. Forest Meteorol. 123: 159–176.

Radzevich, L.F. (1991): Torfianoi fond Belorusskoy SSR i osnovnye napravleniia ego okhrany i ratsionalnogo ispolzovaniia. [The peat fund of the Belarusian SSR and main directions of its rational use] BelNIINTI, seriia 87.51.15. Minsk. [in Russian]

Ramenski, L.G., Tsatsenkin, I.A., Chizhikov, O.N. & Antipin, N.A. (1956): Ekologicheskaia otsenka kormovykh ugodii po rastitelnom pokrova. [Ecological evaluation of grazed lands by their vegetation.] Gosudarstvennoe izdatelstvo selskokhozaystvennoy literatury, Moscow. [in Russian]

Reddy, K.R. & DeLaune, R.D. (2008): Biogeochemistry of wetlands. CRC Press, Boca Raton, p. 111–476.

Richert, M., Dietrich, O., Koppisch, D. & Roth, S. (2000): The influence of rewetting on vegetation development and decomposition in a degraded fen. Restorat. Ecol. 8: 186–195.

Roberge, J.M. & Angelstam, P. (2004): Usefulness of the Umbrella Species Concept as a Conservation Tool. Cons. Biol. 18: 76–85.

Robert, E. C., Rochefort, L. & Gareau, M. (1999): Natural revegetation of two block-cut mined peatlands in eastern Canada. Canad. J. Bot. 77: 447–459.

Robertson, G.P. & Groffmann, P.M. (2007): Nitrogen transformation. In: Paul, E.A. (ed.): Soil Microbiology, Biochemistry, and Ecology. Springer, New York, p. 341–364.

Rochette, P. & Eriksen-Hamel, N.S (2008): Chamber measurements of soil nitrous oxide flux: are absolute values reliable? Soil Sci. Soc. Am. J. 72: 331–342.

Rode, M., Scheider, C., Ketelhake, G. & Reißhauer, D. (2005): Naturschutzverträgliche Erzeugung und Nutzung von Biomasse zur Wärme- und Stromgewinnung, BfN-Skript 136, Bonn – Bad Godesberg, 183 S.

Rodewald-Rodescu, L. (1974): Das Schilfrohr. Die Binnengewässer, 27. Schweizerbart, Stuttgart. 302 S.

Roll, H.-J. & K. Hedden (1994): Vergasung von grob gemahlenem Schilfgras im Flugstrom. Symposium Miscanthus. Biomassebereitstellung, energetische und stoffliche Nutzung. Schriftenreihe Nachwachsende Rohstoffe. FNR. Band 4. Landwirtschaftsverlag Münster, p. 159–172.

Roobroeck, D., Butterbach-Bahl, K., Brüggemann, N. & Boeckx, P. (2010) Dinitrogen and nitrous oxide exchanges from an undrained monolith fen: short-term responses following nitrate addition. Eur. J. Soil Sci. 61: 662–670.

Rosenthal, G. (2003): Selecting target species to evaluate the success of wet grassland restoration. Agriculture, Ecosystems and Environment 98: 227–246.

Rosenthal, G. (2010): Secondary succession in a fallow central European wet grassland. Flora 205: 153–160.

Roulet, N.T., Lafleur, P.M., Richard, P.J.H., Moore, T.R., Humphreys, E.R. & Bubier,J. (2007): Contemporary carbon balance and late Holocene carbon accumulation in a northern peatland. Glob. Change Biol. 13: 397–411.

Rydin, H. & Jeglum, J.K. (2006): The biology of peatlands. Oxford University Press.

Saarnio, S., Alm, J., Silvola, J., Lohila, A.-L., Nykänen, H., Martikainen, P.J. (1997): Seasonal variation in CH4 emissions and production and oxidation potentials in microsites of an oligotrophic pine fen. Oecologia 110: 414–422.

Sachs, T. (2009): Land-atmosphere interactions on different scales. The Exchange of Methane between wet Arctic Tundra and the Atmosphere at the Lena River Delta, Northern Siberia. PhD thesis, AWI für Polar- und Meeresforschung, Universität Potsdam. 181 p.

Sachs, T., Giebels, M., Boike, J. & Kutzbach, L. (2010): Environmental controls on $CH_4$ emission from polygonal tundra on the microsite scale in the Lena river delta, Siberia. Global Change Biol. 16: 3096–3110. DOI: 10.1111/j.1365-2486.2010.02232.x

Safford, L. & Maltby; E. (eds) (1998): Guidelines for integrated planning and management of tropical lowland peatlands with special reference to Southeast Asia. IUCN Commission on Ecosystem Management / Tropical Peatlands Expert Group (TROPEG). IUCN. Gland.

Salonen, V. (1990): Early plant succession in two abandoned cut-over peatland areas. Ecography 13: 217–223.

Salonen, V. (1992): Effects of artificial plant cover on plant colonization of a bare peat surface. J. Veg. Sci. 3: 109–112.

Samaritani, E., Siegenthaler, A., Yli-Petäys, M., Buttler, A., Christin, P.-A. & Mitchell, E.A.D. (2010): Seasonal net ecosystem carbon exchange of a regenerating cutaway bog: how long does it take to restore the C-sequestration function? Restorat. Ecol., doi: 10.1111/j.1526-100X.2010.00662.x.

Sanderson, E.W., Redford, K.H., Vedder, A., Coppolillo, P.B. & Ward, S.E. (2002): A conceptiual model for conservation planning based on landscape species requirements. Landscape and Urban Planning. 58: 41–56.

Sarkkola, S. (ed.) (2008):. Greenhouse impacts of the use of peat and peatlands in Finland. Research Programme Final Report. Ministry of Agriculture and Forestry, Helsinki, 72 p.

Scamoni, A. (1960): Waldgesellschaften und Waldstandorte. 3rd ed. Akademie-Verlag, Berlin.

Schäfer, A. (1999): Schilfrohrkultur auf Niedermoor – Rentabilität des Anbaus und der Ernte von Phragmites australis. Archiv für Naturschutz und Landschaftsforschung 38: 193–216.

Schäfer, A. & Joosten, H. (2005): Erlenaufforstung auf wiedervernässten Niedermooren. – 68 p; Greifswald (Institut für Dauerhaft Umweltgerechte Entwicklung der Naturräume der Erde (DUENE) e.V.).

Schaefer, H.M., Naef-Daenzer, B., Leisler, B., Schmidt, V., Müller, J.K. & Schulze-Hagen, K. (2000): Spatial behaviour of the Aquatic Warbler Acrocephalus palucidola during mating and breeding. J. Ornithol. 141: 418–424.

Scheel, R. (1937): Anbau von Flachs und Hanf. Deutsches Forschungsinstitut für Bastfasern, Sorau. 39 p.

Schmatzler, E. & Hoyer, W. (1994): Entwicklungs- und Pflegemaßnahmen nach industriellem Torfabbau am Beispiel des Neustädter Moores in der Diepholzer Moorniederung. Telma 24: 229–243.

Schmidt, M.H., Lefebvre, G., Poulin, B., Tscharntke, T. (2005): Reed cutting affects arthropod communities, potentially reducing food for passerine birds. Biol. Conserv. 121:157–166.

Schneebeli, M. (1991): Hydrologie und Dynamik der Hochmoorentwicklung. PhD- thesis, ETH Zürich, 133 p., appendix.

Schopp-Guth, A. (1999): Renaturierung von Moorlandschaften. Schriftenreihe für Landschaftspflege und Naturschutz 57: 1–219.

Schouwenaars, J.M. (1992): Hydrologic characteristics of bog relicts in the Engbertsdijksvenen after peat-cutting and rewetting. In: Bragg, O.M., Hume, P.D., Ingram, H.A.P., Robertson, R.A. (eds) Peatland ecosystems and Man: An impact assessment. Department of Biological Sciences, University of Dundee, and International Peat Society. p. 125–132.

Schrier-Uijl, A.P., Veenendaal, A.M., Leffelaar, P.A., van Huissteden, J.C. & Berendse, F. (2008): Spatial and temporal variation of methane emissions in drained eutrophic peat agro-ecosystems: drainage ditches as emission hotspots. Biogeosciences Discussions 5: 1237–1261.

Schroeder, F.-G. (1998): Lehrbuch der Pflanzengeographie. UTB Quelle & Meyer, Wiesbaden.

Schulz, K. (2005): Vegetations- und Standortentwicklung des wiedervernässten Grünlandes im Anklamer Stadtbruch (Mecklenburg-Vorpommern). Institute for Botany and Landscape Ecology. Greifswald, Greifswald University: 168 p .

Schulze-Hagen, K. (1991): Acrocephalus paludicola (Vieillot 1817) – Seggenrohrsänger. In: Glutz von Blotzheim, U.N. (Ed.): Handbuch der Vögel Mitteleuropas. Aula, Wiesbaden, p. 252–291 .

Schulze-Hagen, K., Leisler, B., Schäfer, H.M. & Schmidt, V. (1999): The breeding system of the Aquatic Warbler Acrocephalus paludicola – a review of new results. Vogelwelt 120: 87–96.

Schumann, M. & Joosten, H. (2008): Global peatland restoration manual. International Mire Conservation Group. URL: http://www.imcg.net/pages/publications/papers.php?lang=EN

Seiberling, S. & Klußmann, M. (2010): De-embankment in the non-tidal Baltic Sea: Impact on vegetation and standing crop. Naturschutz und Biologische Vielfalt 91: 67–87.

Sellin, D. & Schirmeister, B. (2004): Durchzug und Brut der Weißbart-Seeschwalbe im Jahr 2003 im Peenetal bei Anklam. Ornithologischer Rundbrief für Mecklenburg-Vorpommern 45: 39–44.

Shannon, R.D. & White, J.R. (1994): A three-year study of controls on methane emissions from two Michigan peatlands. Biogeochemistry 27: 35–60.

Shimova, O.C., Lopachuk, O.N. & Baychorov, V.M. (2010): Ekonomicheskaya effektivnost meropriyatii po sokhraneniyu biologicheskogo raznoobraziya [Economical efficiency of measures for the protection of biodiversity]. Belarusskaya Navuka, Minsk. 123 p. [in Russian]

Shotyk, W. (1989): An overview of the geochemistry of methane dynamics in mires. Int. Peat J. 3: 25–44.

Simberloff, D. (1998): Flagships, umbrellas, keystones: is single-species management passé in the landscape era? Biol. Cons. 81: 247–257.

Simonetti, S. & De Witt Wijnen, R. (2009): International Emissions Trading and Green Investment Schemes. In: Freestone, D. & Streck, C. (eds): Legal aspects of carbon trading. Kyoto, Copenhagen, and beyond. Oxford University Press, Oxford, p. 157–175.

Sjörs, H. (1950): On the relation between vegetation and electrolytes in north Swedish mire waters. Oikos 2: 241–258.

Skog, K.E. & Nicholson, G.A. (2000): Carbon sequestration in wood and paper products. Chapter 5, USDA Forest Service Gen. Tech. Rep. RMRS-GTR-59. pp 79–88 http://www.fpl.fs.fed.us/documnts/pdf2000/skog00b.pdf

Skoropanov, S.G.(1961): Osvoenie i ispolzovanie torfiano-bolotnykh pochv. Izdatestvo ASKhN BSSR, Minsk, 250 p. [in Russian]

Sliva, J. (1997): Renaturierung von industriell abgetorften Hochmooren am Beispiel der Kendlmühlfilzen. PhD thesis, Technical University Munich. 221 p.

Smelovskiy, V. E. (1988): Vyrabotannye torfianye mestorozhdeniia i ikh ispolzovanie. [Cutover peatlands and their utilisation] Pod red. S.G. Skoropanova. Nauka i tekhnika, Minsk, 152 p. [in Russian]

Smith, K.A. & Dobbie, K.E. (2001): The impact of sampling frequency and sampling times on chamber-based measurements of $N_2O$ emissions from fertilized soils. Global Change Biol. 7: 933–945.

Smith, P., Gary Lanigan, G., Kutsch, W.L., Buchmann, N. & Eugster, W. (2010): Measurements necessary for assessing the net ecosystem carbon budget of croplands. Agricult. Ecosys. Environ. 139: 302–315.

Steffenhagen, P., Frick, A., Timmermann, T. & Zerbe, S. (2008): Satellitenbildgestützte Vegetationsklassifizierung unter besonderer Berücksichtigung dominanter Pflanzenarten. Phosphor- und Kohlenstoff-Dynamik und Vegetationsentwicklung in wiedervernässten Mooren des Peenetals in Mecklenburg-Vorpommern. In: Gelbrecht, J., Zak, D. & Augustin, J., Berichte des IGB. 26: 143–144.

Steffenhagen, P., Zerbe, S., Frick, A., Schulz, K. & Timmermann, T. (2010): Wiederherstellung von Ökosystemleistungen der Flusstalmoore in Mecklenburg-Vorpommern. Naturschutz und Landschaftsplanung 42: 304–311.

Stephens, J.C., Allen, L.H. & Chen, E., (1984): Organic soil subsidence. In: Holzer, T. L. (ed.): Man-induced land subsidence. Geological Society of America; Boulder: 107–122.

Stepniewski, W. & Glinski, J. (1988): Gas exchange and atmospheric properties of flooded soils. In: Hook, D.D., McKee, W.H., Smith, H.K. Jr. et al. (eds): The ecology and management of wetlands. Volume 1: Ecology of wetlands. Croom Helm, London, p. 269–278.

Stroh, P. & Hughes, F. (2010): Practical approaches to wetland monitoring: Guidelines for landscape-scale long-term projects. Anglia Ruskin University; Cambridge.

Stull, R.B. (1988): An Introduction to Boundary Layer Meteorology. Kluwer Academic Publishers. 666 p.

Succow, M. (1988): Landschaftsökologische Moorkunde. Gustav Fischer Verlag, Jena.340 p.

Succow, M. & Joosten, H. (2001): Landschaftsökologische Moorkunde. Schweizerbart, Stuttgart. 2. völlig neu bearbeitete Auflage, 636 S.

Succow, M., Koska & I., Clausnitzer, U. (2001): Vegetationsentwicklunsgreihen. In: Succow, M. & Joosten, H. (Eds.): Landschaftsökologische Moorkunde. Schweizerbart, Stuttgart: 181–184.

Tanfilev, G.I. (1895): Bolota i torfyaniki Polesya.[Mires and peatlands of Polesye] [in Russian]

Tanneberger, F. (2008) The Pomeranian population of the Aquatic Warbler (Acrocephalus paludicola) – habitat selection and management. PhD thesis, Greifswald University.

Tanneberger, F. (2011): The relationship between peatland condition and its biodiversity. Literature study. Unpublished report. 104 p.

Tanneberger, F., Bellebaum, J., Helmecke, A., Fartmann, T., Just, P., Jehle, P. & Sadlik, J. (2008): Rapid deterioration of aquatic warbler Acrocephalus paludicola habitats at the western margin of its breeding range. J. Ornithol. 149:105–115.

Tanneberger, F., Tegetmeyer, C., Dylawerski, M., Flade, M. & Joosten, H. (2009): Slender, sparse, species-rich – winter cut reed as a new and alternative breeding habitat for the globally threatened Aquatic Warbler. Biodiversity and Conservation 18: 1475–1489.

Tanneberger, F., Flade, M., Preiksa, Z. & Schröder, B. (2010a): Habitat selection of the globally threatened Aquatic Warbler at the western margin of the breeding range and implications for management. Ibis 152: 347–358.

Tanneberger, F., Bellebaum, J. & Frick, A. (2010b): NATURA 2000 Species Management Plan for the Aquatic Warbler in Brandenburg. Unpublished report. 223 p.

Tanovitskaya, E.M. & Smelovskiy, V.E. (1972): Osushenie poley dobychi frezernogo torfa drenazhem iz plastmassovykh trubok. [Drainage of milling sites using plastic tubes] Minsk, 52 p. [in Russian]

Tanovitskaya, N.I. & Bambalov, N.N. (2009): Sovremennoe sostoianie i ispolzovanie bolot i torfianykh mestorozhdeniy Belarusi. [Current condition and utilisation of mires and peatlands in Belarus] Prirodopolzovanie 16: 82–89. [in Russian]

Tanovitskaya, N.I. & Ratnikova, O. N. (2010a): Osobennosti stratigrafii torfianykh zalezhei verkhovykh bolot Zapadno-Poozerskoy torfiano-bolotnoy oblasti. [Characteristics of bog stratigraphy in the West-Poozersky peatland district] Prirodopolzovanie 17: 94–104. [in Russian]

Tanovitskaya, N. & Ratnikova, O. (2010b): Changes in the areas of natural and damaged peatlands in Belarus over the last 20 years / Materials of international symposium "Chemical, physical and biological processes in soil" 15–17 June, 2010, Poznan. p. 147–154. [in Russian]

Tanovitskiy, I.G. (1980): Ratsionalnoe ispolzovanie torfianykh mestorozhdeniy i okhrana okruzhaiushchei sredy. [Rational use of peatlands and environmental protection] Minsk, 40 p. [in Russian]

Tanovitskiy, I.G. & Obukhovskiy, I.U.M. (1988) Antropogennye izmeneniia torfiano-bolotnykh kompleksov. [Anthropogenic alteration of peatland complexes] Minsk, 165 p. [in Russian]

The Conservation Measures Partnership (2007): Open Standards for the Practice of Conservation. Version 2.0. Accessible under www.conservationmeasures.org.

Thiele, A., Minaeva, T., Sirin, A., Mischenko, A.; Mikitiuk, I.; Chumachenko, S.; Tanavitskaya, N. Kozulin, A. (2008): Inventory on Area, Situation and Perspectives of Rewetting of Peatlands in Russia, Ukraine & Belarus. Prepared for the seminar 'Market Based Instruments for Rewetting of Peatlands' held during 12.–17.11.2008 at the International Academy of Nature Conservation, Isle of Vilm, Lauterbach. Manfred Hermsen Foundation, RSPB.

Thiele, A., Tanneberger, F., Minke, M., Couwenberg, J., Wichtmann, W., Karpowicz, Z., Fenchuk, V., Kozulin, A. & Joosten, H. (2009): Belarus boosts peatland restoration in Central Europe. Peatlands International 01/2009: 32–24.

Timmermann, T. (1999): Anbau von Schilf (Phragmites australis) als ein Weg zur Sanierung von Niedermooren – Eine Fallstudie zu Etablierungsmethoden, Vegetationsentwicklung und Konsequenzen für die Praxis. Archiv für Naturschutz und Landschaftsforschung 38, 2–4: 111–143.

Timmermann, T. (2003): Nutzungsmöglichkeiten der Röhrichte und Riede nährstoffreicher Moore Mecklenburg-Vorpommerns. Greifswalder Geographische Arbeiten 31: 31–42.

Timmermann, T., Margóczi, K., Takács, G. & Vegelin, K. (2006): Restoring peat forming vegetation by rewetting species-poor fen grasslands: the role of water level for early succession. Appl. Veg. Sci. 9: 241–250.

Tiuremnov, S.N. (1976): Torfianye mestorozhdeniia M. Nedra, 488 p. [in Russian]

Tiuremnov, S.N., Largin, I.F., Efimova, S.F. & Skobeeva, E.I. (1977): Torfianye mestorozhdeniia i ikh razvedka (Rukovodstvo po laboratorno-prakticheskim zaniatiiam). [Peatlands and their exploration] M., Nedra, 264 p. [in Russian]

Tomassen, H.B.M., Smolders, A.J.P., Lamers, L.P.M. & Roelofs, J.G.M. (2004): Development of floating rafts after the rewetting of cut-over bogs: the importance of peat quality. Biogeochemistry 71: 69–87.

Toogood, S.E. & Joyce, C.B. (2009): Effects of raised water levels on wet grassland plant communities. Appl. Veg. Sci. 12: 283–294.

Traxler, A. (1997): Handbuch des vegetationsökologischen Monitorings. Methoden, Praxis, angewandte Projekte. Teil A: Methoden. Monographien Band 98A. Umweltbundesamt, Wien. 391 p.

Trommer, C. (1853): Die Bonitirung des Bodens vermittelst wildwachsender Pflanzen. Ein Leitfaden für Boniteure, Landwirthe, Forstmänner und Gärtner. C. A. Koch, Greifswald.

Trumper, K., Bertzky, M., Dickson, B., van der Heijden, G., Jenkins, M., Manning, P. (2009): The Natural Fix? The role of ecosystems in climate mitigation. A UNEP rapid response assessment. United Nations Environment Programme, UNEPWCMC, Cambridge, UK. http://www.unep.org/pdf/BioseqRRA_scr.pdf.

Truskavetskiy, R. (2010): Torfovi grunty i torfovishcha Ukraini. [Peat soils and peatlands of the Ukraine] Miskdruk, Kharkiv, 278 p. [in Ukrainian]

Tucker, G.M. & Evans, M.I. (1997): Habitats for Birds in Europe: A Conservation Strategy for the Wider Environment. Birdlife International; Cambridge.

Tuittila, E.-S., Komulainen, V.M., Vasander, H. & Laine, J. (1999): Restored cut-away peatland as a sink for atmospheric $CO_2$. Oecologia 120: 563–574.

Tuittila, E.-S., Komulainen, V.M., Vasander, H., Nykänen, H., Martikainen, P.J. & Laine, J. (2000): Methane dynamics of a restored cut-away peatland. Global Change Biol. 6: 569–581.

Tuittila, E.S., Vasander, H. & Laine, J. (2004): Sensitivity of C sequestration in reintroduced Sphagnum to water-level variation in a peat extraction peatland. Restor. Ecol.12: 483–493.

Turner, W., Spector, S., Gardiner, N., Fladeland M., Sterling, W. & Steininger, M. (2003): Remote Sensing for biodiversity science and conservation. Trends in Ecology and Evolution. 18 (6): 306–314.

van de Akker, J.J.H., Kuikman, P.J., de Vries, F., Hoving, I., Pleijter, M., Hendriks, R.F.A., Wolleswinkel, R.J., Simoes, R.T.L & Kwakernaak, C. (2008): Emission of $CO_2$ from agricultural peat soils in the Netherlands and ways to limit this emission. In: Farrell, C. & J. Feehan (eds), Proceedings of the 13th International Peat Congress After Wise Use – The Future of Peatlands, Vol. 1 Oral Presentations, Tullamore, Ireland, 8–13 June 2008. International Peat Society, Jyväskylä, p. 645–648.

van den Pol-van Dasselaar, A., van Beusichem, M.L. & Oenema, O. (1999): Methane emissions from wet grasslands on peat soil in a nature reserve. Biogeochemistry 44: 205–220.

van der Peijl, M.J. & Verhoeven, J.T.A. (2000): Carbon, nitrogen and phosphorus cycling in river marginal wetland, a model examination of landscape geochemical flows. Biogeochemistry 50: 45–71.

van Diggelen, R., Middleton, B., Bakker, J., Grootjans, A. & Wassen, M. (2006): Fens and floodplains of the temperate zone: Present status, threats, conservation and restoration. Appl. Veg. Sci. 9: 157–162.

van Duinen, G.-J., Brock, A.M.T., Kuper, J.T., Leuven R.S.W.E., Peeters T.M.J., Roelofs, J.G.M., van der Velde, G., Verberk, W.C.E.P. and Esselink, H. (2003): Do restoration measures rehabilitate fauna diversity in raised bogs? A comparative study on aquatic macroinvertebrates. Wetlands Ecol. Managem. 11: 447–459.

van Duinen, G.-J., Dees, A. & Esselink, H. (2004): Importance of permanent and temporary water bodies for aquatic beetles in the raised bog remnant Wierdense Veld. Proc. Neth. Entomol. Soc. 15: 15–20.

van Huissteden, J., van den Bos, R. & Marticorena Alvarez, I. (2006): Modelling the effect of water-table management on $CO_2$ and $CH_4$ fluxes from peat soils. Geologie en Mijnbouw 85: 3–18.

VCS (2011): Verified Carbon Standard Version 3. http://www.v-c-s.org/docs/VCS Standard – v3.0.pdf. Accessed 30 May 2011.

Veen, P. J. (1992): De Nieuwe Venen. Natuurontwikkeling tussen Nieuwkoop en Botshol. Natuurmonumenten. Unpublished report.

Veenendaal, E.M., Kolle, O., Leffelaar, P.A., Schrier-Uijl, A.P., van Huissteden, J., van Walsem, J., Möller, F. & Berendse, F. (2007): $CO_2$ exchange and carbon balance in two grassland sites on eutrophic drained peat soils. Biogeosciences 4: 1027–1040.

Vegelin, K., Schulz, K., Olsthoorn, G. & Wachlin, V. (2009): Erfolgskontrolle Polder Randow-Rustow 2008. Gebietszustand im 9. Jahr der geregelten Wiedervernässung (2008). Unpublished report. 142 p.

Verberk, W.C.E.P., Leuwen, R.S.E.W., van Duinen, G.A. & Esselink, H. (2010): Loss of environmental heterogeneity and aquatic macroinvertebrate diversity following large-scale restoration management. Basic Appl. Ecol. 11: 440–449.

Verhagen, A., van den Akker, J.J.H., Blok, C., Diemont, W.H., Joosten, J.H.J., Schouten, M.A., Schrijver, R.A.M., den Uyl, R.M., Verweij, P.A. & Wösten, J.H.M. (2009): Peatlands and carbon flows. Outlook and importance for the Netherlands. Report WAB 500102 027, Netherlands Environmental Assessment Agency PBL, Bilthoven.

Verzhitskaya, I.N. (2010): Razrabotka kratkosrochnykh indikatorov vosstanovleniya naruzhennykh bolot razlichnykh tipov i analiz i sistematizaziya tekhnicheskikh rezhenii vosstanovleniya naruzhennykh bolot [Elaboration of short-term indicators for the restoration of various types of degraded peatlands and analysis and systematics of technical solutions for the restoration of degraded peatlands]. Final report, National Academy of Sciences. 86 p. [in Russian]

Vitt, D.H. (2006): Function characteristics and indicators of boreal peatlands. In: Wieder, R.K. & Vitt, D.H. (eds) Boreal Peatland Ecosystems. Springer, New York, p. 9–24.

Vogel, T. & Ahlhaus, M. (2009): Nutzung von Landschaftspflegematerial in Biogasanlagen. 3. Rostocker Bioenergieforum. 14./15. Okt. 2009, Universität Rostock, Institut für Umweltingenieurwesen, Band 23, p. 237–250.

Wagner, A. & Wagner, I. (2005): Leitfaden der Niedermoorrenaturierung in Bayern. Bayerisches Landesamt für Umwelt. 140 p.

Wagner, C. (1994): Zur Ökologie der Moorbirke *Betula pubescens* EHRH. In Hochmooren Schleswig-Holsteins unter besonderer Berücksichtigung von Regenerationsprozessen in Torfstichen. Mttlg. AG Geobotanik Schleswig-Holstein & Hamburg, H. 47, 184 p., app.

Wagner, C. (2006): "Grenzen des Entkusselns" oder: Zum Einfluß der Moorbirke (*Betula pubescens*) auf Regenerationsprozesse in Hochmooren. Arch. Naturschutz & Landschaftsforschung 45 (2), p. 71–85.

Warner, B.G. & Chmielewski, J.G. (1992): Testate amoebae (Protozoa) as indicators of drainage in a forested mire, Northern Ontario, Canada. Archiv fuer Protistenkunde 141: 179–183.

Wassen, M.J., Olde Venterink, H., Lapshina, E.D. & Tanneberger, F. (2005): Endangered plants persist under phosphorus limitation. Nature 437 (7058): 547–551.

Weber, C.A. (1902): On the Vegetation and Development of the Raised Bog of Augstumal in the Memel Delta. In: Couwenberg, J. & Joosten, H. (eds; 2002): C.A. Weber and the Raised Bog of Augstumal. International Mire Conservation Group/PPE Grif & K, Tula: 52–270.

Weihe, P.E. & Neely, R.K. (1997): The effects of shading on competition between purple loosestrife and broad-leaved cattail. Aquatic Bot. 59: 127–138.

Weiss, V. (2001): Derzeitige und künftig zu erwartende emissionsbegrenzende Anforderungen der TA Luft an Feuerungsanlagen für Stroh oder ähnliche pflanzliche Stoffe. In: Gülzower Fachgespräche: Energetische Nutzung von Stroh, Ganzpflanzengetreide und weiterer halmgutartiger Biomasse. Stand der Technik und Perspektiven für den ländlichen Raum, FNR 2001, p. 17–35.

Wendel, D. (1992): Untersuchungen zum aktuellen Zustand und zur Sukzession im Naturschutzgebiet "Mothäuser Heide" (Erzgebirge). Dipl.-thesis TU Dresden, 62 p., 21 app.

Wendel, D. (2010): Autogene Regenerationserscheinungen in erzgebirgischen Moorwäldern und deren Bedeutung für Schutz und Entwicklung der Moore. PhD TU Dresden, Dept. of Forest Sciences, 248 p.

Wetlands International & University of Greifswald (2010): Q&A on AFOLU, 'wetland management' and the road to land-based accounting. Questions and answers. Wetlands International, Ede, 24 p. http://www.wetlands.org/LinkClick.aspx?fileticket=zKxgBiimpvs%3d&tabid=1911.

Whiting, G.J. and Chanton, J.P. (2001): Greenhouse carbon balance of wetlands: methane emission versus carbon sequestration. Tellus 53B: 521–528.

Whiting, G.J., Bartlett, D.S., Fan, S., Bakwin, P.S. & Wofsy, S.C. (1992): Biosphere/atmosphere $CO_2$ exchange in tundra ecosystems: community characteristics and relationships with multispectral surface reflectance. J. Geophys. Res. 97: 16671–16680.

Wichmann, S. & Wichtmann, W. (Eds.) (2009): Bericht zum Forschungs- und Entwicklungsprojekt Energiebiomasse aus Niedermooren (ENIM). Universität Greifswald und DUENE e.V. Abschlussbericht an die DBU, 190 S.

Wichtmann, W. (1999a): Nutzung von Schilf (*Phragmites australis*). Archiv für Naturschutz und Landschaftsforschung 38(2–4): 217–232.

Wichtmann, W. (1999b): Schilfanbau als Alternative zur Nutzungsauflassung von Niedermooren. Archiv für Naturschutz und Landschaftsforschung 38(2–4): 97–110.

Wichtmann, W. & Joosten, H. (2007): Paludiculture: peat formation and renewable resources from re-wetted peatlands. IMCG-Newsletter, issue 2007/3: 24–28.

Wichtmann, W. & Koppisch, D. (1998): Nutzungsalternativen für Niedermoore am Beispiel Nordostdeutschlands. Z. Kulturtechnik Landentwickl. 4: 162–168.

Wichtmann, W. & Succow, M. (2001): Nachwachsende Rohstoffe. In: Kratz, R. & Pfadenhauer, J. (Eds.): Ökosystemmanagement für Niedermoore – Strategien und Verfahren zur Renaturierung. Eugen Ulmer Verlag, Stuttgart. p. 177–184.

Wichtmann, W. & Tanneberger, F. (2009): Feasibility of the use of biomass from re-wetted peatlands for climate and biodiversity protection in Belarus. Report to the Project: 'Restoring Peatlands and applying Concepts for Sustainable Management in Belarus – Climate Change Mitigation with Economic and Biodiversity Benefits' Michael Succow Stiftung zum Schutz der Natur. 112 p.

Wichtmann, W. & Wichmann, S. (2011a): Environmental, Social and Economic Aspects of a Sustainable Biomass Production. J. Sust. Energy Environ., Special Issue: 77–83.

Wichtmann, W. & Wichmann, S. (2011b): Paludikultur: standortgerechte Bewirtschaftung wiedervernässter Moore. Telma, in press.

Wichtmann, W., Knapp, M. & Joosten, H. (2000): Verwertung der Biomasse aus der Offenhaltung von Niedermooren (Utilisation of biomas from fen peatlands) . Z. f. Kulturtechnik und Landentwicklung 41: 32–36.

Wichtmann, W., J. Couwenberg & A. Kowatsch (2009): Standortgerechte Landnutzung auf wiedervernässten Niedermooren – Klimaschutz durch Schilfanbau. Ökologisches Wirtschaften 1. 2009: 25–27.

Wichtmann, W., Tanneberger, F., Wichmann, S. & Joosten, H. (2010a): Paludiculture is paludifuture: Climate, biodiversity and economic benefits from agriculture and forestry on rewetted peatland. Peatlands International 1: 48–51.

Wichtmann, W. Wichmann, S. & Tanneberger, F. (2010b): Paludikultur – Nutzung nasser Moore: Perspektiven der energetischen Verwertung von Niedermoorbiomasse. Naturschutz und Landschaftspflege in Brandenburg 19 (3, 4): 211–218.

Wicke, B., Dornburg, V., Junginger, M. & Faaij, M. (2008): Different palm oil production systems for energy purposes and their greenhouse gas implications. Biomass Bioenergy 32: 1322–1337.

Wiegleb, G. & Krawczyniski, R. (2010): Biodiversity management by water buffalos in restored wetlands. Waldökologie, Landschaftsforschung und Naturschutz 10: 17– 22.

Wild, U., Kamp, T., Lenz, A., Heinz, S. & Pfadenhauer, J. (2001): Cultivation of Thypa spp. in constructed wetlands for peatland restoration. Ecological engineering 17: 49–54.

Wilson, D., Tuittila, E.-S., Alm, J., Laine, J., Farrell, E.P. & Byrne, K.A. (2007): Carbon dioxide dynamics of a restored maritime peatland. Ecosci. 14: 71–80.

Wilson, F., Armstrong, A., Alm, J., Laine, J., Farne, E.P. & Byne, K.A. (2008): Rewetting of cut-away peatlands: Are we creating hot-spots of methane emission? Restorat. Ecol., doi: 10.1111/j.1526-100X.2008.00416.x

Wulder, M.A., Hall, R.J., Coops, N.C. & Franklin, S.E. (2004): High spatial resolution remotely sensed data for ecosystem characterization. Biosci. 64: 511–521.

Yli-Petäys, M., Laine, J., Vasander, H. & Tuittila, E.-S. (2007): Carbon gas exchange of a re-vegetated cut-away peatland five decades after abandonment. Boreal Environm. Res. 12: 177–190.

Zak, D., Augustin, J. & Gelbrecht, J. (2008): Biogeochemische Prozesse in Niedermooren und deren Wirkung auf Gewässer und Klima. In: Gelbrecht, J., Zak, D. & Augustin, J. (Eds.): Phosphor- und Kohlenstoff-Dynamik und Vegetationsentwicklung in wiedervernässten Mooren des Peenetals in Mecklenburg Vorpommern. Berichte des IGB, Leibniz-Institut für Gewässerökologie und Binnenfischerei, Berlin. Heft 26: 12–19.

Zhuravlev, D. (1999): Izmenenie ornitofauny v rezultate sukzessii na vtorichno zabolochnenykh territoriakh PGREZ. [Changes in the ornitofauna as a result of succession on rewetted areas PGREZ] In: Belavezhskaya Pushcha na rubezhe tretego tysyacheletiya. Materialy konferenzii 22.–24.12.1999. [Belaveshskaya Pushcha at the edge of the third millenium. Conference proceedings 22.–24.12.1999] Minsk. p. 290–293. [in Russian]

Zhou, L., Zhou, G. & Jia, Q. (2009): Annual cycle of $CO_2$ exchange over a reed (Phragmites australis) wetland in Northeast China. Aquat. Bot. 91: 91–98.

Zinke, P. & Edom, F. (2006): Hydraulische und hydrologische Erklärung von Ökotopstrukturen am Regenmoor Kriegswiese im mittleren Erzgebirge. Arch. Naturschutz & Landschaftsforschung 45: p. 43–60.

Names of Belarusian locations are transliterated using Rule No. 8/14809 (08.08.2006) of the Commitee for Land Resources, Geodesy and Cartography, Belarus. Russian language references are transliterated using ALA-LC (American Library Association & Library of Congress) romanization tables for Slavic alphabets in the latest version published in 1997 (except in case of names such as 'Tanovitskaya' which are spelled with '-aya' instead of 'aia' for the sake of consistency with other publications). Diacritical marks are omitted.

# List of contributors

Jürgen Augustin, Leibniz Centre for Agricultural Landscape Research (ZALF), Müncheberg, Germany.

Susanne Bärisch, Michael Succow Foundation, Greifswald, Germany.

Richard Bradbury, The Royal Society for the Protection of Birds (RSPB), Sandy, UK.

Tatsiana Broska, Institute of Experimental Botany, National Academy of Sciences of Belarus, Minsk, Belarus.

Andrei Burlo, APB BirdLife Belarus, Minsk, Belarus.

Olga Chabrouskaya, APB BirdLife Belarus, Minsk, Belarus.

Hanna Chuvashova, APB BirdLife Belarus, Minsk, Belarus.

Mike Clarke, The Royal Society for the Protection of Birds (RSPB), Sandy, UK.

John Couwenberg, Institute of Botany and Landscape Ecology, Greifswald University, Greifswald, Germany.

Frank Edom, HYDROTELM, Dresden, Germany.

Igino Emmer, Silvestrum, Jisp, The Netherlands.

Viktar Fenchuk, APB BirdLife Belarus, Minsk, Belarus.

Rob Field, The Royal Society for the Protection of Birds (RSPB), Sandy, UK.

Jack Foxall, The Royal Society for the Protection of Birds (RSPB), Sandy, UK.

Greta Gaudig, Institute of Botany and Landscape Ecology, Greifswald University, Greifswald, Germany.

Marek Giergiczny, Department of Economic Sciences, University of Warsaw, Warsaw, Poland.

Alexander Grebenkov, Department of Energy Efficiency of the State Committee of Standardization, Minsk, Belarus.

Hans Joosten, Institute of Botany and Landscape Ecology, Greifswald University, Greifswald, Germany.

Zbigniew Karpowicz, The Royal Society for the Protection of Birds (RSPB), Sandy, UK.

Alexander Kozulin, Centre for Bioresources, National Academy of Sciences of Belarus, Minsk, Belarus.

Nazdeya Liashchinskaya, APB BirdLife Belarus, Minsk, Belarus.

Uladzimir Malashevich, APB BirdLife Belarus, Minsk, Belarus.

Merten Minke, APB BirdLife Belarus, Minsk, Belarus.

Frank Moerschel, KfW Bankengruppe, Frankfurt am Main, Germany.

Robert O'Sullivan, Climate Focus, Washington D.C., USA.

Galyna Poshtarenko, The Ukrainian Society for the Protection of Birds (USPB), Kyiv, Ukraine.

Lydia Pshenitsyna, APB BirdLife Belarus, Minsk, Belarus.

Norbert Schäffer, The Royal Society for the Protection of Birds (RSPB), Sandy, UK.

Sebastian Schmidt, Michael Succow Foundation, Greifswald, Germany.

Jenny Schulz, Michael Succow Foundation, Greifswald, Germany.

Andrei Shunko, GEOPLAN, Maladziecna, Belarus.

Andrea Strauss, International Academy for Nature Conservation, Federal Agency for Nature Conservation (BfN), Vilm, Germany.

Franziska Tanneberger, Michael Succow Foundation and Institute of Botany and Landscape Ecology, Greifswald University, Greifswald, Germany.

Nina Tanovitskaya, Institute for Nature Management, National Academy of Sciences of Belarus, Minsk, Belarus.

# List of contributors

Annett Thiele, APB Birdlife Belarus, Minsk, Belarus.

Sviataslau Valasiuk, APB BirdLife Belarus, Minsk, Belarus.

Irina Viarshitskaya, Institute of Experimental Botany, National Academy of Sciences of Belarus, Minsk, Belarus.

Irina Voitekhovitch, APB BirdLife Belarus, Minsk, Minsk.

Sabine Wichmann, Institute of Botany and Landscape Ecology, Greifswald University, Greifswald, Germany.

Wendelin Wichtmann, Michael Succow Foundation, Greifswald, Germany.

Jens Wunderlich, Michael Succow Foundation, Greifswald, Germany.

Tatsiana Yarmashuk, APB BirdLife Belarus, Minsk, Minsk, Belarus.

Dmitry Zhuravlev, Centre for Bioresources, National Academy of Sciences of Belarus, Minsk, Belarus.

Tomasz Zylicz, Department of Economic Sciences, University of Warsaw, Warsaw, Poland.

# Index

abandoned (see also abandonment) 9, 20, 39, 43ff., 49f., 69, 72f., 84, 86, 109, 117, 122, 127, 130, 133, 138, 166, 169, 178f., 191f., 195
abandonment 42f., 49f., 52, 57, 65, 70ff., 86, 112f., 138ff., 146, 154, 174, 178, 192
*Acorus calamus* 112
*Acrocephalus paludicola* 58, 61, 63, 65, 130f., 134, 137, 145f., 151, 165, 171, 194
*Acrocephalus schoenobaenus* 72f., 75, 131
activity shifting 99, 138
additionality 104, 138, 155
Afforestation, Reforestation and Revegetation (ARR) 97, 99f., 143
agricultural (land) use 6-9, 23, 50, 107, 130, 166, 176
Agricultural Land Management (ALM) 97, 143
Agriculture, Forestry and other Land Use (AFOLU) 92, 97, 98f., 143
*Alcedo atthis* 72, 176
*Alnus glutinosa* 16, 83, 110, 115
*Anas* 65f.
*Andromeda polifolia* 36, 76, 173
Anthropogenic (GHG) emissions 20, 91, 129
*Anthus pratensis* 67, 73
APB-BirdLife Belarus 1, 10, 61, 93, 113, 126, 133, 135, 142, 153f., 185f., 197, 215f.
*Aquila clanga* 58, 61, 65, 79f., 134, 137, 145f., 174
*Aquila pomarina* 65, 83, 174, 181f.
ash content **125ff.**
*Asio flammeus* 174
assessment of GHG emissions/ fluxes 136, 141, 143, 152, 189
assessment on Peatlands, Biodiversity and Climate Change 91ff.
assessment tool = GHG assessment tool
Astrauskoje 39, 47f., 51f., 55f., 58, 140, 142, 187
*Aulacomnium palustre* 48
awareness raising 135, 149, 151, 155

Barcianicha 27, **33ff.**, 53, 74ff., 140f.
baseline 41ff., 77, 96f., 99, 106, **136ff.**, 141, 143, 145, 155, 165, 171, 176, 179, 182, 192
baseline scenario 41, 43, 47f., 50, 52, 56, 98f., 137f., 192f.
beaver 48, 72, 102, 138, 155, 169ff., 180, 182
Belarusian Socialist Soviet Republic 6f.
*Betula humilis* 171, 173
*Betula nana* 187
*Betula pendula* 43, 44, 46, 49, 51, 74, 76, 171, 175
*Betula pubescens* 74, 177, 180
*Bidens frondosa* 50
Biebrza valley 63f.
biodiversity 1f., 10, 13, **61ff., 80ff.**, 89, 91ff., 96f., 107ff., 112, 126, **128ff., 133ff., 144ff.**, 149f., 154ff., 165ff., 171, 173, 176ff., 181f., 186, 189, **192ff.**, 197
biodiversity benefits 1, 89, 129, 131, 133, 147, 150, 155, 171, 173, 176, 179, 182, 186, 189, 193, 194, 197
biodiversity monitoring 77, 146, 154, 194f.
biofuel 102f., 127, 129, 155f.
biogas production 112, 116, 125, 130
biomass 40f., 47, 49, 67, 83, 85ff., 98, **107ff., 113ff., 125ff.**, 131, 135ff., 141, **143ff.**, 150, 152, 155, 166f., 181, 189, 191ff., 195, 197
biomass harvesting 36, 83, 109f, 114ff., 125ff., 130f., 135ff., 147, 155f., 176, 190f.
biotope type 10, 66f.
bird monitoring 145, 176, 180
Black Alder = *Alnus glutinosa* 16, 83, 110, 115
BMU-ICI project 1ff., 10, 32, 35, 51, 70, 80, 86, 97, 115, 125, 127, 129, **133ff.**, 166, 169, 189ff.
bog 3ff., 7f., 18f., 21, 27, 35, 39ff., 43, 45ff., 51, 61, 66f., 70f., 73ff., 77, 79, 83f., 86, **109ff.**,133, 139f., 145ff., 153, 163f., **169ff.**, **177ff., 184ff.**, 192

*Bombina bombina* 76, 79, 84
*Botaurus stellaris* 65, 68, 72f., 145, 176, 179, 182
breeding site 62, 64, 69, 79, 86, 131, 146, 192, 194f.
briquettes (from biomass) 9, 116, 125ff., 129, 131, 144, 177
*Bubalus bubalis* 113, 118
*Bubo bubo* 174
*Calamagrostis epigeios* 45, 48, 158, 161
calibration 32, 35f., 41, 81, 141f., 153, **189ff.**
*Calliergon cordifolium* 46
*Calliergonella cuspidata* 46, 182
*Calluna vulgaris* 43, 46ff., 74, 76, 163, 173, 187
capacity building 1, 134, 136, 152f., 155
*Carabus menetriesi* 79
carbon (C) cycle 13, 15
carbon credits 1, 13, **93ff.**, 99, 101, 103f., 129, **137ff.**, 144, 152, 154ff., 166f., 194
carbon dioxide = $CO_2$
carbon stock 29, 93, 98ff., 143f., 147f., 193
*Carex* 16, 36, 45f., 50f., 66, 71, 74, 79, 107, 115, 171, 177, 187, 190f.
*Carex acutiformis* 70
*Carex canescens* 46, 48
*Carex disperma* 171
*Carex elata* 36., 46
*Carex nigra* 36, 46, 48, 173
*Carex paupercula* 171
*Carex riparia* 70, 116
*Carex rostrata* 36., 46, 177
*Carex vesicaria* 46, 48
Cattail = *Typha*
CCBA validation 145f.
$CH_4$ **13ff., 30ff.**,37, **39ff.**, 49, 52f., 99, 116, 141, 150f., 190ff.
*Chamaedaphne calyculata* 36, 48
Chernobyl 61, 72f., 136, 142
*Chlidonias hybridus* 72
*Chlidonias niger* 65, 72
choice experiment 66f.
*Ciconia nigra* 65, 75, 174, 176, 182
CIM (Centre for International Migration) 1, 121, 142, 144, 152f., 155
*Circus aeruginosus* 72, 79

# Index

*Circus pygargus* 65, 73
Clean Development Mechanism (CDM) 42, 103f., 144
climate benefits 129f., 171, 173, 175, 179f., 182, 186, 189
Climate Convention = United Nations Framework Convention on Climate Change
climate effect 14, 18f., 136f., 150
Climate, Community and Biodiversity (CCB) Standards 97, 146f., 193ff.
closed chamber 30ff., 34ff., 53
$CO_2$ **13ff.**, **29ff.**, 34, 36f., **39ff.**, 49, 53f., 92f., 99f., 104, 107, 109, 116, 129, 133, 139, 141, 148, **152ff.**, 189
$CO_2$ equivalent 14, 30, 33, 39f., 42, **47ff.**, 129f., 139, 141, 144, 166, 171, 173, 175f., 179, 182, 190f
colonization 46, 51, 69, 70, 77f., 84, 86, 186
combustion 107, 116, 125, 127ff.
Common Reed = *Phragmites australis*
communication 149, 153, 155, 166
community benefits 147, 195
compliance market 37, 43, 89, 95, 99, 103f., 149
conservation 1f., 6, 8, 10, 61f., **64ff.**, 78ff., 84, 86, 90ff., **96ff.**, 104, 107, 110f., 113, 126f., 129ff., **133ff.**, 141ff., **149ff.**, 165f., 169, 171, 192f.
Convention on Biological Diversity (CBD) 91, 94
Convention on Combating Desertification 91
cooling effect 14, 33
cranberry 110f., 171, 174
*Crex crex* 65, 68, 72f., 81, 145, 176, 180
cropland 20, 29, 40, 42, **99ff.**
cutover peatland (bogs and fens) 7f., 46, 74, 110, 112, 192
*Dactylorhiza maculata* 171
dairy cows/cattle 9, 112, 125, 130
Dakudauskaje 10f., 25, 139f., 163, 169, 181ff.
Dalbeniski 123f., 139f., 145, 147, 169ff., 187
dam 11f., 26ff., 75, 102, 139f., 157, 163, **169ff.**, 178ff., 182f., 185ff., 195
Darwin Initiative 10, 69
decomposed peat 71, 109
decomposition 4, 15f., 39, 85f., 93, 138, 173, 178
degradation 3, 10, 70, 71, 80, 84, 90, 91, 93, 94, 98, 115, 130, 147, 182

degraded peatlands 10, 70, 89ff., 136f., 142, 166
denitrification 16
depleted peat 9ff., 90
*Deschampsia caespitosa* 43
*Dicranum scoparium* 48
domestic project cycle 105f.
drainage 1, 3, **5ff.**, 11, 16f., 20, 29, 41, 47, 48ff., 52, **61ff.**, 66, 70f., **82ff.**, 92f., 98f., 102f., 108, 110, 112f., **128ff.**, 138ff., 147f., 166, 169f., 172f., 176, **178ff.**, 186f., 195
drained peatland 1, 3, **6ff.**, 11, 15, **17ff.**, 23, 35, 37, 41ff., 46, 48, 70f., 78ff., 92ff., 98ff., 102, 112f., 122, 129ff., 133, 137f., 143, 146, 148, 156, 165f., 180, 191f., 194
draining effect 11, 182
*Drosera rotundifolia* 43, 47
draught 6, 9
*Dryopteris filix-mas* 48
dwarf shrub **43ff.**, 48f., 124, 163, 181f., 192
ebullition 16, 31, 34
economic benefits 1, 130, 133, 195, 197
ecosystem respiration 15f., 31, 33, 36
eddy covariance 30, **34ff.**, 55, 191
effectiveness 11, 48, 106, 126, 193f.
Ellenberg indicator values 38, 41, 180, 189
*Elymus repens* (*Agropyron repens*, *Elytrigia repens*) 46, 50, 112, 180
*Emberiza schoeniclus* 67, 72f., 75
emission factors 20, 48, 50, 52, 100, 137
emission reductions purchase agreement (ERPA) 94, 105
emission(s) reduction 1, 30, 35, 37, 41f., 47, 50, 52, 89, 94, **96ff.**, **103ff.**, 129, 133f., **136ff.**, **141ff.**, 146, 148, 152, 154ff., 165ff., 171, 173, 189
*Empetrum nigrum* 46
enclosure time 30ff.
energy 6, 9, 32, 35, 87, 89, 94, 99f., 102f., 107, 110, 113, 115f., 125, 127, 129, 136f., 142, 144, 151, 156, 195
engineering 10, 51, 90, 136, 139, 152, 155, 170, 173, 179
*Eriophorum angustifolium* 36, 43, 45f., 190
*Eriophorum vaginatum* 39, **43ff.**, 49, 76, 171, 182, 186
Estonia 29, 101, 128
*Eupatorium cannabinum* 112

exchange market 94f.
fen **3ff.**, 15, **17ff.**, 35, 40, 45f., 49f., 59, 61, **63ff.**, 70f., 73ff., 77, 79f., 83f., 86f., 112, 114f., 126, 130f., 133, 135f., 139ff., 145ff., 153, 166, 169, **171ff.**, 179, 181f., 191f.
fermentation 107, 125
*Filipendula ulmaria* 112
fire 85f., 170, 172, 175, 178, 182, 184, 189
forward sales 95f.
forward-looking baseline 42, 48, 50, 139, 141, 171, 176, 179, 182
*Funaria hygrometrica* 47
*Gallinago gallinago* 65f., 68, 73, 75
*Gallinago media* 61, 65, 79f., 145
gap filling 35f., 41, 189ff.
gas concentration 15, 30f., 34, 99
GEST approach 35, 37, 39, 41, 141ff., 166, 189, 191f.
GEST-FIRE 189
GEST-FOREST 189, 191
GEST-HERB 189, 191f.
GEST-PREDICT 189, 192
GEST-TRANSIENT 189, 191
GEST-WATER 189, 191
GHG benefits (see also climate benefits) 43, 83
Global Carbon project 92
Global $CO_2$ emissions 13, 20
Global Environment Facility (GEF) 3, 8, 10f., 69f., 74f., 86, 89, 92, 135, 144f., 180, 183, 186
Global Peatland Database 20, 54
Global Peatland Initiative 92
global warming potential (GWP) 14f., 32f., 39, 47, 49, 51, 130
*Glyceria fluitans* 46, 71
*Glyceria lithuanica* 171
*Glyceria maxima* 46, 71, 116
Gold Standard 97, 195
grassland 9, 19, 36, 39f., **46ff.**, 73, 83, 86, 90, 99ff, 109, 111ff., 115ff., 130, 158, 161, 174ff., 191
Greater Spotted Eagle = *Aquila clanga*
greenhouse gas (GHG) emissions 1, 2, 13, **30ff.**, 43, 52, 78, 83, 86, 90f., 93, 98ff., 104f., 107, 1110f., 129, 133f., 136, 141ff., 148, 150ff., 154ff., 166, 171, 179f., 189, 191, 193f.
greenhouse gas (GHG) monitoring 193
greenhouse gas emissions site type (GEST) 35ff., 41f., 47, 49f., 52, 123, 136, 139, 141, **143ff.**, 153, 159f., 162, 165f., 171, 173, 175, 179, 182, 189ff.
gross primary production (GPP) 14ff., **31ff.**

# Index

ground water (groundwater) 1, 4f., 7, 14, 37, 44, 51, 71f., 75, 111, 114, 174ff., 177ff., 181, 184, 186, 194
*Grus grus* 66, 68, 72f., 80, 84, 174, 182, 186
GWP = global warming potential
habitat conditions 70, 85f., 136, 192f.
*Haliaeetus albicilla* 72
harvesting of biomass = biomass harvesting
heating value 125ff., 129
*Hierochloe odorata* 112
high intensity grassland 112, 116
Horeuskaje 70, 139f., 146, 169
Hrycyna-Starobinskaje 23ff., 28, 70, 139f., 145f., 158, 160, 161, 169, 174f.
hummock-hollow complexes 169, 173
hummock-hollow lawn complexes 46, 169, 173
*Huperzia selago* 171, 173, 187
hydromorphological analysis 51
Improved Forest Management (IFM) 97f., 143
indicator species 37, 61, **77ff.**, 82, 194
indicator(s) 37, 77f., 189, 195
Indonesia 20, 29, **91ff.**, 104
integrated management of peatlands 92
International Climate Initiative (BMU-ICI) 1, **133ff.**, 166f.
International Mire Conservation Group (IMCG) 20, 54, 91ff.
International Sacharov Environmental University (ISEU) 127
International Standards Organisation (ISO) 97, 143
Jelnia 10, 21, 47, 51, 140, 153, 163f., 169, **184ff.**
*Juncus articulatus* 50
*Juncus effusus* 50, 191
*Juncus filiformis* 48
Kreditanstalt für Wiederaufbau Entwicklungsbank (KfW) 1, 133, 135, 166
Kyoto Protocol 14, 37, 42, 89f., 93f., **99ff.**, 136, 147, 149
*Lagopus lagopus* 174
land fund 8f.
land use 2, 6f., 9f., 19f., 37f., 41, 43, 51, 63f., 68ff., 82, 86, 89, 91ff., **98ff.**, 104, **197ff.**, **126ff.**, 137, 139, 144, 147, 150, 153, 166f., 170, 172, 174, 178, 181, 184, 195
land use options 2, **107ff.**, **112ff.**, **126ff.**

Land Use, Land-Use Change and Forestry (LULUCF) 93f., **100ff.**
leakage 11, 26, 97, 99, 138ff., 150, 169, 193
*Ledum palustre* 46, 48, 74, 76, 163, 173, 182, 191
legal sources 89
Leibniz Centre for Agricultural Landscape Research (ZALF) 1, 134, 142, 152ff.
Lesser Spotted Eagle = *Aquila pomarina*
*Leucorrhinia dubia* 84
*Limosa limosa* 61, 65
*Linnaea borealis* 173
litter accumulation 39, 131
*Locustella* 79
low intensity grassland 113
*Luscinia svecica* 72, 75
*Lycaena dispar* 79
*Lysimachia thyrsiflora* 46
*Lysimachia vulgaris* 45f.
*Lythrum salicaria* 46
macrorelief 11, 184
management recommendations 146, 171, 174, 176, 180, 183, 187
managing a rewetted fen 83
marketing strategy.96
measurement footprint 35f.
medicinal plants 112
*Menyanthes trifoliata* 22, 36, 112
methane = $CH_4$
methane emissions 18f., 34, 35, 103, 148
methanogenesis 16, 19, 40
*Metroxylon sagu* 112
Michael Succow Foundation (MSF) 1, 127, 133f., 150, 153, 166
microrelief 45
milled peat 6f., 43, 47, 172
milling of bogs 71, 83, 169
*Milvus migrans* 72
*Milvus milvus* 72
mineralization 1, 3, 16f., 107, 110f., 113, 131, 133, 176, 179, 183
Ministry of Natural Resources and Environmental Protection 1, 10, 93, 105f., 108, 133, 136, 185
mire **1ff.**, 8, 10f., 22, 41, 45, **61ff.**, **66ff.**, 74, **77ff.**, 80, 84, 86f., 107ff., 112, 126, 130, **133ff.**, 147, 151, 153, 169, 175, 178, 184f., 187
*Miscanthus* **126ff.**, 132
*Molinia caerulea* 43, 50, 71
monitoring 2, 30, 35, 37, 42f., 64f., 69, 74f., 77f., 80ff., 89, 93, 100, 102, 105f., 126, 130, 134, 136, 139, **143ff.**, 150, 154f., 165, 176., 179f., **189ff.**

monitoring methodology 136f., 141, 143, 165
monitoring scheme 65, 78, 146
monitoring strategies 80, 82
Marocna 11, 26, 111, 140, 183
*Motacilla alba* 75
*Motacilla citreola* 75
*Motacilla flava* 67, 73
*Myriophyllum alternifolium* 46
$N_2$ 13, 16
$N_2O$ **13ff.**, 18f., **30ff.**, 42, 53, 93, 99f., 130, 141, 152ff.
nature conservation/protection fund 6, 8
nature protection = nature conservation 6, 8, 10, 86, 90, 110, 113, 127, 129, 135, 152, 130, 153, 167, 185
N-cycle 13
net ecosystem production (NEP) 16f.
net primary production (NPP) 15, 109
nitrate 16
nitrification 16
nitrogen = $N_2$
nitrous oxide = $N_2O$
Non-governmental Organisation (NGO) 95, 105, 148f., 151, 155, 165
*Numenius arquata* 61, 65, 80, 174, 182, 186
nutrient level / status 41, 64, 82, 86
Obal 26, 139f., 145, 169
opaque chamber 30ff., 53
*Oryza sativa* 16
over-the-counter (otc) market 94f., 97
oxygenation 17
Paliessie **3ff.**, 61, 68, 151, 165f., 174
paludiculture 1, 87, 98, **107ff.**, **127ff.**, 136, 141, 144, 149ff., 156
*Pandion halietus* 72, 174, 179
paper pulp 93, 114
peat accumulation 4, 8, 17, 71, 98, 108
peat briquettes 9, 116, 126, 129, 144
peat extraction 1, 3, **6ff.**, 20, 23ff., 43f., 47, 49f., 71, 74f., 83, 89f., 100f., 109, 111f., 129, **138ff.**, 155f., 163, 166, **169ff.**, 174, 178, **181ff.**
peat fire = peatland fire
peat formation 3f., 6, 43, 109, 182
peat fund 6, 8f.
peat layer **3ff.**, 15f., 19f., 42, 70, 75, 83, 110, 137, 166, 169, 178
peat properties 10, 70

# Index

peat stock 4, 6, 8
peat/peatland cadastre 6, 8
peat-forming vegetation 43, 71
peatland **1ff.**, **23ff.**, **40ff.**, **50ff.**, 59, **61ff.**, **82ff.**, **89ff.**, 104, **107ff.**, 122, **125ff.**, **140ff.**, 165ff., **169ff.**, 176ff., **180ff.**, **186ff.**
peatland distribution 3, 13
peatland district 3ff., 21, 35
peatland fire 20, 92f., 136, 166f.
peatland rewetting 1ff., 9ff., 20, 30, 37, 42, 46, 61, 78, **81ff.**, 89, 91f., 94f., **97ff.**, 110, 129, 135ff., 139, 141ff., **146ff.**, 154, 156, 165, **192ff.**
Peatland Rewetting and Conservation (PRC) 97ff., 136f., 143, 150, 165
pellets (from biomass) 116, 120, 125, 127
percolation mire 70
permanence 104, 165, 183, 193
*Phalaris arundinacea* 19, 45f., 50, 70, 115ff., 120, 125, 190
photosynthesis 15, 30, 43
photosynthetic active radiation (PAR) 31ff., 35, 53
*Phragmites australis* **33ff.**, 46, 50, 71, 74, 83, 110, **113ff.**, 120, 129, 171, 173, 177, 181
*Picea abies* 45
*Pinus sylvestris* 36, 44, 46, 74, 76, 182
*Plathanthera bifolia* 171
*Pleurozium schreberi* 48
*Pluvialis apricaria* 62, 72, 174
*Polytrichum* 39, 45, 47, 56, 74, 76, 170f., 179f., 190
*Polytrichum strictum* 36, 43, 46f., 48, 76, 173, 180, 187, 191
Poplau Moch 60, 83f., 139f., 146, 169
*Populus tremula* 46, 51, 74, 171
*Porzana parva* 73, 77
*Porzana porzana* 65, 72f., 77, 81
*Porzana pusilla* 73, 77
PRC requirements 98, 143
precipitation 4ff., 35, 47, 69, 169f., 172, 174, 178, 180f., 184
prediction tool 154, 192
processing of biomass 114f., 126, 131, 137
Project Design Document (PDD) 105f., 147, 154
Project Document/Description (PD) 98, 104, 143, 139
project scenario 41ff., 46ff., 50ff., 56, 99, 192f.
proxy/proxies 13, 35, 37ff., 41ff., 99, 141, 155f., 189, 191
*Quercus robur* 45

raised bog 47, 51, 84, 111, 133, 139f., 147, 169f., 182, 184
Ramsar Convention 90f.
*Rana arvalis* 60, 74, 76, 84, 176
raw material 107, 109, 112f., 115, 127
Raznianskaje 35f., 53, 140f.
re-colonization 43f., 82, 186
Reducing Emissions from Deforestation and forest Degradation (REDD) **94ff.**, 104, 134, 143, 149
Reed Canary Grass = *Phalaris arundinacea*
reed harvesting 83, 115, 128, 131
reliability 11, 156, 191
renewable 9, 109f., 142
renewable resource 109
research 2. 5ff., 16, 41f., 61, 75, 82, 93, 96, 109ff., 114, 125, 133f., 141ff., 152ff., 166, 175, **189ff.**
respiration 14ff., **30ff.**
restorability 70f.
restoration 1f., 9f., 20, 43, 61, 64, 69ff., 77f., 80ff., 84, 86, **89ff.**, 97, 107, 109f., 115, 125, 129, 135ff., 146f., 149ff., 153ff., 166, **170ff.**, 187, 189, 193
restoration approach 170, 172, 175, 178, 182, 185
rewettability 52
rewetting = peatland rewetting
Rewetting of Drained Peatland (RDP) 42, 98f., 143
rewetting site 11, 74, 135, 140, 146, 150, 156, 166, 169, 171, 173, 175, 179, 183, 191
rewetting strategy 10
rhizodeposition 15f.
*Riparia riparia* 75
Royal Society for the Protection of Birds (RSPB) 1, 9f., 69, 93, 133, 135, 142, 150f., 166, 185
*Rubus chamaemorus* 171, 173, 187
Russia (Russian Federation) 1, 7, 20, 29, 64, 92, 101, 104, 114, 133f., 137, 140, 150, 165, 169
*Sagittaria* 16
*Salix lapponica* 173
*Salix myrtilloides* 171, 173, 182
Scara-Dabramysl 69f., 140
Scarbinski Moch 139f., 162f., 169. 177ff.
*Scheuchzeria palustris* 36
*Schoenoplectus* 16, 45
scientific justification 10, 139, 167
*Scolopax rusticola* 81
S-cycle 13
Sedge Warbler = *Acrocephalus schoenobaenus*

selection of indicator species 61, 78
shunt species 16, 19, 40f., 144, 193
site selection 10, 35, 137ff., 146, 152ff., 156
Social Carbon Standard 97
*Sphagnum* 8, 12, 19, 22, 39, **43ff.**, 66, 71, 74, 76, 78, 84, 109, 129, 171ff., 182, 186f., 193
*Sphagnum angustifolium* 48, 171, 173, 186
*Sphagnum cuspidatum* 36, 48, 182
*Sphagnum magellanicum* 36, 48, 173, 182
Sporauski zakaznik 35, 36, 108, 121, 126, 135, 141, 151, 169
Sporava 10, 62, 79, 126, 140, 144, 146f.
static chamber method 141
streamlines 51, 58
subsidence 17, 40, 48, 51, 99, 138, 187
succession 11, 43, 45ff., 49, 51f., 71, 78, 80, 83, 86, 129, 131, 139, 154, 169, 178, 182, 192
sustainable (land) use 2, 91, 110f., 115, 127, 131, 134, 144, 147, 150, 153, 166, 181
sustainable biomass use 141, 144
target species 80, 82, 84, 86, 129, 145f., 177, 192ff.
*Tetrao tetrix* 66, 84
tourism 108, 110, 112, 195
touristic potential 108
trace gas 14, 30f., 34, 152f.
transgression mire 77
transitional mires / peatlands 3, 5, 21f., 66f., 110
transparent chamber 31f.
*Trichophorum alpinum* 46
*Tringa glareola* 66, 72, 79, 186
*Tringa tetanus* 65
*Triticum aestivum* 36
*Typha* 16, 45f.
*Typha angustifolia* 71, 110, 115
*Typha latifolia* 36, 71, 110, 115f.
Ukraine 1, 29, 62, 64, 101, 103, 122, 133, 135ff., 140f., 149f., 165ff.
United Nations Development Programme (UNDP) 1, 3, 9f., 69, 89, 92f., 133, 135ff.
United Nations Framework Convention on Climate Change (UNFCCC) 14, 20, **89ff.**, 99f., 102, 104, 134, 136, 144, 147ff., 166
*Urtica dioica* 36, 45, 50
*Utricularia intermedia* 173
*Vaccinium macrocarpon* 110
*Vaccinium microcarpum* 173, 182

# Index

*Vaccinium myrtillus* 43
*Vaccinium oxycoccos* (*Oxycoccus palustris*) 43, 110
*Vaccinium uliginosum* 46, 74, 76, 110, 163, 173, 182
*Vaccinim vitis-idaea* 43, 48, 110
*Vanellus vanellus* 65f., 72f., 75
vegetation change 70, 83, 193
vegetation development 13, 41ff., 45ff., 49, 51f., 70, 71, 74, 189, 192f.
vegetation form 37f., 41, 79, 152, 189, 191
vegetation prediction 47, 50, 192
verification 52, 96f., 106, 136, 143, 147, 165, 193ff.
Verified Carbon Standard (VCS) 37, 42, **96ff.**, 104, 106, 136f., 139, 141, 143, 150, 155, 165, 193

Voluntary Carbon Standard (see also Verified Carbon Standard) 143
voluntary emission reduction 104ff., 136, 144
voluntary market (see also voluntary carbon market) 37, 42f., 89, **93ff.**, 104f., 134, 137, 142, 146, 154, 156
Vyhanascanskaje 39f., 42., 49f., 57, 59, 70, 138, 140, 145
Water Buffalo = *Bubalus bubalis*
water flow 4, 11f., 27, 51, 69, 183
water level 4, 7, 10ff., 17ff., 31f., **35ff.**, **69ff.**, 75f., 80, 83ff., 98f., 102f., 107, 109, **111ff.**, 121, 128, 130, 136, 138, **141ff.**, 151ff., 156, 166, **170ff.**, 186f., **189ff.**

wet *Sphagnum* lawn 39, 47ff., 171ff., 182
wetland 4, 38, 46, 61ff., 74, 80, 84, **90ff.**, 95, 99, 101f., 110, 116, 125, 127, 131, 144, **146ff.**, 151, 153, 182, 185, 194
wetland restoration 93f., 147
willingness to pay 66, 97, 111
wooden dams 11, 171
Zada 139ff., 147, 157ff., 169, 172ff., 187
zakaznik 35f., 47, 61, 68, 70, 108, 121, 126f., 135, 141, 151, 169, 181f.
*Zizania aquatica* 112
Zvaniec 10, 22, 28, 61f., 64f., 66, 68f., 79, 140

# This book is also available in Russian

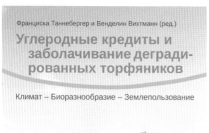

Франциска Таннебергер и Венделин Вихтманн (ред.)

## Углеродные кредиты и заболачивание деградированных торфяников

*Климат – Биоразнообразие – Землепользование*

Теория и практика – уроки реализации пилотного проекта в Беларуси

2011. XII, 224 с., 100 преимуществ. цветн. рис., 41 таб., 30 встав., 27,6 x 21 см.
**ISBN 978-3-510-65272-3**   € 29,90

Информация о книге на русском языке: www.schweizerbart.com/9783510652723

**Углеродные кредиты и заболачивание деградированных торфяников – Идея хорошая, но как осуществить ее практически?**

Осушенные торфяники занимают около 0,3% земной поверхности, однако концентрируют на своей площади непропорциональные 6% от общего объема антропогенных выбросов $CO_2$. Поэтому эта глобальная проблема должна быть решена. Мы знаем места сосредоточения «горячих точек» – в Юго-Восточной Азии, Центральной и Восточной Европе, в некоторых частях США, в Северо-Восточном Китае. Мы знаем, что делать: восстановить уровень грунтовых вод. Но остается еще очень много вопросов:

Как будет влиять повторное заболачивание на потоки парниковых газов? Будет ли выделяться метан и в каком количестве? Можно ли измерить выбросы, свести в отчет и в последующем проверить эти данные? Могут ли полученные сокращения выбросов парниковых газов способствовать выполнению обязательств в рамках Киотского протокола или быть проданы на добровольном углеродном рынке? Как повторное заболачивание будет влиять на биоразнообразие? Могут ли повторно заболоченные торфяники по-прежнему продуктивно использоваться?

Беларусь занимает восьмое место среди стран мира по выбросам парниковых газов от торфяников либо третье – если рассматривать их как удельные выбросы на единицу площади. В последние годы в стране десятки тысяч гектаров осушенных торфяников были заболочены.

Эта книга представляет собой синтез возникших проблем и их соответствующих решений в рамках пилотного проекта, реализованного в Беларуси в период 2008–2011 гг. В ней представлены данные и результаты проекта, а также связанное с ними современное применение основных принципов интеграции науки и политики, экологии и экономики. Его опыт и рекомендации будут вдохновлять практиков, ученых и политиков.

## Содержание

Предисловие Программы защиты окружающей среды Организации Объединенных Наций .. V
Предисловие Министра природных ресурсов и охраны окружающей среды Республики Беларусь ............ VII
Предисловие Фонда Михаэля Отто .......... IX
1 Введение ............ 1
2 Торфяники в Беларуси ............ 3
Распространение и классификация торфяников в республике Беларусь • Исследование и осушение болот • Использование болот и торфа • Повторное заболачивание торфяников
3 Торфяники и климат ............ 13
Торфяники и парниковые газы • Общая картина выделения $CO_2$ с торфяников • Измерение эмиссий парниковых газов с торфяников • Растительность как индикатор потоков парниковых газов – GEST подход • Прогнозирование развития растительности при повторном заболачивании и без заболачивания
4 Торфяники и биоразнообразие ............ 61
Показатели биологического разнообразия белорусских торфяников • Связь между состоянием торфяников и степенью биоразнообразия • Целевые и индикаторные виды • Повторное заболачивание торфяников и управления биоразнообразием
5 Движущие силы и возможности финансирования ............ 89
Установленные законом обязательства по восстановлению нарушенных торфяников Беларуси • Возрастающее внимание к торфяникам со стороны глобальных конвенций об изменении климата • Продажа сокращений выбросов при повторном заболачивании на добровольном углеродном рынке • Продажа сокращений выбросов при повторном заболачивании на обязательном углеродном рынке • Добровольные проекты по сокращению выбросов парниковых газов – как начать работу в Беларуси
6 Варианты землепользования после повторного заболачивания ............ 107
Обзор • Использование земли для пищи и кормов • Использование биомассы в качестве сырья • Использование биомассы для получения энергии • Польза от использования земель на повторно заболоченных торфяниках
7 Проект BMU-ICI ............ 133
Резюме проекта • Выбор территорий и деятельности по повторному заболачиванию • Проектные мероприятия для климата • Проектные мероприятия по улучшению биоразнообразия • Политические проектные мероприятия • Повышение роли обмена информации и осведомлённости • Повышение потенциала • Извлеченные уроки • Проект-близнец BMU-ICI на Украине
8 Практические примеры заболачивания ............ 169
Введение • Долбенишки • Жада • Гричино-Старобинское • Щербинский Мох • Докудовское • Ельня
9 Рекомендуемая научно-исследовательская деятельность и мероприятия по мониторингу ............ 189
Рекомендуемая научно-исследовательская деятельность • Рекомендуемые мероприятия по мониторингу
10 Благодарность ............ 197
Список литературы ............ 199
Специалисты, внесшие вклад в создание книги ............ 215
Алфавитный указатель ............ 217

# Schweizerbart Science Publishers • Stuttgart

Johannesstr. 3a, 70176 Stuttgart, Germany., Tel. +49 (0)711 351456-0, Fax: +49 (0)711 351456-99, order@schweizerbart.de, www.schweizerbart.de